Lecture Notes in Electrical Engineering

Volume 751

The book series *Lecture Notes in Electrical Engineering* (LNEE) publishes the latest developments in Electrical Engineering - quickly, informally and in high quality. While original research reported in proceedings and monographs has traditionally formed the core of LNEE, we also encourage authors to submit books devoted to supporting student education and professional training in the various fields and applications areas of electrical engineering. The series cover classical and emerging topics concerning:

- Communication Engineering, Information Theory and Networks
- Electronics Engineering and Microelectronics
- Signal, Image and Speech Processing
- Wireless and Mobile Communication
- Circuits and Systems
- Energy Systems, Power Electronics and Electrical Machines
- Electro-optical Engineering
- Instrumentation Engineering
- Avionics Engineering
- Control Systems
- Internet-of-Things and Cybersecurity
- Biomedical Devices, MEMS and NEMS

For general information about this book series, comments or suggestions, please contact leontina.dicecco@springer.com.

To submit a proposal or request further information, please contact the Publishing Editor in your country:

China

Jasmine Dou, Editor (jasmine.dou@springer.com)

India, Japan, Rest of Asia

Swati Meherishi, Editorial Director (Swati.Meherishi@springer.com)

Southeast Asia, Australia, New Zealand

Ramesh Nath Premnath, Editor (ramesh.premnath@springernature.com)

USA, Canada:

Michael Luby, Senior Editor (michael.luby@springer.com)

All other Countries:

Leontina Di Cecco, Senior Editor (leontina.dicecco@springer.com)

**** This series is indexed by EI Compendex and Scopus databases. ****

More information about this series at http://www.springer.com/series/7818

Mladen Božanić · Saurabh Sinha

Mobile Communication Networks: 5G and a Vision of 6G

Springer

Mladen Božanić
Faculty of Engineering and Built Environment
University of Johannesburg
Johannesburg, South Africa

Saurabh Sinha
Office of the Deputy Vice-Chancellor
University of Johannesburg
Johannesburg, South Africa

ISSN 1876-1100 ISSN 1876-1119 (electronic)
Lecture Notes in Electrical Engineering
ISBN 978-3-030-69272-8 ISBN 978-3-030-69273-5 (eBook)
https://doi.org/10.1007/978-3-030-69273-5

This Springer imprint is published by the registered company Springer Nature Switzerland AG
The registered company address is: Gewerbestrasse 11, 6330 Cham, Switzerland

Preface

Roughly every 10 years, a new generation of mobile networks appears. The first generation (1G) network, which appeared in the 1980s, was analog in nature and supported only voice calling. Since then, there have been three radical updates, dubbed the second, third, and fourth generation networks (2G, 3G, and 4G, respectively). 2G marked the transition from analog to digital signaling, text messaging and basic Internet services; 3G introduced faster Internet access and video calling; and 4G allowed for much faster data rates, supporting services such as high-definition video streaming, while supporting a large number of users in a single cell. As of 2019, fifth generation (5G) networks have been deployed, carrying the initial premise as the enabler of Internet of everything (IoE) applications. Each time a new wireless generation evolved, added services required faster data rates, which could only be achieved by moving to channels with more bandwidth. While the 5G capacity and data rates are sufficient at present, 5G is going to have to continue to evolve in the same manner as all the previous generations of networks did, to support the increased data traffic. The International Telecommunication Union forecasts that global mobile Internet traffic is going to increase to over 5000 exabytes/month in 2030. 5G networks have enough in their store to keep up with the demand for several years; however, the 5G evolution will not be able to continue indefinitely, as 5G has some fundamental limitations. This requires a fresh perspective to be taken into consideration when imagining the future of communications.

This is where the sixth generation of mobile networking (6G) fits in. To enable 6G networks to catch up where 5G leaves off by 2030, envisioning of 6G has to start in early 2020s, in parallel with 5G network deployment. In fact, this process already started in the late 2010s, with the biggest push forward asserted by Finnish (6G Flagship, University of Oulu) and Chinese researchers. The vision of 6G differs from research group to research group. On the one extreme, one can expect that 6G is going to be an enhancement of conventional mobile communications (enhancement of 5G and thus of all the previous network generations), where the cellular phone remains the main tool of communication. This type of network would remain human-centric, as all previous generations are. On the other extreme,

6G might become something that can be described as ubiquitous wireless intelligence, where communication services are just a single set of services in a framework of services existing in the edge cloud. This scenario would result in 6G being machine-centric, application-centric, or data-centric. Most likely, however, 6G will be defined as something in between. What is common to all envisioned scenarios is that 6G will require transmission of large amounts of data almost instantly, which will only be practical if data transmission speeds are in the region of Tb/s. For this, new and emerging data transmission technologies, including communication via millimeter-waves, terahertz waves, or visible light will be needed, seeing that the existing communication channels typically use frequencies below 6 GHz, and are typically utilized at maximum capacity. Some of the other basic requirements of 6G networks also include low end-to-end latency, in the region of less than 1 ms, which is needed to support the required verticals; high energy efficiency, expressed in energy per transmitted bit, to decrease the overall power consumption brought about by extremely high data rates; coverage that is ubiquitous, and always-on, achieved by integrating terrestrial wireless with non-terrestrial variants (e.g., satellite systems, aerial networks, and underwater communication links); network management that is made more effective by connected intelligence with machine learning capabilities; as well as support for the evolution of old service classes and for new ones.

This book aims to explore the main ideas and the outlook of past and future wireless network trends, respectively, present capabilities of 5G and other wireless systems, technologies that would be the potential enablers of 6G, potential 6G applications and requirements, as well as unique challenges and opportunities that 6G research is going to present over the next decade. The book is divided into 10 chapters. Chapter 1 presents the introduction to the topic of research and places the book in the context of recent 5G and 6G research efforts. 6G vision, as well as many technical aspects of 6G, will be briefly introduced in this chapter. The focus of Chap. 2 will be the history of wireless network generations, with special focus on 4G and even more strongly on 5G, the shortcomings of 5G and predictions in terms of where 6G can help rectify the shortcomings of previous wireless network generations. Historical trends will be used to justify and predict the future of the telecom "explosion".

With Chap. 3, a series of technical chapters begins. First, channel coding theory and channel capacity are discussed as a way of elaborating why higher frequency bands or other communication paradigms have to be explored in 5G and 6G networks. Subsequent discussions cover millimeter-wave frequencies, terahertz frequencies, and modulation. Chapter 4 looks into active device (transistor) technologies that could enable the expansion of communications to channels with higher center frequencies. Some of the basic millimeter-wave and terahertz circuits are also discussed in this chapter. Lastly, this chapter discusses the systems approach to packaging and package optimization for better performance of future wireless network circuits. Chapter 5 looks into visible light communications (VLC) as the main communication technology alternative to millimeter-wave and terahertz radio for establishing ultra-fast data links in 6G. The main advantages and shortcomings of VLC are discussed in detail, together with the physical aspects of VLC and most likely use cases. Whereas VLC technology is nowhere near as mature as radio, it is potentially much better in various scenarios, which include vehicle-to-everything (V2X) communications, underwater communications, or communications where low interference or high security are needed. In Chap. 6, the concept of 6G as the "green network" is discussed. This involves predominantly the introduction of energy harvesting into the network design, where mobile network users use the energy harvested from light, movement, ambient RF or other sources as a source of power, with or without the aid of batteries and supercapacitors. Furthermore, in view of many techniques for energy saving that have been investigated and are being introduced in 5G networks, specifically on the side of base stations, ongoing research to achieve even better energy saving and conservation techniques for 6G networks is also discussed.

Chapter 7 will cover some of the advanced, possibly even futuristic, technological aspects that are expected to be unique to 6G. Advanced aspects include ultra-massive multiple-input, multiple-output, beamforming, and channel control with programmable (reflective) metasurfaces. Futuristic concepts include holographic radio, quantum and molecular communication, nanophotonics, aerial networks established with the help of drones and satellites, as well as the use of blockchain technologies in network security. Chapter 8 considers another set of advanced topics, this time related to intelligence within the 6G network design. Concepts such as pervasive artificial intelligence (AI), machine learning, and edge intelligence are discussed. This chapter also discusses the architecture that 6G should take to make full use of AI principles, as well as the controversial topic of ethics in AI.

Chapter 9 steps away from technical discussions and discusses the health issues associated with underlying technologies of 5G and 6G. This chapter is mostly dedicated to the investigation of the influence of the short-term and long-term influence of different types of electromagnetic radiation on various human health aspects. Both the physics of wave propagation and medical research studies are discussed. Potential health issues associated with VLC are also discussed in this chapter. Finally, Chap. 10 concludes the book by discussing 6G as a fusion of communications, imaging, sensing, localization, and many other 6G verticals. These include the concepts of robotics, autonomous systems, V2X, the internet of things, IoE, Industry 4.0, extended reality, and others.

Peer-review acknowledgement: The authors would like to recognize the support of technical reviewers, as well as language and graphic editors, who have all contributed to the book-compiling process. We value the system of independent scholarly review and the value that this adds to the production of research text that adds to the body of scientific knowledge.

Johannesburg, South Africa
January 2021

Mladen Božanić
Saurabh Sinha

Contents

About the Authors

Mladen Božanić, Ph.D. (Eng), SMIEEE obtained his B.Eng. (with distinction), B.Eng. (Hons) (with distinction), and Ph.D. degrees in Electronic Engineering from the University of Pretoria (UP) in 2006, 2008, and 2011, respectively. In 2008, he joined Azoteq, a fabless IC design company originating in South Africa with where he has been responsible for the silicon-level design, simulation, characterization and design for testability of various analog, RF, digital, and mixed-mode circuits. While actively working in the industry, he also participates in research activities, currently with the University of Johannesburg (UJ) where he is serving as a senior research associate. Since 2011, Dr. Božanić has been fulfilling the role of specialist editor of the South African Institute of Electrical Engineers. He is a recipient of the SAMES Award and CEFIM Fellowship Award, and an author or co-author of over 25 peer-reviewed journal and conference articles, one book chapter and five books.

Saurabh Sinha, Ph.D. (Eng), Pr. Eng., SMIEEE, FSAIEE, FSAAE obtained his B.Eng., M. Eng. and Ph.D. degrees in Electronic Engineering from the University of Pretoria (UP). As a published researcher, he has authored or co-authored over 130 publications in peer-reviewed journals and at international conferences. Professor Sinha served UP for over a decade, his last service being as Director of the Carl and Emily Fuchs Institute for Microelectronics, Department of Electrical, Electronic and Computer Engineering. On October 1, 2013, Prof. Sinha was appointed as Executive Dean of the Faculty of Engineering and the Built Environment at UJ. As of December 1, 2017, Prof. Sinha has been the UJ Deputy Vice-Chancellor: Research and Internationalisation. Among other leading roles, Prof. Saurabh Sinha also served the IEEE as a member of the Board of Directors and IEEE Vice-President: Educational Activities.

Chapter 1
Leap to 6G?

Abstract This chapter presents the introduction to the topic of research of this book and places the chapter and the rest of the book in the context of recent 5G and 6G research efforts. 6G vision and many technical aspects of 6G are briefly introduced.

Some 40 years ago, humanity entered the mobile communication age. The penetration of cellular telephones into the general population was initially slow and limited to only a few countries. However, at one point, in late 1990s, it just exploded. Older generations will still recall arranging meetings at the exact spot and exact time, because it was not possible to change meeting arrangements on the fly without cellphones; newer generations are born into the world where cellular telephones are omnipresent and probably cannot picture the world without them (and may laugh at classic movies where the main hero drives halfway across town to find a phone booth). In essence, in this short time span, within the duration of one human lifetime, we went from having virtually no wireless communication (two-way radio being the most notable exception) to being able to phone from the street, then from being able to send text messages on the fly to being able to check sports scores when not close to the television and finally to watching videos in the doctor's waiting room. One may recall riding in the subway not even 20 years ago where people were reading daily papers to catch up on last evening's news; now virtually everybody's phones are in their hands, some reading the news in real time, others listening to streaming music and some catching up on the latest episode of their favorite show.

This became possible with the mobile network evolution. Roughly every 10 years, a new generation of mobile networks appeared [1]. The first generation (1G, although not initially called that), appeared in 1980s, was analog in nature and supported only voice calling. Since then, there have been three radical updates dubbed the second, third and fourth generation networks (2G, 3G and 4G respectively) [2]. 2G or global system for mobile communications (GSM [3]), introduced in the 1990s, marked the transition from analog to digital signaling, essentially paving the way to having more available services than just calling, eventually introducing text messaging and basic internet services as part of its evolution. At the turn of the century, 3G brought in faster internet access and video calling. 4G, introduced in the 2010s and still widespread

M. Božanić and S. Sinha, *Mobile Communication Networks: 5G and a Vision of 6G*, Lecture Notes in Electrical Engineering 751,
https://doi.org/10.1007/978-3-030-69273-5_1

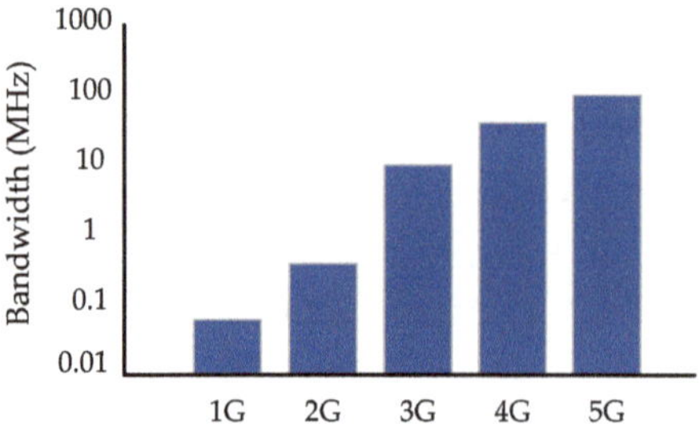

Fig. 1.1 Bandwidth required by various wireless generations. Adapted from [5]

today, allows for much faster data rates, enabling services such as high-definition (HD) video streaming, while supporting a large number of users in a single cell. By 2019, the fifth generation (5G) networks were being deployed, carrying the initial premise as the enabler of the internet of everything (IoE) applications [4]. Each time a new wireless generation evolved, added services required faster data rates, which could only be achieved by moving to channels with more bandwidth. This trend is shown in Fig. 1.1 [5].

The question, at this point, is whether the increased data demand is going to continue. After all, everybody already has a cellphone, which already has all the features that one can imagine are needed: large screen capable of HD picture quality, still and video cameras, multiple installed applications, global positioning system (GPS) capability, multiple sensors, health tracking and many other features. Despite each person still really needing one device in principle, somehow, more and more users are connected all the time. This is because in the IoE concept and its predecessor, the internet of things (IoT), various devices are now also internet users, complementing humans and their phones, tablets or computers. Meanwhile, increased data availability keeps promoting the finding of new ways in which these data can be utilized, thus resulting in a further increase in traffic demands. There are no indications that this positive feedback is going to stop soon; without any doubt, while the 5G capacity and data rates are sufficient at present, 5G is going to have to continue to evolve in the same manner as all the previous generations of networks did to support the increased data traffic. The International Telecommunication Union (ITU) forecasts that global mobile internet traffic is going to increase by 55% annually, from 57 exabytes/month in 2020 to over 5 000 exabytes/month (5 zettabytes/per month) in 2030 (a 100-fold increase in only 10 years) [6, 7]. This trend is shown in Fig. 1.2. It is, of course, expected, by virtue of design, that 5G networks will have enough in store to keep up with the demand for several years, as they gradually evolve into beyond-5G (B5G) networks. The B5G evolution, however, will not be able to continue indefinitely, as 5G has some fundamental limitations, requiring that a fresh perspective once more be taken into consideration when imagining the future of communications [8].

This is where the sixth generation of mobile networking (6G) fits in. 6G is probably not going to appear naturally as a merely renamed B5G, although the terms B5G and 6G are sometimes used interchangeably [9]. It will rather pick up where B5G leaves off [10]. For this to happen, the trend of network evolution, depicted in Fig. 1.3, has to continue. According to this trend, the development of every new generation can only

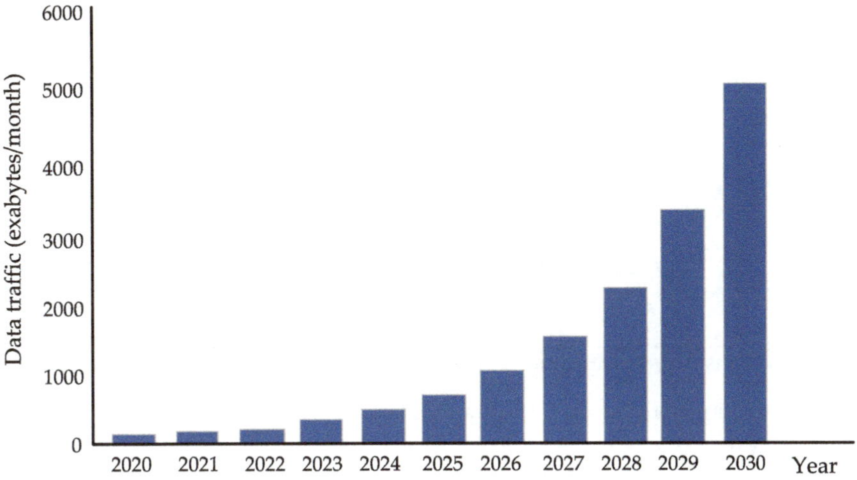

Fig. 1.2 ITU forecast of global mobile data traffic increase between 2020 and 2030 [6, 7]

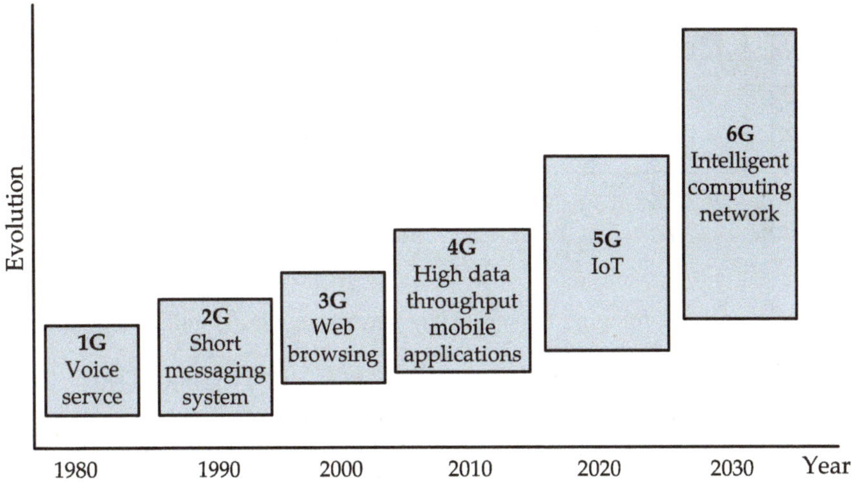

Fig. 1.3 Evolution of mobile networks, from 1G to envisioned 6G

take a maximum of 10 years on average, and this is only if development is started roughly at the time of commercial deployment of the previous generation. Note that the maturity of the network is only reached about ten years into the deployment of each generation, as seen in Fig. 1.4 [11]. Thus, in order to catch the correct wave and have 6G ready by 2030, the envisioning of 6G has to start in early 2020s, as 5G networks are deployed. In fact, this process has started already in late 2010s, with the biggest push forward asserted by Finnish (6G Flagship, University of Oulu) [12] and Chinese [6] researchers. The first 6G wireless summit was held in Finland in March

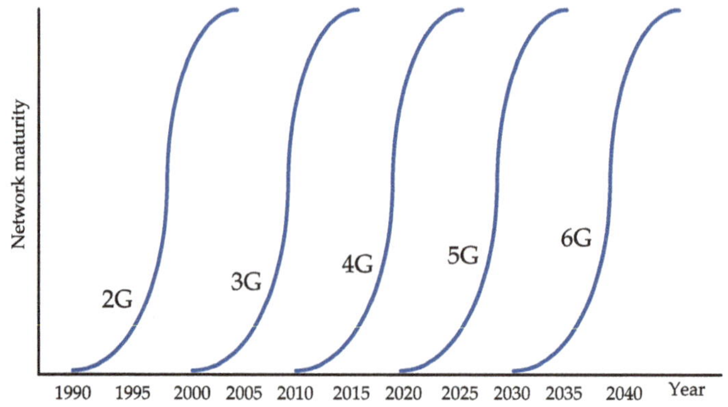

Fig. 1.4 Maturity level of wireless networks. Adapted from [11]

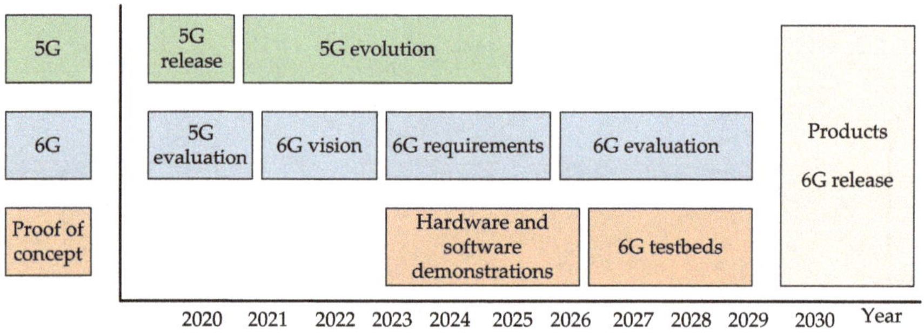

Fig. 1.5 Roadmap enabling 6G deployment by 2030 [13]

2019 as a means of bringing the major key players together to initiate 6G thinking on a larger scale.

Envisioning alone, naturally, is not enough and to achieve the goal of 6G deployment by 2030, 6G research should rather loosely follow a set roadmap, such as the roadmap depicted in Fig. 1.5, as proposed in [13]. According to this roadmap, collaborative efforts will focus on 5G evaluation and 5G evolution for the first five years, while setting the specifications of 6G. The first 6G testbeds can be expected to appear only post-2025. Defining a roadmap, however, is not a massively daunting task, compared to the task of determining how 6G should look. On the one extreme, one can expect that 6G is going to be an enhancement of conventional mobile communications (enhancement of 5G and thus of all the previous network generations), where the cellular phone remains the main tool of communication [2]. This type of network would remain human-centric, as all previous generations are. On the other extreme, 6G might become something that can be described as ubiquitous wireless intelligence, where communication services are just a single set of services in a framework of services existing in the edge cloud (part of the network consisting of user devices and their links to the network itself) [12]. This scenario would result in 6G being machine-centric, application-centric or data-centric. Most likely, however, 6G will be defined as something in between. What is common to all envisioned scenarios is that 6G will require transmission of large amounts of data almost instantly, which will only be practical if data transmission speeds are in the region of Tb/s. For this, new and emerging data transmission technologies, including communication via millimeter-waves, terahertz waves or visible light will be needed, seeing that the existing communication channels typically use frequencies below 6 GHz and are typically utilized at maximum capacity [14].

This book aims to explore the main ideas and the outlook of the past and future wireless network trends and present capabilities of 5G and other wireless systems, technologies that would be potential enablers of 6G, potential 6G applications and requirements, as well as unique challenges and opportunities that 6G research is going to present over the next decade.

The vision of 6G is of primary importance. As mentioned earlier in this introduction, it will take some time before the clear path towards 6G is identified. At present, at the start of the new decade, there are numerous research ideas that have to be considered as a part of the 6G vision. Secondly, the vision is, naturally, driven by the prospective services, termed verticals, which include not only the communications and their need for increased speed and bandwidth, but also other applications such as imagining, global and local positioning, machine communications, education, manufacturing, health, energy management and many others. Lastly, the means of achieving this vision (or rather, technologies with the potential to make 6G happen) should not be disregarded. In this chapter will thus expand on these three topics while introducing other important considerations of 6G. The remainder of the book will, subsequently, dissect these 6G topics into a multitude of smaller research aspects that will be handled in separate, yet related, sections and chapters.

The extent of the 6G research opportunity will become clear shortly. As the chapter progresses, the list of 6G-related topics will grow, allowing for the outline of the book to be detailed in the last section of the chapter.

1.1 The 6G Vision

To understand the 6G vision, one must go back and briefly investigate the shortcomings of 5G that is currently being deployed worldwide. Note that Chap. 2 will look into the history of the development of wireless network communications, including 5G, in more detail.

The first 5G standard was released in December 2017 communications, whereas the first deployments were seen in 2018 [15]. What must be noted straight away is that there is no ground-breaking technology in the 5G concept [2]. 5G inherits the strategy of the performance enhancement that was started with 3G and continued with 4G. The operators are deploying more hardware and taking up more of the same spectrum in order to achieve performance gains. Like preceding generations, 5G is expected to evolve continuously. In most cases, the first deployment strategy is based on the third-generation partnership project (3GPP) new radio standard for dense areas and still utilizes the spectrum between 2 and 6 GHz. Other strategies that will be deployed include utilizing the lower frequency bands in the millimeter-wave spectrum, the massive multiple-input and multiple-output (mMIMO) technologies, beamforming, holographic radio and small cell densification. Lastly, 5G allows for three service options: enhanced mobile broadband (eMBB), ultra-low reliable low-latency communications (URLLC) and massive machine-type communication (mMTC), with targets of improved data rate, transmission latency, and connectivity performance respectively [10]. Note that these targets cannot be achieved simultaneously, and only a compromise between them can be sought. At the same time, the initial premise of 5G as the carrier of IoE services has not yet been achieved. At present, the most important key performance indicators (KPIs) of 5G are the data rate in the order of 1 Gb/s, end-to-end delay of 5 ms and processing delay of 100 ns. B5G networks should be able to improve these KPIs to achieve a data rate of 100 Gb/s, end-to-end delay of 1 ms and processing delay of 50 ns [4].

The goal of the 6G initiative is to revolutionize the wireless evolution. This means that the approach has to be changed from having "connected things" to having "connected intelligence" [16]. 6G should be able to integrate applications ranging from autonomous driving to extended reality (XR) [4]. However, first and foremost, 6G must introduce sustainability in mobile development, and remove the digital divide that still exists in the world, by offering everybody access to digital information. In other words, 6G must succeed where 5G failed. In order to achieve all of the above, high data rates, high reliability and low latency across heterogeneous devices have to be achievable practically simultaneously, unlike in 5G. Some of the basic requirements of 6G networks can thus be summarized as follows [6, 16, 17]:

- 6G will require very high data rates, in the region of 1 Tb/s.
- Low end-to-end latency, in the region of less than 1 ms, is needed to support required verticals.
- Channel capacity theory dictates that higher frequency bands with wider bandwidth channels are utilized alongside the frequency bands utilized in the past. This means moving to channels that lie above 6 GHz utilized by networks prior to 5G, and even beyond 86 GHz, identified as the maximum frequency of 5G (e.g., 73–140 GHz and 1–3 THz), or looking into other ways to establish communication, e.g. utilizing the concept of visible light communication (VLC).
- The high data rate compels consideration of energy efficiency, expressed in energy per transmitted bit.
- 6G must be a ubiquitous, 3D and always-on broadband global network, where wide coverage is achieved by integrating terrestrial wireless technology with non-terrestrial variants (e.g. satellite systems, aerial networks and underwater communication links).
- Because of the complexity of the network, the previous requirement can only be achieved if 6G appears in the form of connected intelligence with machine learning capability.

- Finally, the eMBB, URLLC, and mMTC services are no longer sufficient, and new service classes will be required.

Expectedly, opinions on the number and type of service classes that can be anticipated in 6G differ from research group to research group [2–4, 10, 16]. Service classes indicate trade-offs between factors such as the achievable distance, speed, latency, power consumption, and others. In general, there is an overlap in ideas, where some services are expected to be human-centric and others to be machine-centric. According to Saad et al. [4] (6G Flagship), there should be four services in principle:

1. *Mobile broadband reliable low latency communication (MBRLLC).* MBRLLC should introduce stringent rate-reliability latency requirements together with energy efficiency and low rate-reliability latency in mobile environments. Examples of applications in this service include virtual reality (VR), augmented reality (AR), mixed reality (MR) and XR, autonomous vehicles and drones, as well as legacy eMBB and URLLC services.
2. *Massive URLLC (mURLLC).* mURLLC combines ultra-high reliability with massive connectivity, massive reliability and scalable URLLC. Examples of applications in this group of services include IoT, sensing, robotics and blockchain.
3. *Human-centric services.* These services include capturing the data that improve the quality of physical experience (QoPE) for humans.
4. *Multi-purpose 3CLS and energy services (MPS)*, where 3CLS stands for convergence of communication, computing, control, localization and sensing. These services have to join high control stability, low computing latency, high localization accuracy, high sensing and mapping accuracy, low latency and high reliability in communications within target energy guidance for transfer in information. Examples of applications of these services include connected robotics and autonomous systems (CRAS), telemedicine, environmental imaging or mapping and some special cases of XR services.

One of the 6G requirements that is easy to overlook is the requirement for 6G networks to be extremely energy-efficient. For every bit of data that is transmitted, a certain amount of energy is required. With data rates in Tb/s, as opposed to Gb/s in 5G, the power consumption figures of 6G devices will be several orders of magnitude higher than those of 5G devices. This is why 6G is envisioned as a green network, where energy harvesting and the use of new materials are introduced to improve the overall system efficiency and add to the sustainability of the network [6]. Ultra-long battery life or even battery-free communication would be the ultimate goal of 6G networks, and communication efficiency in the order of 1 pJ/b should be one of the targets [13].

A more detailed breakdown of 6G requirements and KPIs, and how they compare with what is achievable with 5G and B5G, can be seen by analyzing data in Table 1.1. At this point, it should already be clear to the reader that 6G cannot achieve all these requirements by continuing a brute force approach of further improvement of previous generations of networks and that 6G is indeed going to have to be a completely new concept.

Table 1.1 Comparison of requirements and KPIs of 5G, B5G and 6G. From [4, 7, 13]

Requirement/KPI	5G	B5G	6G
Application types	eMBB, URLLC, mMTC	Reliable eMBB, URLLC, mMTC, hybrid (URLLC + eMBB)	MBRLLC, mURLLC Human-centric services, MPS
Device type	Smartphones, sensors, drones	Same as 5G plus XR equipment	Same as B5G plus CRAS and wireless brain-computer interactions
Data rate	1 Gb/s	100 Gb/s	1 Tb/s
End-to-end delay	5 ms	1 ms	<1 ms
Processing delay	100 ns	50 ns	10 ns
Reliability	99.999%	99.9999%	99.99999%
Power efficiency	–	–	1 pJ/b
Spectral efficiency	30 b/s/Hz	30 b/s/Hz	100 b/s/Hz
Jitter	–	–	1 μs
Location precision	20 cm (2D)	20 cm (2D)	1 cm (3D)

It is impossible to say exactly how this new concept is going to look. It is, however, possible to observe some trends in industry and technology development that indicate some possible directions 6G research could take. At the end of 2019, Qi Bi from China Telecom identified ten such trends [5], which are summarized in the list below.

- *Trend 1: 6G will continue to move to higher frequencies and more bandwidth.* With 5G already reaching the capacity of channels situated in lower parts of the spectrum, high data demands of 6G will require transmission to be moved to frequencies above 24 GHz, above 60 GHz or even above 300 GHz. This means, without a doubt, that millimeter-waves and terahertz will have to be used. This concept will be expanded upon in Sect. 1.3.1.
- *Trend 2: mMIMO will remain as a key technology in 6G.* The capabilities of mMIMO, a system that uses a great number of antennas (hundreds in each transmitter and receiver), have already been demonstrated in 5G. Field tests showed that peak spectral density of 100 b/s/Hz was possible. Although it may not be possible to replicate such peak densities in real-world operation, it does indicate the usefulness of the mMIMO approach. Average spectral density of just under 10 b/s/Hz seems to be a more realistic figure for commercial deployments, but some sources indicate that a figure of 30 b/s/Hz will be achieved in 5G already (see Table 1.1). 6G will continue to strive to triple this efficiency figure.
- *Trend 3: Data rate and spectral efficiency will remain in the 6G focus.* To achieve transmission over long distances, high power and highly efficient transmissions are needed. However, according to Shannon's formula (described in Chap. 3), a system cannot be optimized simultaneously for both spectral efficiency and power efficiency. Contemporary wireless systems use adaptive techniques to decide

whether the system should operate at high power or high spectral efficiency, depending on the channel condition. In 6G, however, transmission will have to be shifted into the realm in which compromise between these two efficiencies will still result in satisfactory figures for both metrics.

- *Trend 4: 6G will continue to push the device technologies to smaller process nodes.* In order to support higher frequencies, higher bandwidths, faster processing and lower heat dissipation, 6G will have to use smaller process nodes, which will also unlock higher chip densities. In typical 3G chipsets, technologies had the minimum dimension of 65 nm. This decreased to 28 nm for 4G, while 5G uses the 7 nm processes. To make it feasible, 6G is going to have to use 3 nm processes or smaller. It becomes increasingly difficult to engineer smaller process nodes, and device scaling continues to be the focus of many researchers, as will be discussed in Chap. 4.
- *Trend 5: Cloud services in 6G will be taken onto another level.* With 6G, data transmission rates in mobile devices will be so fast that the impression when saving any kind of data onto cloud services will be the same as saving the same data on the hard drive or solid-state drive (SSD) locally. This could potentially result in fast expansion of cloud services. Having all the data available on the cloud will have another advantage: it will also be possible to do all the computation directly on the cloud, whereas the computation on the terminal device can be saved for presentation of information. This will make services such VR, AR, MR and XR more impressive and artificial intelligence (AI) possible.
- *Trend 6: 6G will continue to integrate mode transmission nodes.* This trend already started in 4G networks when the 3GPP standard incorporated Wi-Fi and Bluetooth operation into cellular operation. In 5G, different hardware systems will support eMBB, mMTC and URLLC. 6G is likely to add even more nodes. The most promising node for 6G is satellite communications. If 6G is to be ubiquitous, then rural areas also need to be covered, where almost half of the world's population lives. Instead of bringing in the signals to rural areas via terrestrial backhaul, this can be done via a satellite network. Satellite phones have been available for a long time already, but the satellite capability was never properly integrated into cellular phones owing to the cost of this technology. Bringing satellites into the picture will also require international collaboration, as satellites seldom belong to just one authority and satellite signals cross borders much more easily than terrestrial signals.
- *Trend 7: Grant-free transmissions will be more prominent in 6G.* In past generations of cellular systems, most cellular transmissions were grant-oriented, meaning that the power level and the scheduling of transmissions was coordinated by the system centrally via a base station with the aim to minimize channel interference and maximize channel resources. In 3G and newer generations, some grant-free transmissions are possible, but the number of different types of grant-free transmissions will have to grow for optimum performance of the 6G network to be achieved.

- *Trend 8: Moving from unified to personalized service.* The design architecture of old generations of wireless systems was driven by cost. This means that all the services using older networks could expect the same performance qualities as any other services, irrespective of factors such as the importance and the urgency of the service. However, with the increasing number of different types of vertical industries, end-to-end performance will have to be varied so that special capabilities could be offered to some application categories, as well as personalized services to consumers.
- *Trend 9: mMTC will need to take shape in an older generation before it can succeed in 6G.* In 5G, high expectations were set for the success of mMTC services. Unlike communication between people, which can be interrupted and subsequently resumed, mMTC can only succeed in reality if high coverage, low cost and a long battery life are offered simultaneously. 6G is unlikely to offer all three simultaneously initially, meaning that mMTC will first have to gain traction in 5G, which will be considered an older generation at the time of 6G deployment, before it could be offered in 6G.
- *Trend 10: 6G will emerge as a computing network, rather than just a transmission network.* As mentioned earlier (trend 5), most memory and computing in 6G is going to move from terminals to the cloud. The question on where the cloud resides, however, has not been discussed. In order to reduce the latency of the network, it makes sense to put the data center (cloud) at the edge of the network, which is the idea behind the concept of edge computing. Shifting the data to and from the cloud edge as well as inside the cloud while performing the actual computation may require involvement of AI and in particular machine learning for it to become feasible. Extracting information in real time that machine learning will offer also ties in with delivering personalized service for different users, as discussed as part of trend 8. The role of AI and machine learning, however, does not need to be limited to only edge computing. AI could, for example, manage harmonious transmission between users by providing optimized transmission link configurations based on user's needs, or it could be deployed to assist in big data analysis, which will be discussed later in Sect. 1.4.1.

The vision of 6G as an intelligent computing network with transmission capability that emerged in the analysis of the predicted 6G trends only makes sense if there are services that will require such network architecture. To elaborate further, some of the services or verticals that are expected to use 6G networks are thus discussed next.

1.2 6G Vertical Industries

Applications and services that 6G will be able to support are normally referred to as vertical industries, or verticals. Whereas older generations of wireless networks mostly supported voice, text video and internet capability, with the increased data rate, 5G and 6G are becoming increasingly capable of supporting many other services. This section looks at the services that will most likely become integral parts of 6G.

The first thing to realize is that although 6G is not going to be only about faster communication, conventional, yet enhanced, communications are nevertheless expected to remain one of the most important services in 6G. However, many new communication paradigms will be introduced, including holographic and tactile communication and communication between new devices, including wearables and implantable sensors. Completely new services will include merging of VR and AR into XR, robotics, autonomous systems including autonomous driving, accurate indoor positioning, expanding IoT into IoE, as well services that belong to smart industries (Industry 4.0 and subsequent Industry 5.0).

1.2.1 Enhanced Conventional Communications

Conventional communications refer to traditional, human-centric communications, with the aid of devices such as phones, tablets and computers [2]. In 6G, these types of communications will be enhanced in many ways. The concept of ubiquitous connectivity has already been mentioned in this chapter; it is worth reiterating that network connectivity will exist in all three dimensions; that is, the network will not only be accessible on the ground but also, for example, in a flying plane, even without on-board Wi-Fi service. Communications will also be available in remote or isolated areas (with network coverage possibly provided via satellite), and there are indications that some forms of communication may even exist underwater. Faster data rates and lower latency figures are implied, but 6G will also aim to reduce the cost of mobile communications in general and to extend the battery life of mobile devices.

1.2.2 New Communication Paradigms

At least two new communication paradigms are expected to be enabled by 6G. The first is the communications of wearable devices, integrated interfaces and implantable sensors, which are sometimes collectively dubbed brain-computer interactions [2, 4]. These devices could be used instead of cellphones, tablets and computers, or as a way to complement them (as smart watches do nowadays). Activation and control of these devices could be established through empathic or haptic communications. For these devices to become commonplace, they will have to be lightweight and should be able to communicate strictly at low power levels and precise frequencies. These devices should have reliable power supplies with prolonged battery lifetime, or they should support wireless charging.

A somewhat different concept is holographic communications [2, 13]. With data rates in excess of 1 Tb/s, the data throughput will be high enough to allow for 3D image projection, for scenarios such as virtual in-person meetings. Projecting the image alone in real-time is not the only requirement. Transmitting the stereo audio and some other sensory information (e.g. tactile) is also important. The feedback to the person or persons being projected and their perception of the holographic communication experience is also important, thus the concept of holographic communication is on the border line of what can be understood as virtual teleportation.

1.2.3 Multisensory Extended Reality Applications

Multisensory XR is envisioned to be a natural point where VR and AR will converge [4]. 6G will allow for a full, immersive XR experience, which can only be achieved if low-latency data transmission is possible. Development and deployment of XR, on its own, will not be straightforward and will require multidisciplinary collaboration between engineering and social disciplines. Involvement of the latter is needed, as XR involves human senses, cognition and psychology.

1.2.4 Robotics and Autonomous Systems

CRAS is a wide term and includes technologies such as drone-delivery systems, autonomous cars and other vehicles, as well as autonomous robotics [4]. Once again, these services require high reliability systems, high bandwidths and data rates, as well as low latency, to perform in a satisfactory manner.

Autonomous vehicles that are in the process of field testing today use numerous interrelated systems, including radars, lidars and cameras, to observe their environment and plot a course of their movement. This is combined with GPS and various types of wireless connectivity in the form of dedicated short-range communications (DSRC) and vehicle-to-everything (V2X) systems, which are used to communicate to other objects, such as other vehicles on the road. In the ideal scenario, many of these technologies could be replaced by the ever-present 6G or be integrated as a part of the 6G network. In addition, 6G AI and machine learning with edge computing will be the ideal tools for processing the information collected via all different means mentioned above [7].

Similar reasoning applies to robotics and other technologies that are under the umbrella of CRAS.

1.2.5 Accurate Indoor Positioning

Needless to say, GPS has brought about a revolution in outdoor positioning of objects. On the other hand, indoor positioning, which also requires the third dimension to be taken into consideration, is still far from perfect.

There are many obstacles in the way of realizing accurate indoor positioning systems, many of which are related to the complexity of electromagnetic propagation indoors in the presence of obstacles [2]. 6G could assist in the realization of this concept by introducing a blend of radio-frequency-type (RF-type), VLC and other types of communication technologies.

1.2.6 Imaging and Sensing

Many applications of imaging and sensing are expected to form an integral part of 6G. Both active and passive sensing approaches may be useful in future wireless networks.

Active sensing in the form of radar could be used for the detection of small objects, as well as in the concepts of automotive safety, autonomous driving and flying (aircraft and drones) [18]. Radar will also be used for gesture recognition, which may become one of the preferred means of capturing user input in 6G. Passive sensing comes in the form of imaging (spectrometry). There are many potential applications of this technique, including exploring material properties and composition, environmental sensing, inspecting mechanical damage in aircraft and other man-made objects, sensing of foreign particles and contaminants in food research, improving safety and security, as well as using sensing in medicine (as an alternative to X-rays). Another application of imaging is in localization, where imaging can assist in pinpointing the location of an object to centimeter-level accuracy [19].

1.2.7 Internet of Things, Internet of Everything and Smart Cities

The IoT and the related concept of IoE have been identified as playing the main role in achieving global sustainable development goals (SDGs) and improving the quality of life [1, 7, 13, 20]. The IoT has evolved from the concept of radio frequency identification (RFID) and is seen as the enabler of concepts that can be seen as "all things smart", such as smart cities, smart homes and buildings, smart healthcare (e-health), smart environmental monitoring, smart businesses and smart industries.

In smart homes and buildings, the IoT connects multiple sensors, which can be utilized, for example, to optimize energy consumption or reduce the carbon footprint. In smart cities, smart energy distribution systems can be complemented by smart transportation, building automation, environmental monitoring and/or security and surveillance. In smart healthcare, IoT solutions can be deployed to monitor various health indicators, such as blood pressure, glucose levels, the heart rate and temperature of individuals, and automatically report these medical data to the relevant health department if needed. In smart businesses and industries, IoT can help with logistics, tracking of goods, safety, quality management and operational efficiency. This is not an exhaustive list by any means; a more detailed list will be presented in Chap. 10. One should also note that new services are added to the IoT regularly.

Once again, in 6G, the 3D-distributed 6G networks will allow for faster information flow, while edge computing will become paramount to make distributed decisions on the larger sensor arrays.

1.2.8 Industry 4.0 and Industry 5.0

Last among many verticals that 6G will enable are those related to the fourth and fifth industrial revolutions (Industry 4.0 and Industry 5.0 respectively). These concepts are closely related to, if not enabled by, IoT.

At present, the world is on the brink of Industry 4.0 and it appears that it is moving from one industrial revolution to another faster than ever before. The first industrial revolution, which started around the mid-18th century, was brought about by the ability to use water and steam power for mechanizing production [21]. The second industrial revolution started in the mid-19th century and is associated with electric power, which allowed mass production. The third industrial revolution started with the invention of the transistor, which triggered the use of electronics and information technology to automate production. It is in the era of the third industrial revolution that wireless communications came into existence; the 1G, 2G, 3G and even 4G networks were made possible by this revolution.

As we transition from the third to the fourth industrial revolution, we are seeing the shift from 4G to 5G, but the roles are being exchanged. Whereas the third industrial revolution brought inexpensive wireless communications into existence, 5G is expected to promote the development of Industry 4.0 instead. It is predicted that the transition from 5G to 6G will promote Industry 4.0 further and subsequently its transition to Industry 5.0. In fact, Industry 4.0 is the first industrial revolution that was actually announced in advance, before being developed [22]. The interrelation between various wireless networks and the evolution of industry is illustrated Fig. 1.6.

While the third industrial revolution introduced the so-called cyber systems into the picture, the physical and the cyber world remained separate. The paradigm shift brought about by Industry 4.0 brings cyber and physical systems together into what can be considered cyber-physical systems. Some examples include interconnecting machines, robotics and human intervention into agriculture and high-precision manufacturing [2, 13]. In many automatic and high-precision manufacturing processes, human intervention has already been reduced to negligible levels. Better integration

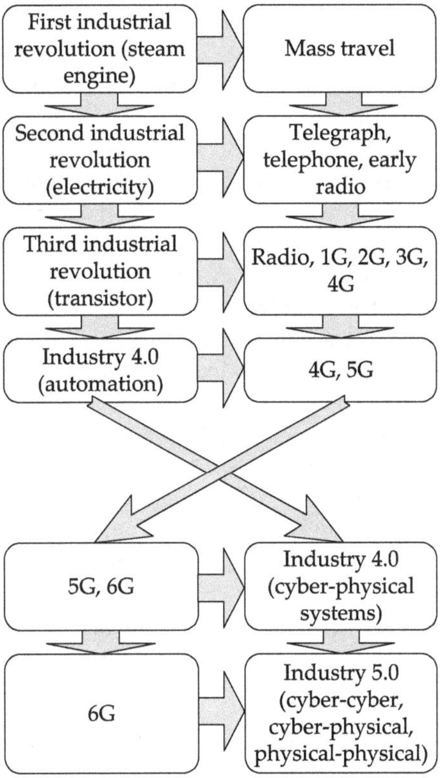

Fig. 1.6 Relation between industrial evolution and the evolution of communication technologies

of the processes will be promoted by the low latency and high reliability that 6G improves upon from previous wireless network generations, potentially marking the start of Industry 5.0. In fact, 6G and Industry 5.0 could bring about a co-existence of systems, including cyber-cyber, cyber-physical and physical-physical systems.

1.3 Technologies Enabling 6G

Many technologies that previous generations of wireless networks utilized will remain in 5G. However, at least two technologies can be considered to become the true enablers of 6G. These are communication utilizing extra-high frequencies (EHF), also termed millimeter-wave frequencies, which are frequencies above 30 GHz, and terahertz frequencies, which are frequencies above 300 GHz, on the one side, and VLC on the other.

1.3.1 Millimeter-Wave and Terahertz Communications

It is believed that millimeter-wave and terahertz communication are the most promising enabling concepts of 6G, offering amounts of bandwidth that would allow transmission of signals at data rates required by 6G. While millimeter-wave frequencies are already being used in 5G, where channels at frequencies as high as 73 GHz are to be available [23], terahertz frequencies have not been used in commercial wireless networks yet and are expected to be piloted in 6G.

In the frequency spectrum, the millimeter-wave and terahertz bands are situated above the RF and microwave bands and below the infrared band in the frequency spectrum [24], as illustrated in Fig. 1.7 [20]. With the frequency increase, the contiguous available bandwidth increases roughly in the same proportion. The largest amount of contiguous bandwidth that is available in millimeter-wave is 36 GHz, which is two orders of magnitude higher than any channel in the part of the spectrum below 6 GHz. Meanwhile, there are channels with over 100 GHz of contiguous bandwidth in the terahertz band, as shown in Table 1.2 [25, 26].

There are several reasons why RF researchers have traditionally been cautious when attempting to utilize frequencies in the order of 10 GHz and above, of which two will be specifically mentioned. The origin of the first reason is the propagation characteristics of waves that are becoming shorter and shorter as the frequency increases. As a result, communication at millimeter-wave and terahertz frequencies is established predominantly with line-of sight (LOS) waves. The second reason is the high attenuation characteristics of channels in these frequency ranges. Attenuation increases with moisture and precipitation, thus limiting the distances over which reliable links can be established, typically to about 1 km for millimeter-wave links [27]. Fortunately, in dry air, there are frequency windows where lower attenuation is experienced, allowing some degrees of freedom in the communication link design [28]. However, these links are usually established with high-power transmitters and extremely directional antennas, both on the transmitting and receiving side. Despite this, it is clear that 6G will have to move from a network of small cells to a network of tiny cells [4]. If it is noted that small cells can be classified into micro cells (range of approximately up to 500 m), pico cells (200 m) and femto cells (100 m) [29], it

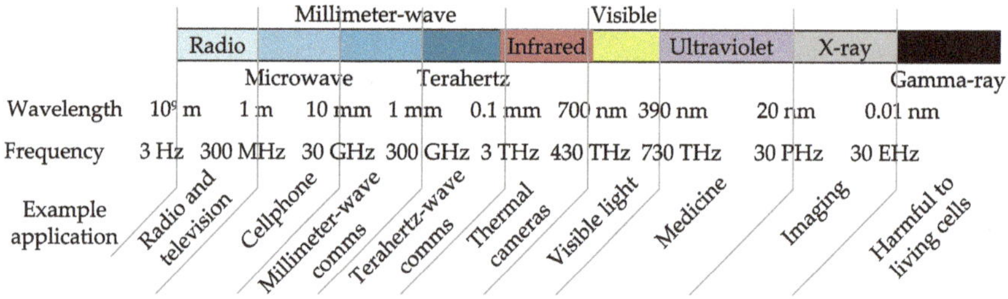

Fig. 1.7 Electromagnetic spectrum and various applications of different frequency ranges. From [20]

Table 1.2 Contiguous bandwidth available in channels in millimeter-wave and terahertz parts of the electromagnetic spectrum

Frequency band	Frequency range	Bandwidth (GHz)
Millimeter-wave (EHF)	31.8–33.4 GHz	1.6
	40.5–43.5 GHz	3
	45.5–47 GHz	1.5
	57–64 GHz	7
	71–76 GHz	5
	81–86 GHz	5
	90–100 GHz	10
	164–200 GHz	36
Submillimeter (terahertz) wave	0.41 THz	65
	0.49 THz	87
	0.66 THz	153
	0.84 THz	142
	0.94 THz	48
	1.03 THz	58
	1.30 THz	38
	1.35 THz	51
	1.49 THz	92
	1.56 THz	29
	1.83 THz	25
	1.98 THz	57

can be deduced that tiny cells are likely to be able to cover only several tens of meters per base station.

Over the last decade, in preparation for 5G and in anticipation of 6G, many research efforts have emerged attempting to solve the link distance effort. Approaches include introducing more powerful transmitters, increasing the sensitivity of receivers, antenna improvements and deploying techniques such as antenna steering and signal repeating [30].

The main advantage of having more powerful transmitters is the improved power amplifier designs, and the receivers are improved with advances in low-noise amplifier (LNA) designs [31]. Research into supporting circuits (oscillators, filters, mixers, etc.) also continues. Most improvements become possible as more advanced transistor technologies are engineered. Technologies that are able to achieve ever increasing frequencies are appearing as a result of technology scaling. The highest frequency operation is achieved with wide-bandgap (WBG) III-V technologies [18], specifically with indium-phosphide (InP) technologies deploying high-electron mobility transistors (HEMTs) or heterojunction bipolar transistors (HBTs), which have reached and surpassed maximum oscillation frequency (f_{max}) values of 1 THz [32]. In addition, gallium-arsenide (GaAs) technologies include metamorphic HEMTs (mHEMTs) that have both f_{max} and the transitional frequency (f_T) of beyond

0.5 THz [33]. The cost of these technologies, however, is prohibiting for mass-scale integrated circuit (IC) development, which remains reserved for silicon-based technologies. Inexpensive complementary metal-oxide semiconductor (CMOS), bipolar CMOS (BiCMOS) and their variations nowadays typically deploy transistors with f_T and f_{max} both between only 200 GHz and 300 GHz. Fortunately, the device size and thus frequency scaling of these technologies continue, with some steam still left, despite indications that Moore's law is soon to come to an end [34]. The fastest reported silicon-based technology at present is a silicon-germanium (SiGe) BiCMOS technology, which uses HBTs with $f_{max} = 0.7$ THz [35]. At least two projects in further silicon-device frequency scaling continue: the European Union TARANTO project [36] and the Defense Advanced Research Projects Agency (DARPA) T-Music project [20]. One of the expected deliverables of the T-Music project is a transistor with $f_T = 750$ GHz. What should be mentioned here is that research into fundamentally different technologies, which could replace traditional transistor-based technologies, also continues. Examples of such technologies include graphene transistors and carbon nanotubes (CNTs) [31].

Millimeter-wave and terahertz antennas must be designed to increase the link between reliability and achievable distance. These antennas have to be highly directional ("pencil beam") to counter the path losses. Fortunately, the gain and directivity of an antenna increase with the square of the frequency for the fixed antenna aperture side. On the other hand, directivity leads to the problem where the transmitter and the receiver have to be almost perfectly aligned for the transmission link to be established in the first place, necessitating the concept of antenna steering to be introduced. Another technique that is used in antenna design is integrating a large number of antennas in both the receiver and the transmitter. These antenna arrays allow for multiple signal propagation paths to be established, leading to the mMIMO concept described earlier. There are some indications that signal coverage could also be promoted by introducing large intelligent surfaces, which alter the wave propagation environment [7], as well as through holographic radio [4].

Questions about which modulation schemes may be suitable for deploying in millimeter-wave and terahertz links have also been raised. It is certain that traditional digital modulation schemes may not be sufficient in these channels. Techniques such as scalable orthogonal frequency division multiplexing (OFDM) and non-orthogonal multiple access were identified at the turn of the decade [14].

6G networks could benefit from another very important concept relating to terahertz frequencies. Terahertz frequencies are ideal frequencies for high-resolution sensing and imaging [18]. The main reason for this is that terahertz (as well as millimeter-wave) radiation is non-ionizing and as such is not harmful to human tissue, as is the case with other types of radiation such as X-ray and gamma-ray radiation shown in Fig. 1.7. The photon energy of terahertz waves is between 0.1 and 12.4 meV, which is more than three orders of magnitude lower than the ionizing photon energy levels [20]. Thus, terahertz waves could be (and are) the vehicle enabling implementation of various sensors, which will become an integral part of 6G networks for applications such as IoE, Industry 4.0, vehicle-to-vehicle (V2V) and V2X communications, and others.

Table 1.3 Advantages and disadvantages of terahertz and visible light communication technologies. Adapted from [6]

Metric (better figure in **bold**)	Terahertz	VLC
Available bandwidth	100 s GHz	**100 s THz**
Propagation mode	**LOS with some non-line of sight (nLOS)**	LOS only
Data rates	**100 Gb/s**	10 Gb/s
Spectrum regulation	Yes	**No**
Signal interreference	High	**None**
Transmission power	High	**Low**
Cost	High	**Low**

1.3.2 Visible Light Communications

It is highly likely that VLC will complement millimeter-wave and terahertz links in 6G networks. VLC are a subset of optical wireless communications (OWC), which also include communication using infrared and ultraviolet communication [6]. OWC, and naturally, VLC, have been made possible with the advent of light-emitting diodes (LEDs) and lasers, which encode the data into optical frequencies [7]. VLC could be established over long distances (e.g. between towers), over short-range links, or even on a micro-scale, e.g. inside transmitter and receiver ICs [31].

VLC has several advantages over radio links. Most important is the fact that VLC can offer ultra-high bandwidths. Secondly, there is no government regulation in the light portion of the spectrum. Other advantages include network security (light can only be received by the receiver at which the light is directed), lack of signal (electromagnetic) interference, low transmitter power and the low cost of deployment, which is due to the relatively inexpensive hardware. Table 1.3 illustrates an attempt to compare the VLC technology with the terahertz technology that was described in the previous section [6]. From this Table 1.3, it appears that VLC has advantages over terahertz communications in most respects, except in the achievable data rate (and the propagation mode, which is self-explanatory). It is, however, expected that data rates of over hundreds of Gb/s or even up to 1 Tb/s would be possible as research into LEDs and LED-related technologies (specifically, micro-LEDs and optical beamforming) continues in the next ten years preceding 6G deployment.

1.3.3 Other 6G Enabling Technologies

Two more communication technologies have recently been attracting the attention of researchers. These are quantum communication and molecular communication [6].

In quantum communication, the data information is encoded using quantum particles (qubits). The main advantage of quantum communication is that information encoded in such a way cannot be tampered with owing to quantum principles, including correlation of entangled particles and inalienable law. The superposition nature of qubits also promotes the high data rates that can be achieved. There are several directions of quantum communication research: quantum teleportation, quantum key distribution, quantum secret sharing and quantum secure direct communication.

Molecular communication would typically be used inside the human body, where it would not be feasible to use radio communication. This type of communication would be established by means of biochemical signals. Carriers of biochemical signals are small particles a few nanometers in diameter, which propagate in aqueous or gaseous medium. Molecular communication consumes very little energy for generation and propagation; however there are still challenges related to the method to interface between the chemical and electrical domains.

1.4 Other 6G Considerations

Several important considerations of 6G networks deserve somewhat more detailed discussion.

1.4.1 Role of Artificial and Ethical Intelligence in 6G

6G is most definitely going beyond a classic mobile internet network and iterative or recursive processes will become inadequate in this context. Instead of these, the network will be empowered by augmented AI [7]. AI and, more specifically, machine learning have already been applied to some concepts in 5G [6]. AI in 5G networks is limited to the optimization of traditional network architecture, but its full potential is not realized because the AI was not included in the network architecture design from the start. There is plenty of time for this mistake to be rectified in 6G.

One of the most powerful techniques in AI at present is deep learning, which is based on a deep neural network (DNN) [7]. While the DNN data for training are readily available, DNN requires the training to be done in a centralized fashion, which may not be feasible in 6G. Instead, AI training in 6G will rather take place at distributed locations (at the network edge), in a process called learning federated learning. If this concept is to succeed, ultra-fast communication with low latency between the central network manager and the distributed clouds becomes extremely important.

AI in 6G will introduce a series of new features, including self-aggregation, context awareness, self-configuration and opportunistic setup, among other benefits [2]. It will also allow for the transformation from cognitive to intelligent radio to take place in 6G. In the latter, the network can in an intelligent way adapt to the conditions of the available hardware, channel and the user density and subsequently choose the protocol that will result in optimum performance [16]. This is also referred to as holistic management of communication [13]. Big data analytics is another major application of AI in 6G and is necessitated by the amount of information that will be available in 6G. There are four types of big data analytics: Descriptive analytics, diagnostic analytics, predictive analytics and prescriptive analytics. Descriptive analytics refers to mining historic data to gain insight into the network performance, channel conditions and traffic. Diagnostic analytics, among other uses, is necessary to detect the faults in the network and improve the network reliability. Predictive analytics refers to using the data to predict future events, including traffic patterns and resource availability. Lastly, prescriptive analytics is used to suggest the decision options for resource allocation, network slicing and virtualization, edge computation, as well as other uses.

One should be aware that the user preference might not always be satisfied with the optimized option produced automatically by AI algorithms [2]. One could thus argue that because of this the free will of human beings will be weakened in 6G. A possible way forward could be to introduce a set of prohibitive clauses, which would ensure that the intelligent services would only be provided within some permissible boundaries.

Permissible boundaries for the AI governing 6G networks should be expanded to be governed with a clear code of ethics [37, 38]. Understanding ethics means understanding the moral agency, or the capacity to make choices and impose those choices on the world.

For a technology that crosses all international and cultural boundaries, developing the code of ethics becomes difficult. Before it can be done, some fundamental questions in ethics must be reconsidered. At least some of these questions are:

- How can a code of ethics for AI be constructed for the global reach?
- How can a code of ethics for AI be constructed so that that it will be largely acceptable internationally?
- How can ethical decision-making and agency be embedded into AI: What ethics is chosen for embedding in the 6G network, given variation in ethics cross- culturally and within different societies?

It is evident that the complexity of ethical intelligence is great and that its evolution will have to be in parallel with the evolution of 5G and 6G networks. It may be false to assume that the people who succeed in developing AI for 6G networks will be equally successful in developing the code of ethics, despite their own intelligence; the effort should involve a much bigger scientific community than just engineers and computer scientists. Fundamentally, a situation where codes of ethics are influenced by a small number of people, potentially consisting only of powerful groups or individuals, should be avoided.

However, the work on AI itself will deal directly with issues that are also of central concern to ethics. Some ethical issues are shared with other rapidly developing technologies. These issues include the problem of prediction, the already mentioned issue of interfacing with social and cultural issues and issues of manipulation of data and the existence of complex systems and responsibility. Thus, the task at hand is actually developing codes of ethics in a situation of uncertainty.

1.4.2 Licensing and Regulation Challenges in 6G

6G necessitates the use of a much wider portion of the electromagnetic spectrum than any of the previous wireless network generations. In fact, 6G will be using all the spectrum allocated to 5G and previous wireless network generations and will add additional spectrum to it. Thus, for 6G to be deployed globally, the spectrum should be properly regulated across multiple countries and territories. In addition, a clear distinction must be made with regard to which portion of the spectrum should become licensed and which portion is to remain unlicensed [3]. The problem arises from the fact that spectrum regulation is done on country level and it may be difficult to synchronize it between counties. Sometimes, there are even different licensing approaches within one country [39]. At present, the biggest issue is standardization of the higher part of the millimeter-wave spectrum and that of terahertz frequencies. Some standardization can, however, be transferred to the likes of the ITU and the Institute of Electrical and Electronic Engineers (IEEE).

Some steps had already been taken by mid-2019, as identified in [6]:

- The IEEE took the first steps in standardization by forming the IEEE 802.15 standardization group to standardize the 250-300 GHz frequency range.
- ITU classified frequencies between 120 and 200 GHz into wireless communication.

On the national or regional level, the following has been done to date:

- The EU research group Horizon 2020 has funded several projects under "Networking research beyond 5G".
- The United States of America (USA) Federal Communication Commission (FCC) started experimental licensing of the part of the spectrum between 95 GHz and 3 THZ.
- Japan has also acted by establishing a research and development program aimed at terahertz frequency bands.

1.4.3 Privacy and Security Concerns

With the large amount of data that will be exchanged (uploaded, cached and transmitted) in 6G, privacy concerns of 6G networks increase [10, 13]. Privacy is maintained by imposing a certain level of security on the network, while the secrecy of information can also contribute to this cause.

In 6G networks, new technologies and techniques will need to be deployed to enhance the privacy, secrecy and security of the users [2]. The utilization of the available AI algorithms would be one means of detecting network anomalies and serve as an early warning mechanism in enhancement of security efforts. 6G should also use more advanced encryption methods, possibly involving quantum key distribution. Furthermore, blockchain technology has been suggested as a way of realizing anonymization and untraceability of users in order to avoid leakage of information. 6G networks should also be structured in such way that user privacy is not sacrificed for the sake of higher network intelligence. Lastly, secure channel coding and channel-based adaptation should be done on the physical level [10].

It should be noted that VLC is also considered to be more secure than electromagnetic types of communication owing to the nature of transmission, as discussed previously.

1.4.4 Energy Harvesting

Energy harvesting is another important concept in 6G. With the increased number of mobile battery-powered devices, all having to transmit high amounts of data and possibly getting involved in AI data processing [7], energy consumption is going to increase and charging of these devices will cease to be trivial. In addition, particular difficulties could be encountered if one has to charge body implants and wearable devices. This is why energy harvesting has been proposed to complement or even replace traditional electrical-socket type charging of batteries.

There are several approaches to energy harvesting [2, 3]. In one of the approaches, it is renewable energy that is harvested. For example, kinetic or vibratory energy (from walking or exercising) could be harvested for wearables, or solar cells could be utilized where possible. In another approach, devices could be charged via radio waves or laser beams originating from base stations, or simply by harvesting the ambient RF energy.

As new materials and surfaces emerge, new ways of energy harvesting may also be found [6].

1.4.5 Non-terrestrial Communications

In order to achieve ubiquitous 3D connectivity, terrestrial network access points will have to be complemented with non-terrestrial access points [2]. Non-terrestrial communications include airborne and satellite communications, as well as underwater communications.

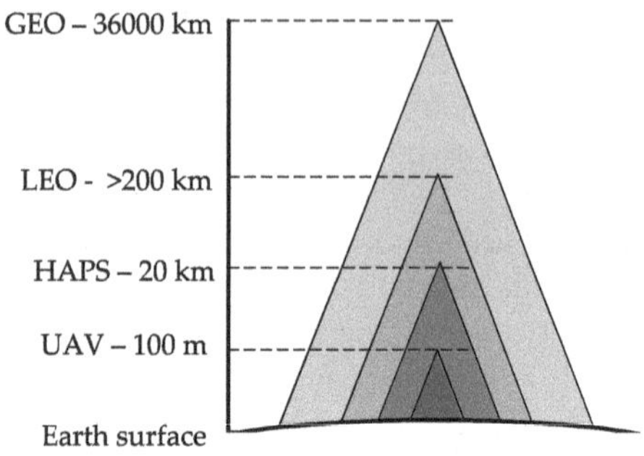

Fig. 1.8 Non-terrestrial base stations—airborne and satellite [40]

Airborne and satellite communication nodes would be used for 3D coverage, and could be deployed at four different altitudes, as illustrated in Fig. 1.8 [40]. Lower base stations would comprise unmanned aerial vehicles (UAVs). UAVs or drones, at present, require a lot of propulsion energy to hover, which is far from being practical; however, they could be deployed on demand, for example in emergency situations. Somewhat more feasible are high-altitude platform stations, which would be stationed in the stratosphere, at an altitude of about 20 km. These could cover an area of hundreds of square kilometers but may face challenges with refueling and stabilization in the air. Finally, two types of satellites are considered in non-terrestrial communications: low earth orbit (LEO) and geostationary earth orbit (GEO) satellites. LEO satellites would orbit at between 200 and 2 000 km and would have lower propagation delay than GEO satellites, but would be non-stationary relative to the earth's surface. On the other hand, GEO, orbiting at 36 000 km, would be stationary relative to the earth's surface and although they can cover a much greater area than LEO satellites, an observable signal propagation delay cannot be avoided. For areal/satellite networks to become a reality, architecture, spectrum, transmission protocol and antenna advancements will be needed, among other requirements.

It may be even more difficult to establish reliable underwater communications [2, 26]. It is a well-known fact that RF signals do not propagate well under water, reaching no further than 10 m, even in the best systems. Two alternatives to electromagnetic communication are communication using acoustic waves and optical communication. Acoustic waves can cover larger distances than electromagnetic waves under water (up to 100 m) but they may not be a good option for 6G because the data rates that can be established are in the region of kb/s. Underwater optical wireless communications (UOWC), which can reach speeds in Tb/s, seem to be a much better alternative. Like terrestrial VLC, UOWC offer low latency figures. Underwater networks, naturally, need to connect to a terrestrial network backbone, and it is most likely that this will be achieved by a combination of surface and areal networks.

1.5 6G Research Directions in the Next Decade

At this point in time, 6G networks can be seen as a great opportunity for humanity. However, it clear that a lot of research is needed to bridge the gap between the reality of current communication systems and the vision of communication systems about a decade from now. According to the Centre for Wireless Communications from Oulu, Finland [9], in order to meet the expected 6G deployment date set for 2030, research should follow the route detailed below. At present (early 2020s),

- Research should focus on further development of vertical solutions for 5G and B5G, which runs in parallel commercial 5G deployment;
- Research on smart networks should start; and
- Research on satellite communication should be expanded.

In three years,

- The 6G community should start collecting feedback from the 5G community.

In five years (2025):

- Research on B5G should reach its peak;
- Optimization of integration of terrestrial 5G and satellite networks should be performed;
- 6G research should increase rapidly;
- Core technologies for 6G should be selected and tested;
- Different 6G algorithms for signal processing and AI and machine learning should be evaluated and compared; and
- Integration work towards experimental 6G should start.

Finally, in ten years,

- Optimized deployment of smart networks should be performed;
- Partial integration of terrestrial and satellite networks should be completed;
- Significant research into other types of communications, such as VLC, quantum communication and molecular communication, should be under way;
- 5G/B5G should reach its full coverage; and finally
- Deployment of commercial 6G networks should start.

This forecast of the 6G future is the vision of the Finnish 6G Flagship program and thus applies in the context of Finland alone. However, it clearly corresponds to the 6G deployment roadmap shown in Fig. 1.5.

1.6 Contribution of This Book to the Body of Knowledge

As seen in this chapter, millimeter-waves are considered one of the most important enabling technologies for 6G. Thus, this book builds on several texts that extensively explored millimeter-waves, as well as terahertz waves to some extent. References [27, 30, 31, 41, 42] detail millimeter-wave power amplifiers, millimeter-wave LNAs, antennas, system-level packaging for millimeter-wave transceivers and methodologies for research, design and innovation in millimeter-wave integrated circuits, respectively.

1.7 Outline of the Book

This chapter presented the introduction to the topic of research of this book and placed this chapter and the rest of the book into the context of recent 5G and 6G research efforts. 6G vision, as well as many technical aspects of 6G, was briefly introduced in this chapter.

The focus of Chap. 2 will be the history of wireless network generations, with special focus on 4G and even more on 5G, shortcomings of 5G and predictions in terms of where 6G can help rectify the shortcomings of previous wireless network generations. Historical trends will be used to justify and predict the future of the telecom "explosion".

In Chap. 3, channel coding theory and channel capacity will be discussed as a way of elaborating why a higher frequency band or other communication paradigms have to be explored in 5G and 6G. Subsequent discussions will cover millimeter-wave frequencies, terahertz frequencies and modulation.

Chapter 4 will look into active device (transistor) technologies that could enable the expansion of communications to channels with higher center frequencies. Some of the basic millimeter-wave and terahertz circuits are also discussed in this chapter. Lastly, this chapter discusses the systems approach to packaging and package optimization for better performance of future wireless network circuits.

Chapter 5 will look into VLC as the main communication technology alternative to millimeter-wave and terahertz radio for establishing ultra-fast data links in 6G. The main advantages and shortcomings of VLC will be discussed in detail, together with the physical aspects of VLC and most likely use cases. Whereas VLC technology is nowhere near as mature as radio, it is potentially much better in various scenarios, which include V2X communications, underwater communication, or communications where low interference or high security is needed.

In Chap. 6, the concept of 6G as the "green network" will be discussed. This, for the large part, involves introduction of energy harvesting into the network design, where mobile network users use the energy harvested from light, movement, ambient RF or other sources as a source of power, with or without the aid of batteries and supercapacitors. Furthermore, with many techniques for energy saving that have been investigated and are being introduced in 5G networks, specifically on the side of base stations, ongoing research for even better energy saving and conservation techniques for 6G networks is discussed.

Chapter 7 will cover some of the advanced, possibly even futuristic, technological aspects that are expected to be unique to 6G. Advanced aspects include ultra-massive MIMO (umMIMO), beamforming and channel control with the programmable (reflective) metasurfaces. Futuristic concepts include holographic radio, quantum and molecular communication, aerial networks established with the help of drones and satellites, as well as the use of blockchain technologies in network security.

Chapter 8 considers another set of advanced topics, this time related to intelligence within the 6G network design. Concepts such as pervasive AI, machine learning and edge intelligence are discussed here. This chapter also discusses the architecture that 6G should take to make full use of AI principles, as well as the controversial topic of ethics in AI.

Chapter 9 discusses the health issues associated with underlying technologies of 5G and 6G. This chapter is mostly dedicated to the investigation of the influence of the short-term and the long-term influence of different types of electromagnetic radiation on various human health aspects. Both the physics of wave propagation and medical research studies are discussed. Potential health issues associated with VLC are also discussed in this chapter.

Chapter 10 concludes the book by discussing 6G as a fusion of communications, imaging, sensing, localization as well as many other 6G verticals. These include the concepts of robotics, autonomous systems, V2X, IoT, IoE, Industry 4.0, XR and others.

References

1. Heath Jr RW (2019) Going toward 6G. IEEE Signal Process Mag 3–4
2. Dang S, Amin O, Shihada B, Alouini M-S (2020) From a human-centric perspective: what might 6G be? Nat Electron 3:20–9
3. David K, Berndt H (2018) 6G Vision and Requirements: Is There Any Need for Beyond 5G? IEEE Veh Technol Mag 13:72–80
4. Saad W, Bennis M, Chen M (2019) A vision of 6G wireless systems: applications, trends, technologies, and open research problems. IEEE Network, 1–9
5. Ten BQ (2019) Trends in the cellular industry and an outlook on 6G. IEEE Commun Mag 57:31–6
6. Huang T, Yang W, Wu J, Ma J, Zhang X, Zhang D (2019) A survey on green 6G network: architecture and technologies. IEEE Access 7:175758–68
7. Tariq F, Khandaker M, Wong K-K, Imran M, Bennis M, Debbah M A speculative study on 6G. IEEE Wireless Commun 27:118–25
8. Zong B, Fan C, Wang X, Duan X, Wang B, Wang J (2019) 6G technologies: key drivers, core requirements, system architectures, and enabling technologies. IEEE Veh Technol Mag 14:18–27
9. Katz M, Pirinen P, Posti H (2019) Towards 6G: getting ready for the next decade. In: 2019 16th international symposium on wireless communication systems (ISWCS), pp 714–8

10. Gui G, Liu M, Kato N, Adachi F, Tang F (2020) 6G: opening new horizons for integration of comfort, security and intelligence. 2020 [cited 2020 Feb 16]; Available from: https://www.techrxiv.org/articles/6G_Opening_New_Horizons_for_Integration_of_Comfort_Security_and_Intelligence/11634669
11. Chen Y, Zhu P, He G, Yan X, Baligh H, Wu J (2020) From connected people, connected things, to connected intelligence. In: 2020 2nd 6G Wireless Summit (6G SUMMIT). 2020. p. 1–7
12. Latva-aho M, Lppanen K (eds) Key drivers and research challenges for 6G ubiquitous wireless intelligence (white paper). 6G Vissions 1. Oulu: Univerisity of Oulu
13. Calvanese Strinati E, Barbarossa S, Gonzalez-Jimenez JL, Ktenas D, Cassiau N, Maret L et al (2019) 6G: the next frontier: from holographic messaging to artificial intelligence using subterahertz and visible light communication. IEEE Veh Technol Mag 14:42–50
14. Zhang L, Liang Y-C, Niyato D (2019) 6G Visions: mobile ultra-broadband, super internet-of-things, and artificial intelligence. China Communications. 16:1–14
15. Sadhu B, Rexberg L (2019) Phased arrays for 5G millimeter-wave communications. Millimeter-wave circuits 5G radar. Cambridge University Press, Cambridge, p 243
16. Letaief KB, Chen W, Shi Y, Zhang J, Zhang Y-JA (2019) The roadmap to 6G: AI empowered wireless networks. IEEE Commun Mag 57:84–90
17. David K, Elmirghani J, Haas H, You X-H (2019) Defining 6G: challenges and opportunities [from the guest editors]. IEEE Veh Technol Mag 14:14–6
18. Božanić M, Sinha S (2019) Emerging transistor technologies capable of terahertz amplification: a way to re-engineer terahertz radar sensors. Sensors 19:2454
19. Sarieddeen H, Saeed N, Al-Naffouri TY, Alouini M-S (2020) Next generation terahertz communications: a rendezvous of sensing, imaging, and localization. IEEE Commun Mag 58:69–75
20. Rappaport TS, Xing Y, Kanhere O, Ju S, Madanayake A, Mandal S et al (2019) Wireless communications and applications above 100 ghz: opportunities and challenges for 6G and beyond. IEEE Access 7:78729–57
21. Schwab S (2020) The fourth industrial revolution: what it means and how to respond [Internet]. World Economic Forum. [cited 2020 Mar 20]. Available from: https://www.weforum.org/agenda/2016/01/the-fourth-industrial-revolution-what-it-means-and-how-to-respond/
22. Rojko A (2017) Industry 4.0 concept: background and overview. Int J Interact Mobile Technol (iJIM) 11:77–90
23. Abbas T, Qamar F, Ahmed I, Dimyati K, Majed MB (2017) Propagation channel characterization for 28 and 73 GHz millimeter-wave 5G frequency band. In: 2017 IEEE 15th student conference on research and development (SCOReD), pp 297–302
24. Božanić M, Sinha S (2020) Methodologies for design of millimeter-wave and terahertz integrated circuits: an LNA case study. In: 2020 30th international conference radioelektronika (RADIOELEKTRONIKA). Bratislava, pp 1–6
25. Mumtaz S, Jornet JM, Aulin J, Gerstacker WH, Dong X, Ai B (2017) Terahertz communication for vehicular networks. IEEE Trans Veh Technol 66:5617–5625
26. Huang K, Wang Z (2011) Terahertz Terabit wireless communication. IEEE Microwave Mag 12:108–16
27. Božanić M, Sinha S (2017) Millimeter-wave low noise amplifiers. Springer, Cham
28. Kleine-Ostmann T, Nagatsuma T (2011) A review on terahertz communications research. J Infrared Milli Terahz Waves. 32:143–71
29. Abrol A, Jha RK (2016) Power optimization in 5G Networks: a step towards GrEEn communication. IEEE Access. 4:1355–74
30. du Preez J, Sinha S (2016) Millimeter-wave antennas: configurations and applications. Springer, Cham
31. Božanić M, Sinha S (2020) Millimeter-wave integrated circuits: methodologies for research. Design and Innovation, Springer Nature
32. Urteaga M, Griffith Z, Seo M, Hacker J, Rodwell MJW (2017) InP HBT technologies for thz integrated circuits. Proc IEEE 105:1051–67

33. Samoska LA (2011) An overview of solid-state integrated circuit amplifiers in the submillimeter-wave and thz regime. IEEE Trans Terahertz Sci Technol 1:9–24
34. Lambrechts W, Sinha S, Abdallah JA, Prinsloo J (2018) Extending Moore's law through advanced semiconductor design and processing techniques. CRC Press, Boca Raton
35. Pawlak A, Lehmann S, Sakalas P, Krause J, Aufinger K, Ardouin B, et al (2015) SiGe HBT modeling for mm-wave circuit design. In: 2015 IEEE Bipolar/BiCMOS circuits and technology meeting—BCTM. Boston, MA, pp 149–56
36. Chevalier P, Liebl W, Rucker H, Gauthier A, Manger D, Heinemann B, et al (2019) SiGe BiCMOS current status and future trends in Europe. In: 2018 IEEE BiCMOS and compound semiconductor integrated circuits and technology symposium (BCICTS) [Internet]. IEEE, San Diego, CA [cited 2019 Apr 23], pp 64–71. Available from: https://ieeexplore.ieee.org/document/8550963/
37. Boddington P (2017) Towards a code of ethics for artificial intelligence. Springer
38. Etzioni A, Etzioni O (2017) Incorporating ethics into artificial intelligence. J Ethics 21:403–18
39. Matinmikko-Blue M, Yrjölä S, Ahokangas P (2020) Spectrum management in the 6G era: the role of regulation and spectrum sharing. In: 2020 2nd 6G Wireless Summit (6G SUMMIT), pp 1–5
40. Giordani M, Zorzi M (2019) Non-terrestrial communication in the 6G era: challenges and opportunities. arXiv preprint arXiv:191210226
41. Preez J du, Sinha S (2017) Millimeter-wave power amplifiers. Springer
42. Božanić M, Sinha S (2019) Systems-level packaging for millimeter-wave transceivers. Springer, Cham

Chapter 2
The Past, Present and Future of Telecommunications Expansion: A Historical Perspective

Abstract This chapter will look into the detail of the technologies, new at the time, that made every step in the evolution of communications and later wireless networks possible, the way in which these networks were commercially implemented and deployed and how they evolved over time, as well the services that each step in the evolution enabled. Some of the key aspects that brought about the vision of 6G already introduced in Chap. 1 will be elaborated on, based on these historical findings. Note that 5G networks, although already in the process of deployment at the time of writing of this book, are still expected to evolve over the next ten years and thus some predictions and assumptions regarding 5G will also be made.

In 1926, the great Serbian–American scientist and inventor Nikola Tesla (1891–1943, pictured in Fig. 2.1), in an interview with Colliers Magazines [1], gave his vision of the future. He predicted that the whole earth would be converted into a huge brain, and that people would be able to communicate with one another instantly, irrespective of distance. He went on to say that through television and telephony people would see and hear one another as perfectly as though they were face to face, and that the instruments through which they would be able to do this would be amazingly simple compared with the telephone used in his time. He anticipated that a person would be able to carry such a device in his/her vest pocket.

What he actually predicted was the world of videophones, broadcast TV, and mobile communication [2], as it exists roughly a century later, in the era of 4G and 5G. Today, the wide coverage of mobile wireless networks allows everybody, mostly irrespective of their income, to carry a smartphone "in the vest pocket", thereby forming nodes of the predicted great interconnected brain. One could argue that this prediction classifies Nikola Tesla as one of the best prophets that ever walked the earth.

However, in the 2020s, it is time to predict the future again. To be able to do that, one should first look at the history of communications, starting from before Tesla's first prediction and moving to the more recent history of wireless communication, which started with the introduction of 1G. That would undoubtedly allow for some

M. Božanić and S. Sinha, *Mobile Communication Networks: 5G and a Vision of 6G*, Lecture Notes in Electrical Engineering 751,
https://doi.org/10.1007/978-3-030-69273-5_2

Fig. 2.1 Nikola Tesla in 1896

historical trends to be observed, which could possibly hold in the near future as well. Something else is possible when it comes to technology, and that is to invent the future instead. This is another legacy left to humanity by Tesla, the great inventor, often considered the person who invented the twentieth century. Using history to predict the future is the theme of this chapter. This chapter will, thus, go to the trouble of looking into the detail of the technologies, new at the time, that made every step in the evolution of communications and later wireless networks possible, the way in which these networks were commercially implemented, deployed and how they evolved over time, as well the services that each step in the evolution enabled. Thereafter, some of the key aspects that brought about the vision of 6G already introduced in Chap. 1 will be elaborated on, based on these historical findings. Note that 5G networks, although already in the process of deployment at the time of writing of this book, are still expected to evolve over the next ten years and thus some predictions and assumptions regarding 5G will also have to be made.

The outline of the history of telecommunications shown in Fig. 2.2 indicates the rough flow of this chapter.

2.1 The First and the Second Industrial Revolutions and the Beginning of Telecommunications

With the first industrial revolution that started in the mid-eighteenth century, which resulted in a faster pace of life, came a need for faster communication with people spaced apart across the world. The word telecommunication, coined in the nineteenth century, means communication across a distance ("tele" = distance) [3].

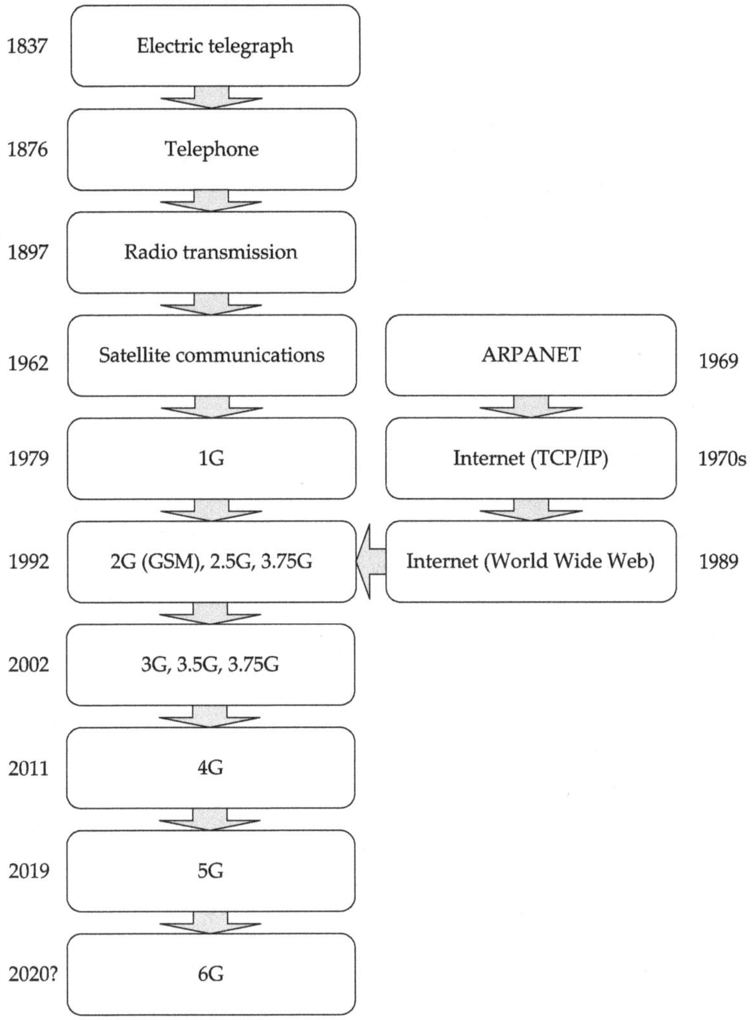

Fig. 2.2 Outline of the history of telecommunications and the rough flow of this chapter

2.1.1 Optical Telegraph

It is believed that the era of telecommunication began in 1794 with the first successful optical telegraph transmission in France [3]. Before electricity, the only feasible way to gather information was to observe what was happening some distance away, at first with the naked eye and later with optical instruments. The optical telegraph relied on observation, but the structure to transmit and observe the information was the main innovation in this concept. For a message to be relayed, somebody had to transmit the signal and somebody had to observe the signal. On the one side, the message would be transmitted using a large mechanical structure consisting of a regulator and two indicators, which could be positioned in a number of configurations, each corresponding to a specific signal in a dictionary. On the other side, an operator would observe the message using a telescope and relay the message further via an own telegraph apparatus. On a relatively flat terrain, a long chain of such "transmitters" and "receivers" could be deployed to form a "telegraph line". Transmitters were usually built on existing buildings, such as churches that already had tall towers. Nevertheless, expanding the optical telegraph network remained a very costly affair and a better way to transmit information was sought.

2.1.2 Electrical Telegraph

Most people are familiar with the electric telegraph concept. Electromagnetism was discovered at the end of the eighteenth century and understood in the nineteenth century, which eventually triggered the second industrial revolution. Although most scientists involved in early electromagnetism experimentations were from Europe, the era of electric telegraphy started almost simultaneously in Great Britain and the USA around 1837.

The easiest way to describe the telegraph is using the concept of the telegraph demonstrated by Carl von Steinheil, illustrated in Fig. 2.3. The telegraph signal is transmitted between station A and station B as a series of current pulses via a copper wire, with the ground as the return conductor. If a telegraph key is pressed at station A, a current will be generated and will shift the telegraph key in Station B, and vice versa. The electric telegraph was improved significantly by the painter and inventor Samuel Morse, who introduced an efficient way of encoding alphabet characters, numbers and punctuation into a series of symbols (dots and dashes), now known as the original Morse code. He also invented a way to print the received messages automatically on

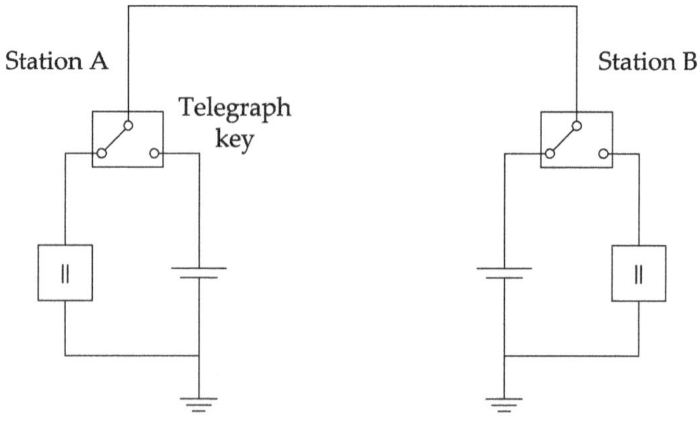

Fig. 2.3 Electric telegraph of Carl August von Steinheil

the side of the receiver. Later, a second telegraph wire was introduced so that signals could be transmitted in both directions simultaneously.

One of the limitations of the electrical telegraph, similar to the optical telegraph, was that messages had to be relayed to cross large distances. This was not possible when crossing large bodies of water via freshly introduced submarine cables. At that point in history, the USA and Europe greatly relied on each other and it was important to find a way to send messages across the Atlantic Ocean. For this, a different approach was needed. This involved transmitting stronger signals across the cable—a signal from a voltage source of 50 V would result in a signal of only a few millivolts on the receiver side. High voltage required the use of large batteries, requiring dots and dashes to be replaced by a system where negative and positive voltages were used. The resultant small positive and negative voltage pulses received on the other side of the telegraph line were used to drive the galvanometer coil to which a pen was mounted either up or down on a moving tape, allowing for the transmitted message to be recorded on paper.

2.1.3 Telephone

The next step in the evolution of telecommunications was the invention of the telephone. Strictly speaking, the first telephone was a telegraph that was adapted for transmission of voice in two directions over the same line. Its invention is attributed to the Scottish-American inventor Alexander Bell around 1876. In his patent, different telegraph instruments could exist on the same line because they could be tuned to different resonant frequencies.

The adoption of the telephone was much quicker in the USA than in Europe, with just under 2000 telephones operating in Europe and just under 50,000 in the USA in 1880. The first telephone communication in Europe was between phones installed in Switzerland and Italy. By 1900, there were 2.2 million telephones operating around the globe, as shown in Table 2.1. Originally, the telephone used the existing telegraph wire network, but it was quickly realized that a separate telephone network had to be established.

The telephone network, like the telegraph network, required switching between subscribers for a call to a specific location to be made. Original telephone exchanges deployed manual switching, with operators, usually women, manually connecting different lines by cables. With the increase in the number of telephone subscribers

Table 2.1 Beginning of telephony: number of subscribers in various regions in the world [3]

Year	Europe	USA	Rest of the world	Total
1880	1900	47,900	–	49,800
1890	227,000	177,000	31,500	435,500
1900	1,355,000	800,000	100,000	22,550,000

across the world, manual switching became complex and the idea of automatic switching came into existence. Before the end of the nineteenth century, most telephone exchanges introduced direct step-by-step systems, where the dialing telephone sent impulses that directed the selector at the telephone exchange to step to the connection leading to the desired subscriber identified by a number. In the first half of the twentieth century, most of the systems were electromechanical systems, evolving from indirect-control systems to crossbar switching systems. In the second half of the last century, these systems were first replaced by fully electrical switching systems, and finally by digital systems still in use for fixed-line telephony today.

Two inventions increased the distances that could be reached by telephone cables. The first was the Pupin coil, named after another Serbian–American, Mihajlo Pupin. The coil was used to introduce an inductive load into the telephone line with the purpose of decreasing the distortion in long transmission lines. The second contribution was the invention of various methods for signal amplification, first by triode vacuum tubes (from 1907) and later by transistors (from 1960s).

2.2 Radio Communications

Perhaps the most radical point in the history of telecommunication was the introduction of radio communications. With the invention of radio, for the first time, communication could be established not only by two parties at two stationary locations, but also by parties in motion. Radio also made communication on water (ships) and in the air (aircraft) possible.

2.2.1 Demonstration of Electromagnetism and Early Radio

The very first steps toward radio communications can be traced to the early nineteenth century when the first attempts to describe the theory of electromagnetic waves were made. While British scientist Michael Faraday envisaged the existence of lines of magnetic force and thus challenged existing theories that magnetism acts instantaneously across a distance, Scottish physicist James Maxwell formulated the theory that electromagnetic waves behave in the same way as light does, and thus have the same set of characteristics as light, including the speed of propagation, polarization, refraction and reflection. The only difference between light and electromagnetic waves was narrowed down to their different wavelengths. The Faraday-Maxwell theory was later experimentally proven by German scientist Heinrich Hertz. These discoveries were still far away from practical solutions for radio communication, or radio telegraph, as the technology was called at that time. The first attempts to implement radio practically came shortly afterwards, most notably from the British scientist Oliver Lodge and the Russian scientist Alexander Popov. The aim was clear:

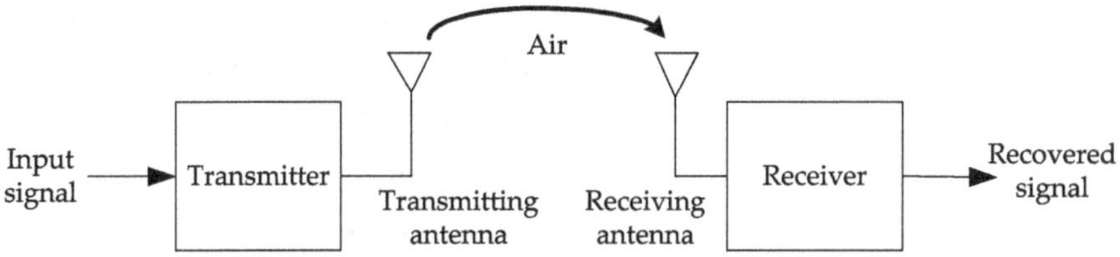

Fig. 2.4 The radio concept

to replace the wire conductor between two stations, called the transmitter and the receiver, with the air as the transmission medium, as shown in Fig. 2.4.

History, however, remembers one person as the person who invented the radio: Guglielmo Marconi. The Italian inventor is, indeed, most famous for being granted the British patent for inventing this technology in 1897 [4] and some other patents later. It is, however, less known that the radio was, in fact, invented by Nikola Tesla himself, who started experimenting with radio waves in 1892 and demonstrated the remote-controlled boat in 1898. He also submitted a basic patent demonstrating this technology in 1897 [5]. The origin of Marconi's wide recognition, even in the text by Huuderman [3], on which several sections of this chapter are heavily based, lies in the fact that he collected several related ideas from a number of scientists and was able to commercialize them successfully [6]. Tesla disputed his achievements and in 1943, the United States Supreme Court recognized Nikola Tesla as the true inventor of radio, which formed the basis of overturning many of Marconi's patents. This recognition came too late, seeing that the name of Marconi was already recorded in history books, and Tesla was already deceased by that time.

2.2.2 Evolution of Radio

The discovery of the triode not only had an influence on telephony, as already stated, but also on radiotelegraphy [3]. In 1912, it was discovered that triodes could equally well be used to generate continuous electromagnetic wafers, as they were capable of detecting and amplifying signals. In other words, an oscillator circuit was invented. World War I held up further development of radio, but in the years between the two world wars, radio penetration occurred widely into ocean ship communication, with the improvements in vacuum tubes in turn incrementally improving upon the radio signal quality and reliability. This was achieved by increasing the transmitter output power and receiver sensitivity.

In the 1920s, shortwave radio systems substantially decreased the cost of building radio stations, while the long-wave radio, utilizing antennas several tens of meters long, was seen as the solution for intercontinental communication. Because of atmospheric, ionospheric and sunspot conditions, both types of radio transmissions suffered disruptions or downtimes during a portion of the day, which became more prominent with increasing distances between radio stations. This problem was solved, on land, with the introduction of radio-relay technology, developed in 1934. In the radio-relay era, radio signals could be received, amplified and retransmitted, covering hundreds or even thousands of kilometers in distance, using relay stations that were, depending on the ground topology, spaced about 50 km apart.

The invention of a portable radiotelephone came in 1943, when Motorola introduced the Handy Walkie, which weighed 16 kg. Although difficult to carry around in one's hand, such a system could be installed in vehicles, including cars and trains. The first commercial mobile telephone network started operating in Missouri in 1946 and utilized the 150 MHz band. Other mobile radio networks started operating in the world in the 1950s and 1960s, but were mostly limited to Europe and North America. Operation of these networks was in manual mode, meaning that the operator had to connect to radio lines or a radio and a fixed line together at the exchange.

The further evolution of the radio was triggered by the invention of semiconductors, specifically transistors, which would later replace the vacuum tubes in amplifiers and oscillators. The rise of solid-state electronics triggered the third industrial revolution. The first transistor that was invented was the point-contact transistor, built in 1948. The invention of a metal-oxide semiconductor (MOS) transistor (1960) had to wait for the discovery of single-crystal silicon. Integrated circuits were first reported in 1958; soon it was discovered that the speed of operation and the number of devices could be doubled every 12–18 months, allowing Gordon Moore to coin the famous Moore's law in 1965 [7].

2.2.3 Satellite Communications

The era of satellite communications started in the 1960s [3]. Soon after World War II, amid the Cold War between the West and the East, the space race began. While both the USA and the Soviet Union had the end goal of sending humans into space, they had to start by developing rockets that could launch man-made objects into space. This speeded up the deployment of early satellites. The first-ever satellite to be launched into orbit was the Russian *Sputnik I*, which sent space telemetry for three weeks in October 1957 before losing orbit and burning up in the atmosphere of the earth.

Soon, satellites were identified as an opportunity to establish transcontinental communication by offering a means for relaying multichannel two-way transmission of telephony, data, and television signals between two stations on earth that did not have a LOS. Before satellites, the only intercontinental connections were made with the aid of maritime telegraph and telephone cables. The first commercial satellite was *Telstar I*, launched in 1962. By 1965, there were three commercial satellites in the earth's orbit, and that number had risen to over 70,000 by 1990 (Table 2.2). A satellite remains in orbit owing to the precise balance of the gravitational attraction of earth and other bodies (such as the moon) and the centrifugal force due to the rotation of the satellite. Telecommunication satellites use four different orbits: LEO,

Table 2.2 Number of satellites in orbit from 1965 to 1990 [3]

Year	1965	1970	1975	1980	1985	1990
No. satellites	3	45	123	2995	8996	76,586

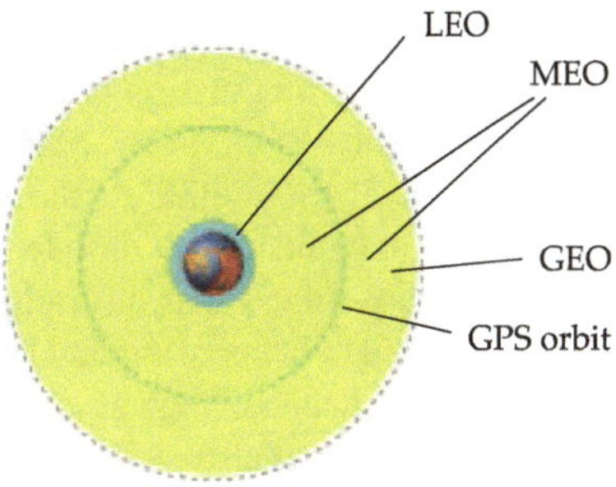

Fig. 2.5 Different satellite orbits (not to scale). Adapted from a free image

medium earth orbit (MEO), GEO and highly elliptical orbit (HEO). Satellites in LEO are launched to an altitude between 500 and 2000 km and satellites in MEO to an altitude between 5000 and 15,000 km. For telecommunications, however, the most important satellites are those in the GEO, which lies at exactly 35,786 km above the surface of the earth. This allows the transmitters and receivers on earth always to point at the same spot in the sky and for the single satellite to be visible at any point in time, which is not the case with other satellite orbits. This orbit is also called the Clarke orbit. Lastly, HEO satellites are launched above the altitude associated with GEO. Different orbits are shown in Fig. 2.5. This figure also shows the orbit of GPS satellites, which circle the earth at an altitude of about 20,000 km.

The unique advantage of satellite communications lies in the fact that even the remotest of areas can be reached by satellite, whether on land, ocean or in the air. On the other hand, satellites offer limited amounts of bandwidth and it thus remains necessary to continue to expand terrestrial communication networks, which can provide much higher bandwidths, especially nowadays when fiber-optic network technologies are available (Sect. 2.5.1).

2.3 Beginning of Cellular Telecommunications

The era of cellular telecommunications started with 1G in the 1980s in most of the developed world. Prior to 1G, only a number of isolated mobile networks existed, mostly available in certain urban areas in Europe and North America. These networks are sometimes referred to as 0G.

2.3.1 Mobile Telecommunications Prior to 1G

By the 1970s, telecommunications had evolved to a point where a very comprehensive landline telephone network existed, with mobile networks already in existence for about 25 years. However, the mobile networks of that time were scattered in urban pockets around the world, were voice-centric and used simple analog systems with single-side-band modulation [3, 8]. Other mobile radio networks were push-to-talk systems, used by the military, police and a limited number of private users. At that time, whenever the mobile connection was established, a portion of the spectrum had to be allocated to the mobile user. With tens or even hundreds of thousands of mobile subscribers sharing a small portion of the spectrum in densely populated areas, it started to become evident that a radical change in how mobile networks were organized was needed. Such a network not only needed to introduce better utilization of the spectrum, but also to standardize the coexistence of land and mobile systems.

The solution was presented in 1979, which marks the year in which the first generation of cellular telephone networks (1G) was released. At that point, 2G was still far away and the term 1G had not been coined yet, so the network was simply known as cellular radio. The official name later adopted in USA was Advanced Mobile Phone Service (AMPS).

2.3.2 1G

The first cellular 1G network ever deployed was the so-called Mobile Control Station network in Japan in 1979. This network was still a voice-only network, and it operated at frequencies around 850 MHz. Note the sudden change in terminology in this section (from "mobile" to "cellular")—while mobile networks existed prior to 1979, the radical invention was the cellular concept (thus, all cellular networks are also mobile networks, but not all mobile networks are cellular networks). Wide area network coverage was established by seamless overlapping cells in a honeycomb configuration, something like what is illustrated in Fig. 2.6. A mobile user would connect to the public-switching telephone network (PSTN) via the closest cell tower.

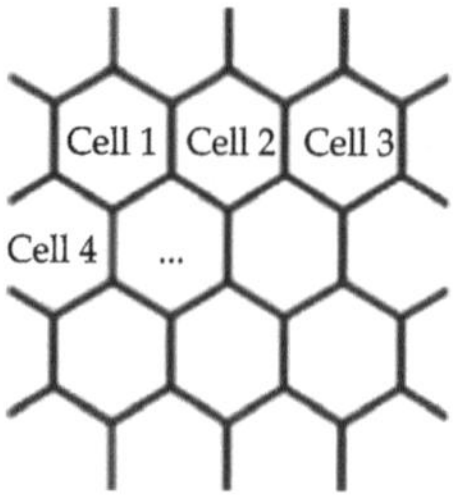

Fig. 2.6 Network cells in honeycomb configuration

This technology concept was radical at the time, and in the early 1980s, many other countries deployed networks operating on a similar principle [9]:

- Nordic Mobile Telephone (NMT) started operating in Nordic countries.
- C-Netz (C-network) started operating in Germany, Portugal and South Africa.
- Total Access Communications System (TACS) started operating in the United Kingdom.
- AMPS, as already mentioned, started operating in the USA and other countries in the Americas.

Although all networks mentioned above used the cellular principle, the technology used for each of the networks was different. This means that specific equipment could only be used in a specific country—i.e. no network roaming was possible [10]. The lack of network standard also meant that signals had low transmission efficiency and no security [11]. The rate of transmission of information (in this case, voice only) was 2.4 kb/s, frequency-division multiple access (FDMA) coding was used to access the spectrum, and transmission channels had 60 kHz of bandwidth [12]. Whenever a call was activated, a portion of the spectrum was allocated to a user, becoming unavailable to other users in the network. This meant that network utilization was based on single-input, single-output architecture (SISO) [13]. Some of the key characteristics of the 1G network are shown in Table 2.3 [13].

The lack of standardization of networks across the world triggered a shift in research efforts: a standardized common network was needed, particularly in Europe where differing standards co-existed in many counties occupying a relatively small geographical area. The need for a pan-European network was identified in 1982

Table 2.3 Detailed comparison from 1G to 6G communications (adapted from [13])

Feature	1G	2G	3G	4G	5G	6G
Period	1980s	1990s	2000s	2010s	2020s	2030 s
Maximum data rate	2.4 kb/s	144 kb/s	2 Mb/s	1 Gb/s	100 Gb/s	1 Tb/s
Maximum frequency	894 MHz	1900 MHz	2100 MHz	6 GHz	100 GHz	>100 GHz
Service type	Voice only	1G + Text	2G + multimedia	3G + HD video	4G + VR/AR	Predicted in Chap. 1
Channel access method	FDMA	FDMA, TDMA	CDMA	OFDM	OFDM	New schemes
Architecture	SISO	SISO	SISO	MIMO	mMIMO	Beyond mMIMO
Backbone network	PSTN	PSTN	Packet N/W	Internet	IoT	IoE
New feature	Mobility	Digitalization	Internet	Streaming	High data rates	Ubiquitous access

already at the European Conference of Postal and Telecommunications Administrations (CEPT) [9]. CEPT is now seen as the first step towards GSM, the predominant standard of 2G.

2.4 Mobile Telecommunications Evolution and the Rise of Internet

In the period between 1990 and 2020 the fastest rate of evolution of telecommunication yet was observed. In the 1990s, the internet became available globally, while mobile telecommunications continued to expand their capability. At the turn of the century, mobile and internet technologies started to fuse.

2.4.1 2G/GSM

The 2G system that had the best overall global reach was GSM, although countries such as the USA and Japan developed their own systems [9]. The ultimate market share of GSM reached 90% [10]. GSM was launched in Europe in 1991. In 2G, for the first time, digital technology was used for both transmission and switching, which greatly improved voice quality. Transmission channels were also given more bandwidth –200 kHz [12]. 2G introduced better spectrum utilization as well by adding time division multiplexing (TDMA) to FDMA that was already used in 1G. GSM, being digital, also allowed for the introduction of sophisticated encryption algorithms to protect the privacy of users [3].

A big win of GSM over 1G was that the same technology was adopted in many countries, which means that it was possible to connect these directly to one another, allowing for the concept of roaming (using a cellular phone outside the native network) to be introduced. Another major innovation was the introduction of the short-messaging service (SMS), for the first time, allowing for sending short text messages between mobile users. This allowed them to avoid making voice calls if they needed to communicate only small amounts of information. 2G/GSM is also associated with ever smaller mobile phone headsets and it is easily the network with the fastest growth in the number of users, from 9 million in 1995 to over 450 million users worldwide in 2000 [3].

2G also had two major upgrades during 1990s, as shown in Table 2.4. The so-called 2.5G was the first network to introduce data packet exchange in the protocol called General Packet Radio Service (GPRS). The speed of data increased to several tens of kpbs, but this service remained slow and costly. Better speeds were achieved in 2.75G or Enhanced Data Rates for GSM Evolution (EDGE), with a theoretical maximum of 1 Mb/s [12]. In reality, these speeds were closer to 200 kb/s. It is interesting to note that although 3G was already launched in the early 2000s, many

Table 2.4 2G evolution

Feature	2G	2.5G	2.75G
Technology	GSM	GPRS	EDGE
Maximum speed	Tens kb/s	Hundreds of kb/s	1 Mb/s (theoretical)
New feature	SMS	Data packet exchange	Internet

later state-of-the art phones such as Nokia N95 and iPhone (both launched in 2007) did not support it; rather, communication was established using the EDGE standard.

As just stated, the 2G network (2.5G, to be more precise) was the first network to support the internet-on-the-go. Thus, at this point, it is worth going one step back to look briefly at how the internet has evolved. The origins of the internet are also important for understanding the emerging concepts of IoT and IoE.

2.4.2 *Minitel*

The idea of sharing information other than voice using telephone lines was explored long before the internet and the World Wide Web (WWW—the way we know internet today), came about. One technology that quickly found widespread use was facsimile, or fax, which played an important part in the second part of the twentieth century. By the late 1970s, however, a need for a more interactive approach had arisen. Several countries attempted introducing a system called Videotex, which would form the first step toward the information society, as it allowed for a computer-like terminal to be connected to a network using phone lines [14, 15]. In most countries, regrettably, these systems failed to attract users, possibly, for one, because they required users to have at least a minimum level of computer literacy, and secondly, because of the cost of acquiring suitable terminals.

The situation was considerably different in France. As part of the vision of France's computerized future, a service called Télétel was rolled out in 1983. The major difference between the French Télétel and Videotex in other countries is that the French Post, Telegraph and Telephony company started giving the terminals, called Minitel (which gave the colloquial name to the network), to subscribers for free. One such terminal is shown in Fig. 2.7. The first service that was offered on Minitel terminals was an interactive phone book that replaced traditional "white pages", which, with the rising number of telephone subscribers, would become outdated by the time the printing of each revision was completed. The number of Minitel services increased rapidly. By 1993, when Minitel use peaked, users could access services such as messaging, online retail, airline or train ticket purchasing, information services such as news and databases other than the phonebook. Minitel even allowed playing of computer games. The number of online services Minitel offered topped over 20,000 even before the WWW came into existence.

Fig. 2.7 A Minitel terminal. Reproduced under Attribution 2.0 Generic (CC BY 2.0) license from https://www.flickr.com/photos/63313714@N00/5060440292

The major problem with Minitel was that it was confined to a single country—France. With the rise of the internet as a global network in its present form, Minitel did not stand a chance in competing. The Télétel network was terminated in 2012 and at the time of writing this book it has been largely forgotten. However, its importance cannot be neglected, as it predicted what the main applications and services of the internet would be over a decade before the WWW deployment.

2.4.3 *The Internet*

The origins of the internet go back two decades before Minitel. However, during the golden days of Minitel, the internet was still a network that could predominantly only be accessed by a number of defense and research institutions, mostly in the USA.

The beginning of the internet can be traced back to the concept of packet exchange proposed in the early 1960s by several researchers of the Massachusetts Institute of Technology (MIT) [16, 17]. In 1966, the idea of a network based on packet switching, together with a set of initial network specifications, was taken by the MIT to the Advanced Research Projects Agency (ARPA, today DARPA). It took three years before the predecessor of the internet—called ARPANET (from "ARPA network")—was born. ARPANET was a small network localized to a single institution, but it was pictured as just one of many networks that could even have different architectures, which could be internetworked together (thus coining the term "internet"). One of the first services of ARPANET was electronic mail, or e-mail, but many other services have been conceptualized. From the very beginning, the principle on which the actual networks were built and the ways of utilizing those networks once the architecture was available (i.e. network services), were researched in parallel. This concept is still true today and it even holds for new technologies, such as 5G and 6G on one side and IoT and IoE on the other.

In the early 1970s, there was only a single way of internetworking (federating) different networks, leading to a need for a protocol to be developed that could meet the needs of the open-architecture network. The protocol that would later become known as transmission control protocol/internet protocol (TCP/IP) was developed based on four ground rules:

1. Each new network could stand on its own and no internal changes would be required to connect this network to the internet.
2. Communications would be on the best effort basis, meaning that if the packet did not make it to its destination, it would be retransmitted until it did.
3. Gateways (routers) would be used to interconnect different local networks.
4. There would be no global control at the operations level of the internet.

One of the issues that remained unresolved in TCP/IP was the lack of global addressing, which would be a vehicle towards expansion of the internet from a network of defense, research and operational institutions to a truly global network that everybody could use. This problem was solved with the introduction of the domain name system (DNS), which allowed for seamless mapping of IP addresses into hierarchical host names. This change happened in 1983, roughly a decade after TCP/IP came into being.

The real revolution came in 1989, a year before ARPANET was officially decommissioned. That year marked the invention of the WWW by Tim Bertens-Lee [18]. The WWW is really the internet in its present form. With the WWW came the possibly to pull up any data, irrespective of its type, format or computing platform, from a web server, with the aid of a simple universal resource identifier. Whether data included text, images or multimedia became irrelevant. The WWW was accessible with newly introduced web browsers supporting hypertext markup language (HTML), which could be installed on networked computers anywhere in the world. Internet or web browsing and numerous associated services (many at the time existing in the Minitel network) became the norm, extending the internet capability much further beyond the initial capability for file sharing and mail exchange.

The 1990s brought another major innovation, quite important in the context of this book. Various manufacturers of mobile phones entered into cooperation, which resulted in the development of the wireless access protocol (WAP), released in 1999, which would allow 2.5G cellular phones to connect to the internet [3]. The first WAP internet browsers used a special wireless markup language, which optimized the context for viewing the content on small, monochrome screens.

At this point in time, wireless cellular communications and the internet merged for the first time. In the twenty-first century, this idea was carried over into 3G and expanded in successive network generations, all the way up to the present where the internet is, in fact, predominantly accessed from mobile terminals. According to the ITU, the number of mobile internet subscribers surpassed that of fixed-line subscriptions in 2012. This is evident from Table 2.5, which shows the total number and the number of mobile internet subscribers per 100 people in the world for the period from 1997 to 2014 [19, 20].

Table 2.5 Total number and number of mobile internet subscribers per 100 people [19, 20]

Year	1997	2002	2007	2011	2012	2013	2014
Global internet users	2	10	20.6	32.5	35.5	37.9	40.4
Mobile internet users	–	–	4	16.7	21.7	26.7	32

2.4.4 3G

3G deployment started around the year 2000. The emphasis of 3G was on higher channel bandwidths (up to 5 MHz [12]) and the use of wideband and spread spectrum techniques, in order to achieve higher data rates than 2.5G and 2.75G, thus providing reliable access to the internet and the capability for multimedia messaging and video calling [11]. The original 3G network could only realistically offer speeds of 346 kb/s, whereas the aim was to achieve 2 Mb/s [10]. The major communication system in 3G was the Universal Mobile Telecommunication System (UMTS) and the channel access in 3G was based on code division multiple access (CDMA). Specifically, two variants were used [9]: wideband CDMA and time-division CDMA. 3G also allowed for the use of some other standards, which were typically hybrids of 2.5G and 3G.

Shortly after 3G deployment, the 3GPP standards group was formed, with the role of defining future network technical specifications and continuing the work of defining mobile standards and systems. The group defined specifications for further evolution of 3G, as shown in Table 2.6 [9]. The new technology, considered an evolution of UMTS and dubbed 3.5G, was called high-speed packet access (HSPA), with downlink speeds (HSDPA, "D" stands for downlink) of up to 14.6 Mb/s. While 3G introduced video calling, albeit with resolution of the picture that was initially not particularly good, transition from 3G to 3.5G introduced the possibility of streaming high-quality or even HD video on mobile devices. The second enhancement of UMTS was called long-term evolution (LTE). LTE is mostly associated with 4G. Release 8 of LTE, however, which introduced data rates of 326 Mb/s and lower latency than that seen in 3.5G, was released in 2007 already, before the official deployment of 4G and can thus be seen as 3.75G.

Table 2.6 3G evolution

Feature	3G	3.5G	3.75G
Technology	UMTS	HSDPA	LTE release 8
Maximum speed	2 Mb/s	14.6 Mb/s	326 Mb/s
New feature	Video calling	Streaming	Low latency

2.4.5 4G

Nowadays, LTE is understood as the ultimate winner in the competition involving all mobile technologies [12]. This single technology, finally, in the second decade of the twenty-first century, succeeded in offering a replacement for numerous other technologies and standards across the world. The main aim of 4G/LTE was to offer over 100 Mb/s of peak downlink (DL) data rate for highly mobile access and approximately 1 Gbps for low mobility or stationary access. Spread spectrum techniques used in 3G were completely abandoned in 4G, and the OFDM technique, based on the division of wide bandwidth into smaller bands, was adopted. OFDM allowed for the link quality and the data rate to be improved with the use of MIMO techniques. Some of the other specifications of 4G include [8]:

- Single-way latency of <10 ms;
- Scalable bandwidth up to 40 MHz, later extendable to 100 MHz;
- DL peak spectral efficiency of 15 bit/s/Hz;
- Uplink (UL) peak spectral efficiency of 6.75 bit/s/Hz; and
- Carrier frequencies of up to 6 GHz [21].

For the first time in 4G, DL and UL data rates are different [22], which is the consequence of differing spectral efficiencies, with figures shown in the list above. Spectral efficiencies are different in DL and UL because in DL, spectral efficiency is prioritized to satisfy the data hungry users, while in the UL, it is more important to prioritize the power efficiency in order to minimize the size of mobile devices and maximize the battery power. As a result, single-carrier FDMA is used instead of the OFDM in the UL.

The fact that LTE had become the universally deployed network throughout the world brought about different advantages, ranging from better regulatory approaches to sharing and cutting cost. Moreover, one could argue that dropping of the roaming charges across networks in different European countries was one of the biggest wins of LTE [10].

LTE evolved rapidly and several new releases (A–C, LTE-A initially stood for LTE-Advanced) had been deployed by 2015, as shown in Fig. 2.8. The main features of different LTE 3GPP releases have been discussed by Saghezchi et al. [8] and are reproduced in Table 2.7. Note that the evolution of 4G still remained under the umbrella of 3GPP, despite the fact that the term "3G" was retained in its name.

At the time of writing this book, LTE is the most widespread mobile technology, offering connectivity for a wide range of devices, not only mobile phones. Most

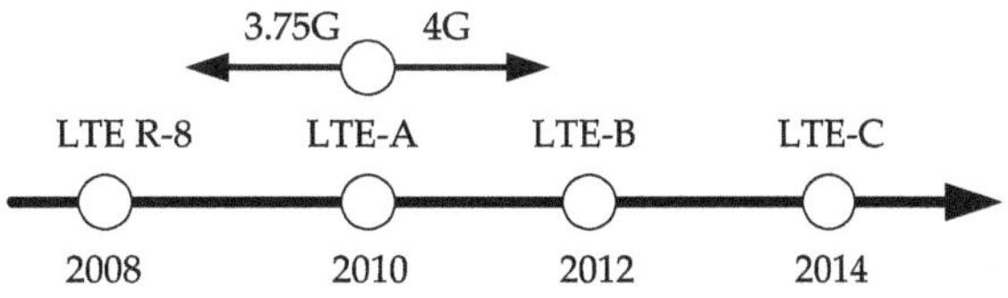

Fig. 2.8 Evolution of 4G and LTE

Table 2.7 Main features of different LTE releases. From [8]

3GPP Release	Features
LTE (Release 8 and 9)	• Supporting both frequency division duplex and time division duplex • Scalable frequency spectrum in six different bandwidths: 1.4, 3, 5, 10, 15 and 20 MHz • OFDM • Supporting up to four-layer spatial multiplexing with single-user MIMO (SU-MIMO) • Achieving 300 Mbps in DL and 75 Mbps in UL • User-plane latency of <20 ms • Multicast and broadcast functionality (R-9)
LTE-A (Release 10 and 11)	• Carrier aggregation to utilize up to 100 MHz bandwidth • Supporting up to eight-layer spatial multiplexing with SU-MIMO • Enhanced multi-user (MU) MIMO • Extended and more flexible reference signal • Relaying functionality • Peak data rate beyond 1 Gbps in DL and 500 Mbps in UL • User-plane latency of <10 ms • Coordinated multipoint (CoMP) transmission and reception (R-11) • Enhanced support for heterogeneous network (HetNet) (R-11)
LTE-B (Release 12 and 13)	• Local area enhancement (soft cell) • Lean carrier • Beamforming enhancement • Enhanced MTC • 3D-MIMO • Enhanced CoMP (R-13) • RAN sharing enhancement • Enhanced self-organizing networks

mobile phones, however, still include support for GSM networks, which are used when placing regular voice calls to other mobile phones and landline telephones on the PSTN, as well as when sending and receiving SMSs. Voice over IP (calling and instant messaging using platforms such as WhatsApp, Viber and WeChat, on the other hand, have been on the rise for the last several years and it is possible that they will render GSM obsolete in the near future. Meanwhile, most operators are in the process of decommissioning the equipment for 3G, as its maintenance is still costly, but most users have already been migrated to 4G.

2.5 The Present

At present, in the 2020s, the world is ready to receive the next generation of networks in the form of 5G. In this era, the backhaul (core) of the global communication network (internet) is a network of high data capacity fiber-optic cables, while "middle-mile" and "last-mile" access is provided with a combination of landline telephone (copper) networks, fiber-to-the-home links, satellites and mobile networks (predominantly 4G but with 5G emerging), as shown in Fig. 2.9 [23].

2.5.1 Fiber-Optics Backhaul and Fiber-to-the-Home

Fiber-optic networks have been around since the 1970s [3], with the first transatlantic cable coming into operation in 1988. Nowadays, the complete telecommunication backbone network uses high-bandwidth fiber-optic cables. The structure of a fiber-optic cable is shown in Fig. 2.10 [23]. Fiber-optic cables are laid both on land and in the ocean. In fact, the low attenuation of fiber-optic cables means that relatively long distances can be traversed without any repeaters, which has made these cables extremely useful in making intercontinental connections, as well as in bringing internet to island nations and third world countries (e.g. to countries in Africa). This concept is illustrated in Fig. 2.11, which shows the distribution of fiber-optic backbone cables around Africa and in the Mediterranean as it stood in 2021.

Fiber-optic cabling has also become a new norm for last-mile delivery of internet access in developed countries, providing high-speed internet access to homes and

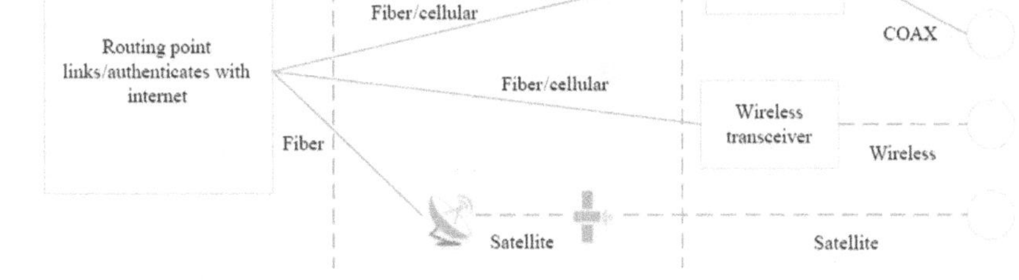

Fig. 2.9 Distribution of global connectivity. Reproduced with permission from [23]

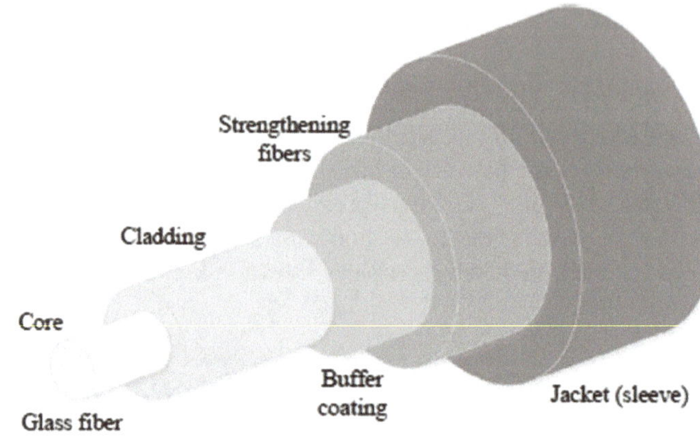

Fig. 2.10 Structure of the fiber-optic cable. Reproduced with permission from [23]

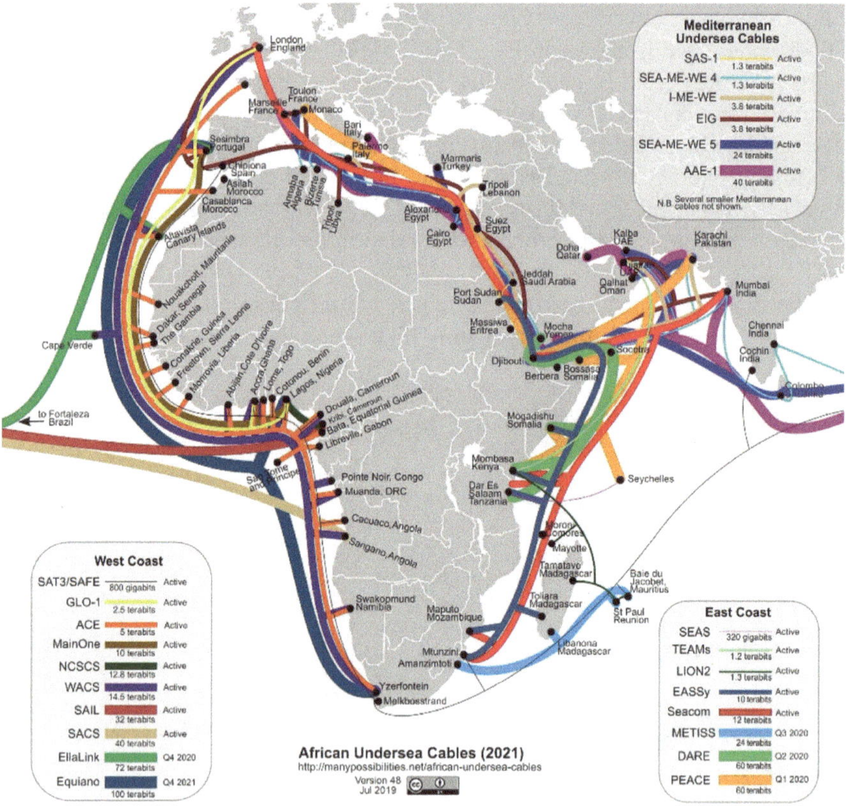

Fig. 2.11 Distribution of African and Mediterranean undersea cables in 2021. Reproduced under Creative Commons license from https://www.flickr.com/photos/ssong/48198601557/

offices. Fiber-optic cables use wavelength division multiplexing to transmit information across multiple wavelengths. A typical fiber-optic cable can carry up to 160 wavelengths and each wavelength can support speeds of 100 Gbit/s. 1 Tbit/s should be achievable in the near future [24]. In addition to being able to provide high-bandwidth access, fiber-optic cables also provide immunity to electromagnetic interference, they do not succumb to corrosion as copper wires do, are immune to intersection and are lightweight, which allows for simple installation. There are, however, many scenarios where fiber-optic last-mile delivery is not an option, for example in the case of remote rural areas where laying fiber-optic cables is not feasible from the perspective of the number of potential users or in dense urban environments where it is difficult to dig up streets and pavements to reach every building. Developing countries also have a challenge of bringing fiber-optic cables to low-income populations, as this is not feasible from the cost perspective. In these cases, access via a high data capacity mobile network, such as 5G, discussed next, is the best solution. These users add to an already large pool of mobile, maritime and airborne users, whose only alternative to broadband internet access and media is via satellite.

2.5.2 5G

All previous generations of cellular networks retained the status of the "newest network" for about a decade. Coincidentally, these decades largely overlapped with calendar decades. The period during which the network was considered the state-of-the-art should not be confused with the lifetime of each network. Most networks survive for a least a decade after the new generation has been deployed. GSM is the best example, still being used some 30 years since the time the first GSM call was recorded. Based on the historical data, it makes sense to assume that after 1G dominated the 1980s, 2G dominated the 1990s, 3G dominated the 2000s and 4G dominated the 2010s, 5G will be in the spotlight for about a decade from 2020. The deployment of 5G, in fact, started in 2019.

Technical Detail of 5G Networks

Some specifications of 5G include [10, 11]:

- Bit rates much faster than those achieved by LTE (~20 Gb/s);
- Significantly lower latencies, going down to 1 ms, which are guaranteed more than 99% of the time;
- Much higher capacity than seen in 4G;
- Device-to-device (D2D) connectivity (direct communication between users without using any network infrastructure);
- Massive machine-to-machine communications as an add-on to the advanced platform of IoT;
- Increased security;
- Increased network reliability; and

- Improved energy efficiency.

Moreover, 5G uses new parts of the spectrum in the microwave band (3.3–4.2 GHz) and it is the first generation of cellular networks that uses millimeter-wave technology. In addition to more spectrum and higher spectral efficiency (bits per Hertz per cell), 5G requires greater cell densification, meaning that more small cells per kilometer squared are required to cover the same area that older generations of networks would cover with a single base station (cell tower) [25].

To illustrate how the performance of the 5G network differs from that of previous generations of networks, one can look at a practical example of the network usage. Fast and reliable streaming of video and downloading of large files are some of the applications that will bring about a massive improvement in 5G. To download an 8-GB movie file, 70 min was needed on 3G networks and seven minutes was needed on 4G networks [26]. The same file can be downloaded in under 7 s on 5G. This performance will be achieved by bringing together many emerging technologies [11]:

- mMIMO for capacity increase;
- Software defined networks (SDN) for flexibility in the network;
- D2D for spectral efficiency;
- Information-centric networking for reduction in network traffic; and
- Network slicing.

The network slicing is an innovation introduced in 5G networks, which is a feature that is not often mentioned when 5G is discussed [27]. It is the ability to partition off segments of the 5G network to provide specific latency, bandwidth and quality-of-service guarantees, for the quick deployment of various services. D2D is another important feature, which allows not only for increased spectral efficiency as mentioned above, but also for improved typical user data rate and capacity per area, extended coverage, reduced latency, and enhanced cost and power efficiency [28].

5G system architecture as envisioned in [8] is shown in Fig. 2.12.

Three major usage scenarios are to be utilized in 5G: eMBB, URLLC and mMTC. The major KPIs of 5G and the definition of these terms have already been discussed in Chap. 1 and listed in Table 1.1 as part of the opening discussion on 6G.

5G Spectrum

The spectrum that 5G uses is divided into three cellular frequency bands: low, median and high [29, 30]. Each of these bands is purposed for a different type of data transfer. The frequencies that will be allocated to 5G networks in different countries differ based on local spectrum regulations, but the approximate frequency distribution of these bands is as follows:

- The low band uses frequencies below 1 GHz, e.g. 700 MHz;
- The mid-band uses frequencies between 1 and 6 GHz, e.g. 3.5–3.8 GHz range; and
- The high- or millimeter-wave band uses frequencies between 24 and 86 GHz, e.g. 24–29 and 37–43 GHz ranges.

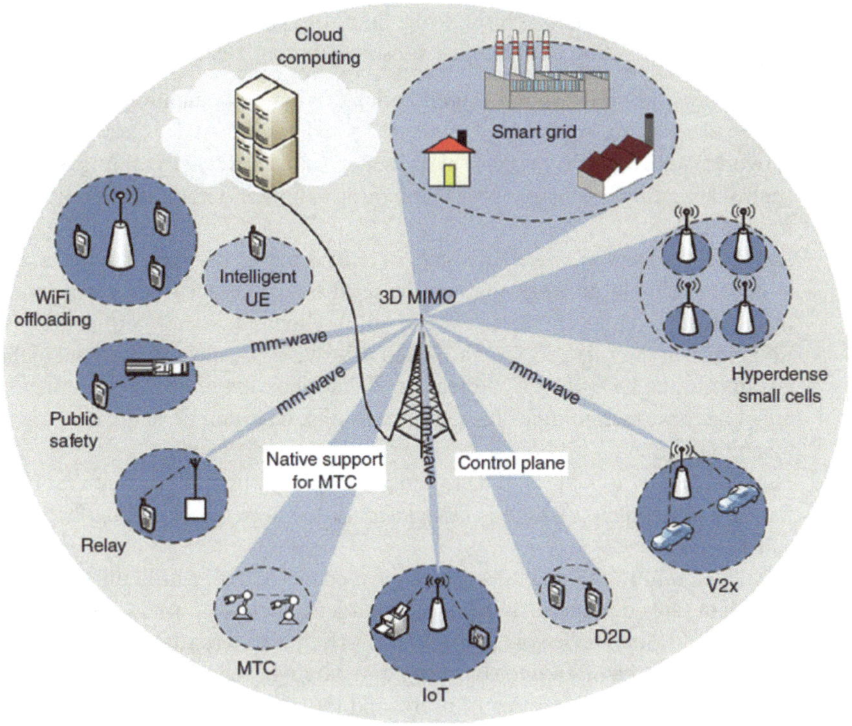

Fig. 2.12 5G system architecture. Reproduced with permission from [8]

Frequencies below 1 GHz will typically be used for indoor coverage and for broader coverage areas. The former is because lower frequencies propagate better through solid objects (such as windows and walls) and the latter is because lower frequency signals are able to propagate further away from the base station in open areas. Frequencies above 6 GHz are also referred to as the super-data layer, because channels in these bands allow for high data bandwidths, which will be discussed in more detail in Chap. 3. Frequencies between 1 and 6 GHz belong to the same range of frequencies used by 3G and 4G (LTE) and offer a compromise between coverage and capacity. With 3G typically phased out, for the foreseeable future, this part of the spectrum will have to be shared with LTE [31]. Initially, mobile devices will predominantly make use of the low- and mid-range frequencies, while the millimeter-wave channels with their much higher bandwidths will be reserved for stationary users. This is likely to change with future network evolution.

5G Verticals

Craven [31] identified several case uses (verticals) of 5G, which are applicable to various economic sectors, including the agriculture, automotive, construction, energy, finance, health, manufacturing, media, public safety, retail and consumer and transport sectors. Some of these case uses are summarized as follows:

1. *Autonomous vehicle control.* Autonomous driving, also mentioned as one of the verticals of 6G in Chap. 1, will be rooted in 5G. At the very least, 5G networks will be able to assist with better traffic safety, by aiding vehicle control signaling, exchanging of video information or networking of vehicles for fleet driving.
2. *Emergency communication.* In a state of emergency or disaster, it is necessary to have a reliable network that can be used for emergency communication. In situations such as earthquakes, 5G will facilitate the location of survivors. Reliability of internet access, via 5G and otherwise, has already been put to the test during the COVID-19 outbreak in 2020, when the majority of humanity relied on these technologies for remote working, entertainment and keeping in touch with others.
3. *Factory automation.* 5G will have a major influence in implementation of the Industry 4.0 concept, which is associated with a high level of manufacturing automation.
4. *Communication while traveling at high speed.* It is more difficult to supply high data rates continuously to moving targets, such as to users traveling on high-speed trains. LTE was the first network that differentiated between the speeds that can be offered to stationary and moving users. 5G will improve the data rates for both categories of users, and thus users moving at high speeds will for the first time be able to enjoy services such as high quality video despite the signal degradation necessitated by the fast movement.
5. *Communication at large events.* At events such as concerts and sports games, a large number of network users are concentrated in small areas. A highly dense 5G network would have the capability to minimize the impact of the number of users, since a large amount of bandwidth is available in 5G channels for sharing. Under these conditions, users should still be able to make video calls and share audio and video clips without noticing the drop in spectrum availability.
6. *IoT.* 5G will be able to provide fast and low-latency connections to the large number of geographically spread devices, including devices used for surveillance, monitoring, environmental sensing, etc.
7. *Media on demand.* As the amount of available bandwidth increases, the cost of data transfer will decrease, thus allowing high-quality media to be accessed readily from any location, at any time though 5G networks. This type of media access is still limited to stationary users, e.g. users using local area networks (LANs) or WiFi in many countries.
8. *Remote healthcare (E-health).* Critical healthcare services can be provided over large distances if communications are reliable. The services can range from virtual doctor consultations and examinations to remote surgeries in extreme situations.

Table 2.8 Details of 3GPP Release 15

3GPP Release	Feature
Release 15	NR
	5G system—Phase 1
	mMTC and IoT
	V2X Phase 2
	Mission-critical internetworking with legacy systems
	Wireless LAN (WLAN) and unlicensed spectrum use
	Slicing—logical and end-to-end networks
	Third party access to 5G services
	Service-based architecture
	Further LTE improvements
	Mobile communications system for railways (FRMCS)

9. *Indoor navigation*. While GPS is sufficient for accurate outdoor navigation, 5G networks with smaller cells deployed in large indoor areas such as shopping malls can be used for indoor navigation.
10. *Smart cities*. 5G will aid the development of many aspects of smart city living, including "smart homes", "smart offices", "smart buildings", "smart grid" and "smart traffic control".
11. *VR and AR*. A very high data rate and tight latency are required for enabling VR and AR. As a result, these concepts are at present mostly reserved for indoor settings where fast data rates can be sustained only on LANs. 5G will gradually allow mobile access to these types of services.

5G Deployment

5G deployment was initiated through the 5G New Radio (NR) specification by the 3GPP technical specification group in 2017 [32]. The NR specification was fast-tracked to ensure that 5G would be ready for trials starting in 2018 and for the start of commercial deployment in 2019 [33]. Initial 5G deployments were non-stand-alone networks. In essence, this means that the 5G network uses 4G infrastructure. It is envisioned that later 5G releases will be stand-alone (SA) networks. 5G NR used different access schemes, including cyclic prefix (CP)-OFDM for both UL and DL, as well as frequency division duplex and time division duplex schemes. Modulation schemes utilized include quadrature phase shift keying (QPSK) and various dimensions of quadrature amplitude modulation (QAM) schemes: 16-QAM, 64-QAM and 256-QAM. QPSK was also used for the physical control of the downlink channel.

The deployment of 5G NR started with what is called the 3GPP Release 15. The main features of this standard are listed in Table 2.8 [34]. The countries that deployed the first 5G networks were:

- China and South Korea in Asia;
- The United Kingdom and Germany in Europe; and
- The USA.

The next set of deployments in Europe encompassed Switzerland and the Nordic countries (Finland, Denmark, Norway, Sweden and Iceland), before expanding to other countries around the world. In 2020, about one fifth of the world's countries were believed to have launched 5G services.

It must be noted, however, that the deployment of 5G did not go smoothly everywhere. For example, in March 2020 it was reported that the government of Slovenia had temporarily halted the introduction of the 5G technology [35], because of lack of scientific evidence on whether the effects of 5G are potentially harmful to human health. However, resistance to the 5G network deployment was to some extent expected, given that the deployment of every generation of wireless networks has been met with some resistance associated with health concerns and this trend is not likely to stop in future generations either. This is why the legality of health claims on 5G and 6G networks and the associated technologies (millimeter and terahertz waves) will also be discussed later in this book.

2.6 The Future—5G Evolution (B5G) and 6G

Originally, 5G was seen as the ultimate network architecture serving all industry sectors. However, it is becoming apparent that this status will be reserved for future networks (in other words, 6G), although it cannot be denied that 5G is still heading in the right direction. Upon completion of the world-wide deployment of the 5G NR standard, the evolution of 5G networks will begin, and the B5G roadmap for the next several years has already been outlined. Thus, the networks will transition from 5G, via B5G, to the intelligent computing network—6G.

2.6.1 5G Evolution

With the continued increase in the number of users in the network that will be involved by adding machine connectivity, future 5G networks will introduce the concept of mass connectivity. Thus, 5G will become the network that will be more than simply a network offering faster data transfer. B5G networks will therefore become too complex to be run by human operators, which will require the introduction of AI, the concept that is expected to be dominant in 6G. How this transition will be achieved remains to be seen.

On the other hand, at the time of writing, two more 5G 3GPP releases have been defined. In 2020, 3GPP Release 16 is ready to be deployed. It is expected that Release

17 will be ready by 2022. The main features of both of these releases are shown in Table 2.9 [36, 37].

It is evident from analysis of the 5G 3GPP Release 17 features that some of the features proposed for 6G in Chap. 1 are already starting to be phased in this 5G release. 6G is thus discussed next.

2.6.2 6G

As in the case of every wireless network generation in the past, where the development of the next generation of networks started at the same time as the deployment of the current network generation, it is now time to start thinking about 6G. As discussed in Chap. 1, with the release of 5G in 2019, the 6G flagship program also started in that year.

Lessons from history and those that are being learned at present (Sects. 2.1, 2.2, 2.3, 2.4 and 2.5) are clear. The availability of every new communications technology (be it in the pre-internet or the current information era) has barely managed to satisfy the hunger for information for a short period. This timeframe was usually just long enough to open new horizons and allow new ideas about possible new use cases to emerge. In the era of wireless cellular telecommunications, this evolution translates into a constant need for more bandwidth and higher data rates. Although the 6G is envisioned as an intelligent computing network, human-to-human communication remains important and can thus still be used as an example. Throughout history, human-centric communication evolved, from transmitting simple Morse-coded messages to voice transmission, and from video to HD and 3D video transmission. In future, 6G will have the data capability to establish holographic links, for example (whether something like this will be practical remains to be seen). It is unclear what the next step in evolution after holographic communication could be, but what is clear is that with each of these types of communication, data traffic and data transfer speeds need to increase and latencies need to decrease, and there is still no indication that the rapid change in these figures is going to slow down soon. A similar trend is observed in industry—the world went from the first to the second to the fourth industrial revolution (emerging at present) and there is no indication that the fifth revolution will be last one. Internet is on the same path—it started as a local network, turned into the WWW, then into IoT, which will eventually turn into IoE. It is difficult even to imagine what could replace "everything" in the next convenient acronym.

It is thus certain that the 6G will be the inevitable replacement of 5G and B5G. Going by the history, this is likely to happen around the year 2030 (recall Fig. 1.4). Whether 6G will satisfy all the future needs and researchers will just keep improving on the existing 6G technology (call it beyond-6G or B6G), like Microsoft aims for Windows 10 to be the last ever Microsoft Windows version, or will need something else, in the realm of 7G by 2040, cannot be predicted at the moment.

Table 2.9 Main features of 3GPP Releases 16 and 17 [36, 37]

3GGP Release	Feature
3GPP Release 16	5G system—phase 2
	V2X phase 3: platooning, extended sensors, automated and remote driving
	Industrial IoT
	URLLC enhancements
	NR-based access to unlicensed spectrum
	5G efficiency: Interference mitigation, self-organizing network (SON), exalt MIMO (eMIMO), location and positioning, power consumption, dual connectivity, device capabilities exchange, mobility enhancements
	Enhancements for common application programming interface (API) framework for 3GPP northbound APIs
	FRMCS Phase 2
3GPP Release 17	NR MIMO
	NR sideling enhancements
	52.6–71 GHz frequency utilization with existing waveforms
	Dynamic spectrum-sharing enhancements
	Industrial IoT/URLLC enhancement
	NR over non-terrestrial networks
	NR positioning enhancements
	Low-complexity NR devices
	Power saving
	NR coverage enhancements
	NR XR
	NB-IoT and LTE-MTC enhancements
	5G multicast broadcast
	Integrated access and backhaul enhancements
	NR sidelink relay
	SON enhancements
	Satellite components in the 5G architecture
	Network automation for 5G—phase 2
	Edge computing
	Proximity services
	Network slicing—phase 2
	V2X enhancements
	Advanced interactive services
	Unmanned aerial systems
	Multimedia priority service
	5G wireless and wireline convergence
	Other features as stipulated in [37]

Thus, one must now limit the predictions only to the aspects of 5G that need to be improved or reinvented in 6G. Note that these go beyond KPIs identified in Chap. 1 and those listed in Table 2.3. These quantitative improvements were conveniently illustrated by Dang et al. [13] in the picture reproduced in Fig. 2.13 to be, in order of the possible improvement gain,

- Security, secrecy and privacy;
- Affordability and customization;
- Energy efficiency;
- Intelligence; and
- Spectral efficiency.

These qualitative improvements are bundled with the inevitable advance of various technologies that will be discussed in the remainder of this book.

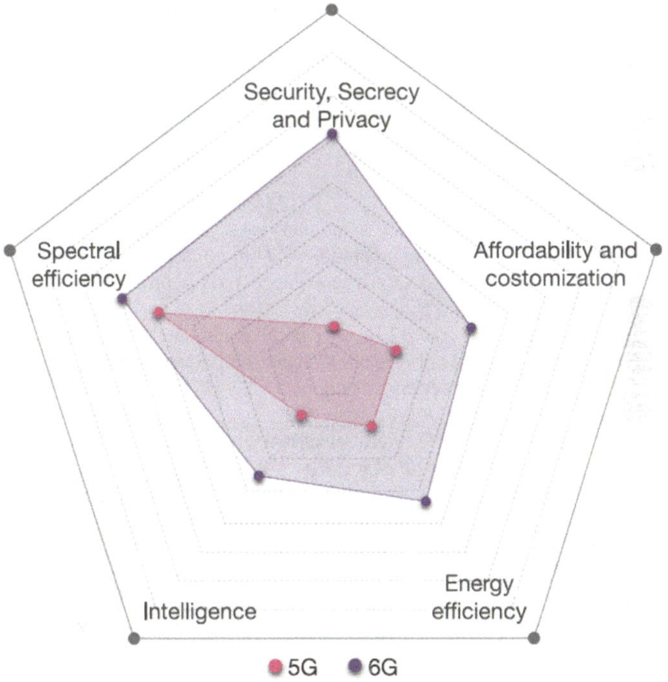

Fig. 2.13 Possible quantitative improvements of 6G over 5G. Reproduced with permission from [13]

2.7 Conclusion

This chapter discussed the history, the present and the future of telecommunications. In the past, telecommunications evolved from very rudimentary concepts, to cable links and then to wireless communications. In 2012, the number of wireless (mobile) users surpassed those that accessed the global network, the internet, via wired or optical links. This indicates that the future is wireless and that neither the present technology—5G—nor 5G evolution networks—B5G—will be able to feed the upcoming hunger for connectivity and new services that the increased connectivity will unlock, for too long. This opens up a great opportunity for 6G to emerge and evolve.

The technologies that are candidates to make 6G possible were briefly introduced in Chap. 1. The discussion of these will be taken forward in various chapters in the remainder of this book. Specifically, Chap. 3 will look into concepts of channel coding theory, channel capacity, millimeter-wave frequencies, terahertz frequencies and modulation.

References

1. Kennedy JB (1926) When woman is boss: an interview with Nikola Tesla. Colliers, Seattle
2. Novak M. Nikola Tesla's Incredible predictions for our connected world [Internet]. Paleofuture. https://paleofuture.gizmodo.com/nikola-teslas-incredible-predictions-for-our-connected-1661107313. Cited 21 Mar 2020
3. Huurdeman AA (2003) The worldwide history of telecommunications. Wiley
4. Frenzel LE (2007) Marconi did not invent radio [Internet]. Electronic Design. https://www.electronicdesign.com/technologies/communications/article/21759673/marconi-did-not-invent-radio. Cited 28 Mar 2020
5. Tesla N (1900) System of transmission of electrical energy [Internet]. https://patents.google.com/patent/US645576A/en. Cited 28 Mar 2020
6. Brenner P (2009) Tesla against Marconi: the dispute for the radio patent paternity. IEEE EUROCON 2009:1035–1042
7. Lambrechts W, Sinha S, Abdallah JA, Prinsloo J (2018) Extending Moore's Law through advanced semiconductor design and processing techniques. CRC Press, Boca Raton
8. Saghezchi FB, Rodriguez J, Mumtaz S, Radwan A, Lee WC, Ai B et al (2015) Drivers for 5G: the 'pervasive connected world'. Fundamentals of 5G mobile networks. Wiley, pp 1–27
9. Osseiran A, Monserrat JF, Marsch P (2016) 5G mobile and wireless communications technology. Cambridge University Press
10. David K, Berndt H (2018) 6G vision and requirements: is there any need for beyond 5G? IEEE Veh Technol Mag 13:72–80
11. Huang T, Yang W, Wu J, Ma J, Zhang X, Zhang D (2019) A survey on green 6G network: architecture and technologies. IEEE Access. 7:175758–175768
12. Hueber G, Niknejad AM (2019) Millimeter-wave circuits for 5G and radar. Cambridge University Press
13. Dang S, Amin O, Shihada B, Alouini M-S (2020) What should 6G be? Nat Electr 3:20–29
14. Cats-Baril WL, Jelassi T (1994) The French videotex system Minitel: a successful implementation of a National Information Technology Infrastructure. MIS Quarterly. Management Information Systems Research Center, University of Minnesota, vol 18, pp 1–20

15. Mailland J, Driscoll K (2017) The French connection machine. IEEE Spectr 54:32–37
16. Leiner BM, Cerf VG, Clark DD, Kahn RE, Klenrock L, Lynch DC et al (2009) A brief history of the internet. ACM SIGCOMM Comput Commun Rev [Internet]. ACM. https://dl.acm.org/doi/abs/10.1145/1629607.1629613. Cited 2 Apr 2020
17. Ryan J (2010) A history of the Internet and the digital future. Reaktion Books
18. McPherson SS (2009) Tim Berners-Lee: inventor of the World Wide Web. Twenty-First Century Books
19. ICT Statistics [Internet]. http://www.itu.int/ITU-D/ict/statistics/ict/. Cited 2 Apr 2020
20. ITU Key 2005–2014 ICT data [Internet]. http://www.itu.int/en/ITU-D/Statistics/Documents/statistics/2014/ITU_Key_2005-2014_ICT_data.xls. Cited 2 Apr 2020
21. Ten BQ (2019) Trends in the cellular industry and an outlook on 6G. IEEE Commun Mag 57:31–36
22. Ankarali ZE, Peköz B, Arslan H (2017) Flexible radio access beyond 5G: a future projection on waveform, numerology, and frame design principles. IEEE Access 5:18295–18309
23. Lambrechts W, Sinha S (2019) Last mile Internet access for emerging economies. Springer
24. Gati A, Salem FE, Serrano AMG, Marquet D, Masson SL, Rivera T et al (2019) Key technologies to accelerate the ICT green evolution—an operator's point of view. arXiv:190309627 [cs] [Internet]. http://arxiv.org/abs/1903.09627. Cited 10 July 2020
25. Vahid S, Tafazolli R, Filo M (2015) Small cells for 5G mobile networks. fundamentals of 5G mobile networks. Wiley, pp 63–104
26. Thornton S (2019) The impacts of 5G on the future: a new era of connectivity [Internet]. 5G Technology World. https://www.5gtechnologyworld.com/the-impacts-of-5g-on-the-future-a-new-era-of-connectivity/. Cited 4 Apr 2020
27. Higginbotham S (2020) What 5G hype gets wrong [Internet of Everything]. IEEE Spectr 57:22
28. Li Z, Moya FS, Fodor G, da Silva Jr B, Mairton J, Koufos K. Device-to-device (D2D) communications. 5G Mobile and wireless communications technology. Cambridge University Press, pp 107–36
29. G Status South Africa—March 2019 [Internet]. EMFSA. 2019. https://www.emfsa.co.za/news/5g-status-south-africa-march-2019/. Cited 9 Apr 2020
30. Fisher T. 5G spectrum and frequencies: everything you need to know [Internet]. Lifewire. https://www.lifewire.com/5g-spectrum-frequencies-4579825. Cited 9 Apr 2020
31. Craven C (2020) What is the 5G spectrum? Definition [Internet]. SDxCentral. https://www.sdxcentral.com/5g/definitions/what-is-5g-spectrum/. Cited 9 Apr 2020
32. What is 5G NR?—A definition & standards—SDxCentral.com [Internet]. SDxCentral. https://www.sdxcentral.com/5g/definitions/5g-nr/. Cited 10 Apr 2020
33. Craven C (2019) The top countries with 5G deployments and trials [Internet]. SDxCentral. https://www.sdxcentral.com/5g/definitions/the-top-countries-with-5g-deployments-and-trials/. Cited 10 Apr 2020
34. Release 15 [Internet]. https://www.3gpp.org/release-15. Cited 24 Apr 2020
35. Slovenia stops the introduction of 5G technology [Internet]. Eng News 24 h (2020). https://engnews24h.com/slovenia-stops-the-introduction-of-5g-technology/. Cited 10 Apr 2020
36. Release 16 [Internet]. https://www.3gpp.org/release-16. Cited 24 Apr 2020
37. Release 17 [Internet]. https://www.3gpp.org/release-17. Cited 24 Apr 2020

Chapter 3
The Quest for Faster Data Rates: Unlocking Millimeter-Wave and Terahertz Frequencies

Abstract Communication with the aid of electromagnetic waves will remain the predominant mode of communication for some time to come. The aim of this chapter is thus to expand on how the full potential of electromagnetic communication can be unlocked to support extreme data rates. First, the frequency spectrum will be looked at, allowing for a closer look at the available transmission ranges for 5G and 6G. Secondly, the concept of channel capacity will be defined, allowing for the contextualization of the importance of the availability of bandwidth to be made, which is the main driving factor in the transition to higher frequencies. Thereafter, the various aspects of establishing links for millimeter-wave and terahertz communication will be discussed in detail. Finally, the methods for accessing the spectrum and its channels to their full capacity will be discussed from the communication systems engineering perspective.

In Chaps. 1 and 2, it was recognized that future generations of wireless networks will have to support much higher data rates than those that can be reached by the best wireless technologies of the present. The currently widespread wireless technology, 4G LTE, can provide data rates in the order of 1 Gb/s in the DL and 5G, in the process of worldwide deployment since 2019, is aiming to bring this figure up to 100 Gb/s. Moreover, it is expected that 6G technologies would have to deliver DL (and even UL) data rates in excess of 1 Tb/s if they were to be successful in supporting the vertical industries envisioned in Chap. 1. The question is, thus, what are the technologies that will be able to assist in achieving these requirements? The simple answer is that despite research into other alternatives, communication with the aid of electromagnetic waves will remain the predominant mode of communication for some time to come. The aim of this chapter is thus to expand on how the full potential of electromagnetic communication can be unlocked to support extreme data rates.

Electromagnetic communication is established through additive white Gaussian noise (AWGN) channels, which have a defined bandwidth and a center frequency. Although channels can be arbitrarily defined, the collection of all the channels forms the frequency spectrum. The visible light, the invisible light consisting of infrared

M. Božanić and S. Sinha, *Mobile Communication Networks: 5G and a Vision of 6G*, Lecture Notes in Electrical Engineering 751,
https://doi.org/10.1007/978-3-030-69273-5_3

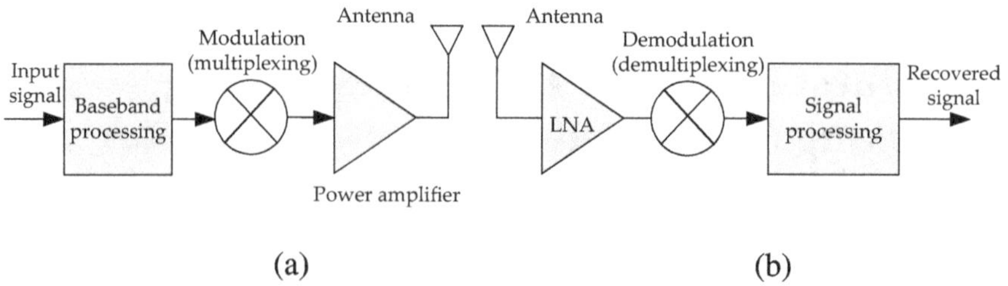

Fig. 3.1 Simplified block diagram of a transmitter (**a**) and receiver (**b**)

and ultraviolet radiation, as well as harmful forms of radiation (X-rays and gamma rays), also take up a portion of the electromagnetic spectrum, as shown in Fig. 1.7. Thus, the complete spectrum is not available for telecommunications. The present expectations are that at least some forms of links could be established in channels with center frequencies up to 3 THz, the limit of which is roughly set by the lower bound on infrared radiation.

The bandwidth of the channel is usually proportional to the center frequency of the channel, meaning that higher frequencies can support higher bandwidth channels. Higher bandwidths are one of the requirements for achieving faster data rates. Meanwhile, most wireless technologies, including cellular technologies such as LTE and WLAN technologies such as WiFi, utilize frequencies up to about 6 GHz, which are some 500 times lower than those allowed in theory. Even the satellite technologies that use frequencies up to about 30 GHz and radar sensing (77 GHz in automotive) extend the utilized range of the spectrum to less than 3% of its total availability. Only the later releases of 5G will begin utilizing frequencies of up to 86 GHz. It is thus easy to present a question, why is this the case? Why are the millimeter-wave frequencies (30–300 GHz) and terahertz frequencies (0.3–3 THz), which can unlock large amounts of bandwidth in the uncongested part of the spectrum, not yet widely utilized for telecommunications?

The answer to these questions lies in understanding of the ease of access of certain frequencies in the spectrum. Here, the problem is twofold: (1) what is the hardware required to access each portion of the spectrum and (2) once this spectrum is accessed, what is the behavior of the electromagnetic waves with certain frequency and characteristics? For the first part, to broadcast the information at a certain frequency, the technology needs to exist to generate the signal at this frequency. Specifically, the technology needs to exist to carry out all required steps in the processing of the digital information, modulating/multiplexing the signal and transmitting the signal with the specific amount of power, as illustrated in Fig. 3.1a. Conversely, technology also needs to exist to receive the signal on the other side of the link, and process the signal in the reverse order by doing low-noise amplification, demodulation and inverse signal processing, as shown in Fig. 3.1b. It is only recently (last 20 years) that transistor technologies capable of accomplishing this started coming into existence. Even at the turn of the century, transistors could amplify and oscillate reliably only up to several gigahertz. For the second part of the problem, the physics needs to be consulted. First, physics dictates that the power of the received signal decreases with the square of the frequency, meaning that the free-space loss increases drastically at millimeter-wave frequencies and even more drastically at terahertz frequencies. Second, the ever-attenuating signal needs to pass through various obstacles in the air, which results in additional atmospheric attenuation. Lastly, the decreasing wavelength of the signal at these frequencies has a major influence on the signal propagation.

This chapter will assume that the first part of the problem associated with the millimeter and terahertz part of the spectrum is or will be solved—that is, that transistor and other technologies capable of supporting transmission and reception of associated signals already exist or will emerge in due course. The reality of this situation is complex and will be dealt with in Chap. 4. This allows for the second part of this problem, which involves analyzing the propagation and associated properties of millimeter and terahertz waves, as well as the engineering aspects of their communication systems, to be tackled on their own in this chapter. Thus, the chapter is divided into several sections. First, the frequency spectrum will be looked at, allowing for a closer look at the available transmission ranges for 5G and 6G. Second, the concept of channel capacity will be defined, allowing for the contextualization of the importance of the availability of bandwidth to be made, which is the main driving factor for the transition to higher frequencies. Thereafter, the various aspects of establishing links for millimeter-wave and terahertz communication will be discussed in detail. Finally, the methods for accessing the spectrum and its channels to their full capacity will be discussed from the communication systems engineering perspective.

3.1 The Electromagnetic Spectrum

Each range and band in the frequency spectrum is defined by its higher and lower frequency bound. Alternatively, frequency bands can also be defined by the wavelength they can carry. The wavelength and frequency are related according to a well-known relation,

$$\lambda = \frac{v}{f}, \tag{3.1}$$

where v is the phase speed of the wave and f is the wave frequency. The phase speed of the electromagnetic wave in free space is the speed of light, which is about 3×10^8 m/s [1].

Frequency bands are defined by the ITU [2]. The portion of the frequency spectrum relevant for 5G and 6G is illustrated in Table 3.1 and encompasses frequencies from 300 MHz to 3 THz. The lowest frequency range in this table is ultra-high frequency (UHF) with frequencies from 300 MHz to 3 GHz (lower frequency ranges are discussed elsewhere; for example, see [3]). In addition to its use in wireless cellular networks, the UHF range is also used for television and radio broadcast,

Table 3.1 Definition of frequency ranges and bands in the frequency spectrum relevant for 5G and 6G

Frequency range	Wavelength range	Range name	Band name	Other names
300–1000 MHz	100–30 cm	UHF	–	RF
1–2 GHz	30–15 cm		*L*-band	
2–3 GHz	15–10 cm		*S*-band	
3–4 GHz	10–7.5 cm	SHF		Microwaves
4–8 GHz	7.5–3.75 cm		*C*-band	
8–12.4 GHz	37.5–24 cm		*X*-band	
12.4–18 GHz	2.4–1.7 cm		*Ku*-band	
18–26.5 GHz	1.7–1.1 cm		*K*-band	
26.5–30 GHz	1.1–1 cm		*Ka*-band	
30–40 GHz	10–1 mm	EHF		Millimeter waves mm-waves E-band (waveguides 60–90 GHz)
40–75 GHz			*V*-band	
75–110 GHz			*W*-band	
110–170 GHz			*D*-band	
170–300 GHz			–	
300 GHz–3 THz	0.1–1 mm	THF	–	THz frequencies Sub-millimeter waves Far infrared

cordless and cellular telephone transmission, as well as for WiFi and Bluetooth®. These bands are also suitable for industrial heating and microwave ovens. The super-high frequency (SHF) range, which includes frequencies from 3 to 30 GHz, and the extremely high frequency (EHF) range, which includes frequencies from 30 to 300 GHz, are mostly used for satellite communication and radar applications, as already discussed. The exceptions are frequencies up to 6 GHz, which are sometimes also utilized by the same applications that utilize UHF.

At a frequency of 30 GHz, the wavelength of a transmitted signal has a value of 1 cm, and above this frequency, it decreases to the order of millimeters. Thus, this value has made it convenient for this point in the spectrum to be taken as the transition between microwave and millimeter-wave ranges. Similarly, the wavelength of 1 mm, equivalent to a frequency of 300 GHz, apart from giving the name the millimeter-wave range, marks the transition from millimeter waves to sub-millimeter waves. Submillimeter waves are better known as terahertz frequencies, but this frequency range is also, although less frequently, referred to as tremendously high frequency (THF).

Further division of frequency ranges into bands is sometimes used for the UHF, SHF and EHF. The *L*-band spans from 1 to 2 GHz, the *S*-band spans from 2 to 4 GHz, the *C*-band spans from 4 to 8 GHz and the *X*-band spans from 8 to 12 GHz. The *K*-band is typically subdivided into three bands, the *Ku*, *K* (proper) and *Ka* bands, where "u" stands for "under" and "a" stands for "above". The *Ku* band spans from 12 to 18 GHz, the *K* band from 18 to 26.5 GHz and the *Ka*-band from 26.5 to 40 GHz. The top part of the *Ku*-band is already in the millimeter range, together with the *V*-band (40–75 GHz) and *W*-band (75–110 GHz).

Note that from the above analysis of frequency bands, each subsequent frequency band has a higher total allocated bandwidth than the previous band. The *L*-band has only 1 GHz of total bandwidth for all the channels in this band. The *S*-band is allowed 2 GHz and the C-band is allowed a total of 4 GHz. However, as the frequency increases even further, the bandwidth becomes abundant. For example, the *V*-band has 35 GHz at its disposal, while the *D*-band has even more, 60 GHz. Yet, the *L*-, *S*-, and *C*-bands are the most frequently utilized bands, which leads to the congestion of this part of the spectrum. This is why, for emerging and future wireless network generations, the millimeter-wave and terahertz frequencies are most appealing, as it becomes possible to support higher data rates, at least, if spectrum availability and bandwidth alone are considered.

One must then argue that communication in the *L*-, *S*-, and *C*-bands has a capacity problem. In fact, in order to achieve the required data rates, LTE already utilizes the allocated portion of the spectrum below 6 GHz to almost the theoretical capacity.

3.2 Sub-6-GHz Communication—The Capacity Problem

The main objective when transmitting information over any communication channel is the reliability of the data transmission. The reliability is measured as the probability of the correct reception, or the reception with the specified bit-error rate (BER), of the signal on the receiver side in the presence of noise. It is always possible to receive signals reliably in the presence of noise; however, the presence of noise sets a limitation on the speed of communication, or in other words, defines the capacity of the channel. This result is known as the channel coding theorem, and was developed by Shannon [4, 5].

In digital communications, the information is sent over band-limited AWGN channels. The bandwidth of a channel is defined as the difference between the upper frequency f_2 and lower frequency f_1, both frequencies marking the points where the magnitude of the signal passing through the channel is typically half of (3 dB below) the maximum gain magnitude [1]:

$$W = f_2 - f_1. \tag{3.2}$$

In millimeter-wave systems, bandwidth is sometimes specified in terms of percentage bandwidth, which is the ratio (in percentage) of the absolute bandwidth to the highest frequency at which the system operates. At 300 GHz, 10% corresponds to 30 GHz bandwidth. It can be shown that the capacity of the channel with noise power spectrum N_0 and signal power P is

$$C = W \log_2\left(1 + \frac{P}{N_0 W}\right) \tag{3.3}$$

This equation is the much celebrated Shannon's formula for capacity, and returns an answer in bits per second (b/s) [4, 5]. Note that the quantity P/N_0W is also the signal-to-noise ratio (SNR) of the signal. Since there is no way to control the noise in the channel, in order to increase the capacity of an arbitrary channel, or the limit on the data rate at which the information can reliably be transmitted over the arbitrary channel, the user is left with the possibility of changing either the power or the bandwidth.

Simple inspection of this equation reveals that if the transmission power increases, whereas the bandwidth is kept constant, the capacity will also increase. For large values of P, SNR >> 1, and C simplifies to

$$C \approx W \log_2 \frac{P}{N_0 W} \tag{3.4}$$

However, P appears as the argument of the logarithmic operation, and the increase in capacity as a function of power becomes negligible if the SNR value is already large. Thus, although in theory any capacity can be reached by increasing the power, the power levels required can quickly become too large for practical applications and the capacity can only be feasibly further increased by increasing W. In this case, the capacity of the channel becomes bandwidth-limited.

As a second way of increasing capacity, one can look at the effect of increasing bandwidth for a fixed power. Bandwidth appears both inside and outside the logarithm in Eq. (3.3). Inside the logarithm, the increase in W decreases the SNR, which would decrease the channel capacity. Outside the logarithm, on the other hand, the increase in W increases the capacity. To determine which W dominates, the one outside the logarithm or the one inside, one can look at the plot of C as a function of W, shown in Fig. 3.2a. From this graph, it appears that the value of C nears an asymptote as W increases to a large value. This means that the increase in bandwidth alone results in an initial capacity increase only, after which it makes no further observable contribution. To find the numeric value of the asymptote, one can use L'Hospital's rule to find the limit on C as W tends to infinity:

$$\lim_{W\to\infty} C = \lim_{W\to\infty} W \log_2\left(1 + \frac{P}{N_0 W}\right) = \frac{1}{\ln 2} \cdot \frac{P}{N_0} = 1.44\frac{P}{N_0} \tag{3.5}$$

At the point of saturation, the capacity enters the power-limited regime, meaning that a further increase in capacity is only possible by increasing the power. Power-limited and bandwidth-limited regimes are illustrated in Fig. 3.2b. The consequence of the existence of these regimes is that an optimal point exists where moderate bandwidth and moderate power will result in a reasonable capacity of the channel, and that that the feasible improvement in capacity can only be achieved if the bandwidth and the transmission power are increased simultaneously.

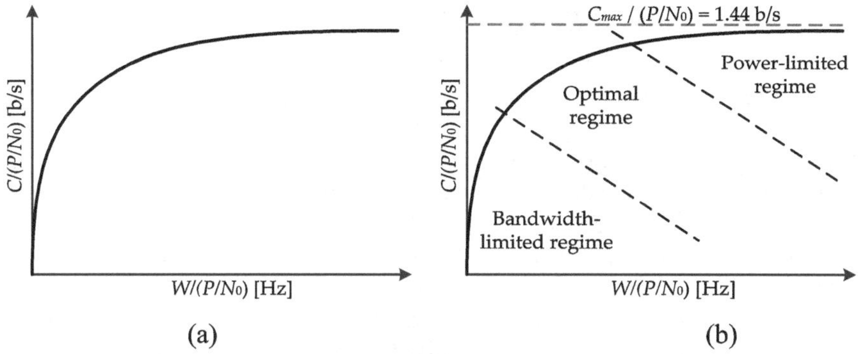

Fig. 3.2 Channel capacity as a function of bandwidth (both quantities normalized by P/N_0) with power kept constant (**a**); the same graph illustrating the infinite-bandwidth limit and the bandwidth-limited, power-limited and optimal regimes (**b**)

Note that the above is true if the transmission of information is done over a single channel. Logic dictates that the maximum data rate can also be increased with parallel transmissions—i.e. transmitting over multiple channels with narrower bandwidths. In practice, for networks such as 5G and 6G, both cases (one single wider channel with higher transmission power or multiple narrower channels with lower transmission powers), require that a lot of bandwidth be made available to achieve the required data rates. Allowing for the transmission to happen only in channels with the upper bound of 6 GHz thus places the limit on the allowed bandwidth, which introduces the capacity problem, which can only be solved (at least when it comes to communication by electromagnetic waves) by moving to millimeter-wave and terahertz bands offering multiple channels with high bandwidths. This will be discussed in the next section.

Note that the capacity of the channel is the bound on the data rate that can be achieved for a specific AWGN channel, but this capacity cannot be achieved with arbitrary signal coding schemes, and practical channel transmission rates are usually lower. The quest for methods to reach the theoretical capacity of the channel forms the basis of the channel coding theory and related research. In 5G, for example, near-capacity is reached by utilizing orthogonal signals. To place this into perspective, the concepts of multiplexing, channel coding, channel access and waveform design will form some of the topics of Sect. 3.4.

3.3 Millimeter-Wave and Terahertz Communication

The millimeter-wave and terahertz portion of the spectrum introduces one main advantage to telecommunication: the abundance of bandwidth. On the other hand, communication in this part of the spectrum has to overcome the challenges of high path loss, high atmospheric attenuation, high sensitivity to blockage and others.

3.3.1 *Abundance of Bandwidth*

The abundance of bandwidth is unlocked by moving to the largely uncongested part of the spectrum, which comprises millimeter-wave and terahertz frequency ranges. The millimeter-wave range encompasses a total bandwidth of 270 GHz, whereas the terahertz range, depending on the definition, typically has about ten times more bandwidth (2.7 THz), as shown in Fig. 3.3 [6, 7].

Apart from physical limitations and technical difficulties associated with access to this part of the spectrum, the spectrum must be allocated to allow its use, which is typically regulated on the government level. In most countries, a small percentage of the total frequency allocation is typically available without the need for licensing, where commercial applications have the incentive to operate. In Fig. 3.3, this spectrum is shown as the unlicensed (free) spectrum. In the millimeter-wave range, two unlicensed bands are available, a smaller one centered at 60 GHz with about 9 GHz of bandwidth, and a larger, less accessible one centered on 180 GHz with about 36 GHz of bandwidth. The remainder of the millimeter-wave range, some 225 GHz worth of bandwidth, can be used only if appropriate (paid) licensing is in place. In the terahertz regime, the licensing and allocation of the spectrum has still not been completed in most countries, as analyzed in Chap. 1.

Spectrum Sharing

Newer 5G releases (B5G), and later 6G, will most likely utilize both the licensed and unlicensed channels, which will have to be to be shared with existing users [8]. These users include satellite and point-to-point communications, as well as radar, sensing and imaging users.

Most present users of the millimeter-wave range are the users in the unlicensed 60-GHz band, despite the fact that this band suffers from large amounts of oxygen attenuation (discussed in Sect. 3.3.3). Europe and Southern Africa allow 9-GHz-wide transmissions between 57 and 66 GHz [9]. This is more than the 7 GHz allowed in the USA, Canada, Korea and Japan, 5 GHz allowed in China [10] and 3 GHz allowed in Australia [6]. Some licensed or reserved allocations can be seen above 60 GHz. Automotive radar and safety applications have already been using frequencies around 77 GHz (76–77 GHz) for some time, after the automotive radar at 77 GHz quickly

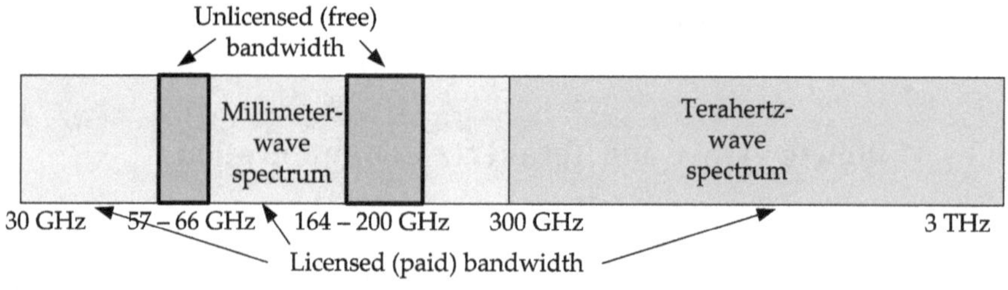

Fig. 3.3 Millimeter-and terahertz-wave bandwidth distribution (adapted from [6, 7])

replaced the previous allocation at 24 GHz because of the small size of the radar antenna, which allowed for seamless integration, as well as because the millimeter-wave propagation characteristics at these frequencies allowed for practical narrow beams [11, 12]. Other radar applications include the ultra-wideband short-range radar, operating at the 77–81 GHz window in Europe [13], as well as radar used for blind landing in aircraft [14]. Higher frequencies (100 GHz and above) are also used for millimeter-wave and terahertz sensing and imagining, as well as for radar for applications such as small target detection, gesture recognition and various automotive applications, including adaptive cruise control, blind spot detection and lane change assistance [7, 15]. However, the preferred frequency for these applications is in windows where oxygen attenuation exhibits local minima. Low-oxygen attenuation ranges below 100 GHz (e.g. 71–76, 81–86 and 92–95 GHz bands) are also used for point-to-point communication.

Terahertz frequencies, specifically, have been proposed for high-speed secure communications [15] which, by their nature, will not be able to form part of the 5G or 6G umbrella. These include WLANs (for fast data transfer without cabling), wireless personal networks (e.g. connection of peripherals, including hard disk drives or SSDs), secure fast data sharing (kiosk downloading or information showers [16]), fast connection of servers, short-range secure military communications and even satellite commutations.

When it comes to the lower end of the millimeter-wave spectrum, frequencies between 30 and 60 GHz are typically used for radio astronomy, satellite communication and imaging. In the USA, the band around 40 GHz is also used for licensed high-speed microwave data links.

In general, 5G, and later 6G, systems will need to support different spectrum authorization modes and sharing scenarios with incumbent services in different frequency bands [10]. Note that the spectrum allocated to the incumbent users will have to be protected [8]. In addition, the spectrum will also have to be shared between different wireless network operators. When operator sharing is introduced, some channel capacity will be lost owing to the limited ability to coordinate between unrelated networks, thus limiting the ability of service providers to maximize the capacity of their deployments individually. Furthermore, it is expected that the dynamic spectrum-sharing techniques will have an important role to play in the deployment of 5G and that the different spectrum authorization regimes will be characterized with different degrees of quality of service (QoS) guarantees.

Achievable Data Rates

Commercial applications in RF and microwave bands used by 4G and older wireless generations are currently able to reach speeds in the range of hundreds of Mb/s. When wireless communication is moved to the millimeter-wave range, operation at 100 Gbps will be possible, but with spectral efficiency of at least 14 bits/s/Hz [17]. However, even this spectral efficiency cannot be easily achieved. In principle, even 1 Tb/s is achievable in the millimeter-wave regime via a single link with the carrier frequency of 250 GHz and a 20% fractional bandwidth [18]. However, it requires a 40-dBi antenna gain at both the transmitter and the receiver, a noise figure of 20 dB and extremely high spectral-efficiency modulation and access schemes to navigate a distance of 10 m.

As a result, speeds surpassing 100 Gb/s and even 1 Tb/s could more realistically be obtained if the operation is moved to above 300 GHz, to the terahertz band, even with spectral efficiency of only a few bit/s/Hz. Consequently, the terahertz band is the only suitable band for ultra-fast communication. This can be justified by looking at Table 1.2 from Chap. 1 again and recalling the capacity equation (Eq. 3.3). The bottom part of Table 1.2 shows the potentially available channels in the terahertz range, which were identified up to 2 THz (as the spectrum above 2 THz remains mostly unexplored), in relation to the channels defined for the millimeter-wave range. While the best channel below 100 GHz offers 10 GHz of contiguous bandwidth and the best channel between 100 and 300 GHz offers 64 GHz of contiguous bandwidth, there are a few channels with over 100 GHz of contiguous bandwidth above 300 GHz. The bandwidth available at terahertz frequencies is thus more than one order of magnitude higher than in commercial millimeter-wave systems [19], which will allow much higher data capacity figures to be reached.

3.3.2 *The Free-Space Path Loss Problem of Millimeter and Terahertz Waves*

One of the main challenges when it comes to millimeter-wave and terahertz communications is the free-space path or the spreading loss. The path loss refers to the attenuation, or the decrease in power density, of the electromagnetic wave as it moves through the obstacle-free line of sight (LOS) path [20]. This is a result of the expansion (or spreading) of the electromagnetic wave in the space and increases quadratically with the operating frequency and the distance between the transmitter and the receiver [21].

For an arbitrary antenna, the power density at the distance r away from the transmitting antenna is

$$p(r) = G_T \frac{P_T}{4\pi r^2}, \tag{3.6}$$

where P_T is the transmitted power and G_T is the gain of the transmitting antenna. According to this equation, the power density decreases quadratically with the distance r away from the transmitter. If this power density is incident on the receiving antenna placed at this distance r, the received power will be the product of the effective aperture of the receiving antenna A_e and this value of power density:

$$P_R = A_e p(r). \tag{3.7}$$

The dimensions of the antenna scale with the signal wavelength λ, which means that the effective aperture of the antenna scales with the square of the wavelength:

$$A_e = \frac{\lambda^2}{4\pi}. \tag{3.8}$$

Furthermore, the receiving antenna can also have a certain amount of gain G_R. Thus, by substituting Eqs. (3.6) and (3.8) into Eq. (3.7) and introducing G_R, one can obtain the formula for the amount of power received by the antenna of the receiver placed in the LOS:

$$P_R = \frac{P_T G_T G_R \lambda^2}{(4\pi r)^2} \tag{3.9}$$

This relationship is known as the Friis law of spreading, and the inverse of the factor $\frac{\lambda^2}{(4\pi r)^2}$, or $\frac{(4\pi r)^2}{\lambda^2}$, is the free-space path or the spreading loss. Hence, the path loss increases not only with the increase in the distance away from the transmitting antenna, but also the decrease in the signal wavelength. Thus, the path loss at 30 GHz is 100 times larger at 30 GHz, where 5G will be operating, than it is at 3 GHz, where 4G operates, at the same distance away from the transmitting antenna. At 300 GHz, where 6G might operate, the path loss would be 10,000 times larger than that in 4G channels. Consequently, the path loss is usually converted into decibels (dB) to simplify calculations. The figures for path loss for several distances at 3, 30, 60, 300 and 600 GHz are shown in Table 3.2.

The high amount of free-space path loss puts a constraint on the radio link power budget. Fortunately, some of the loss can be made up by utilizing high-gaining antennas. Thus, the link budget calculation can be used as an indication of the possibility to use a certain combination of the transmitter and a receiver with specific antennas, to transmit over a predetermined distance at a specified frequency. On the receiver side of the link the received power is

$$P_R = P_T + G_T + G_R - L_0, \tag{3.10}$$

where P_R and P_T are expressed in dBm, G_T and G_R are expressed in dBi and L_0 denotes the free-space path loss. If the predicted received power is higher than the

Table 3.2 Free path loss for a different combination of frequency and distance

Distance/Frequency	3 GHz	30 GHz	60 GHz	300 GHz	600 GHz
3.16 m	51 dB	72 dB	78 dB	92 dB	98 dB
10 m	62 dB	82 dB	88 dB	102 dB	108 dB
100 m	82 dB	102 dB	108 dB	122 dB	128 dB
1 km	102 dB	122 dB	128 dB	142 dB	148 dB

receiver sensitivity, the link is feasible, were there no other losses. The maximum distances that millimeter-waves and terahertz waves could traverse before being submerged into noise would be in the range of only several hundreds of meters to several kilometers, even if there are no obstacles in the path.

The simple format of the link budget equation is convenient for adding additional losses the signal might incur. Typically, the transmitting and the receiving antenna will also have a certain amount of loss (L_T and L_R respectively) due to, for example, the impedance mismatch, pointing loss or the polarization mismatch. Other quantities that have an influence on the link budget are the loss due to the atmospheric attenuation (L_A) and the loss due to the blockage (L_B) [22]. The updated power link-budget equation is then

$$P_R = P_T + G_T + G_R - L_0 - L_A - L_B - L_T - L_R. \tag{3.11}$$

Atmospheric losses and the losses due to blockage are also quite significant in the millimeter-wave and terahertz regimes. These and other losses, as well as propagation of millimeter and terahertz waves in general, are discussed next.

3.3.3 *Propagation, Attenuation and Losses*

For transmission in the millimeter-wave and the terahertz regimes, propagation is possible only by LOS waves [20]. There are two types of LOS waves: direct waves and reflected waves. Figure 3.4 illustrates how a user can be reached by either type in an urban environment. Most transmissions are accomplished by direct waves, because the electromagnetic waves have a weak ability to diffract around obstacles with a size larger than the wavelength [22]. When the wavelength is in the range of several millimeters or several tens of millimeters, humans, animals, furniture and

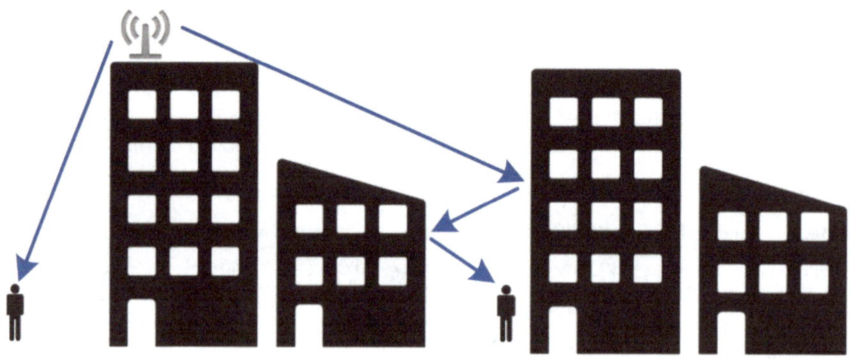

Fig. 3.4 Establishing a 5G connection in urban environments via direct waves (left arrow) and reflected waves (right arrows)

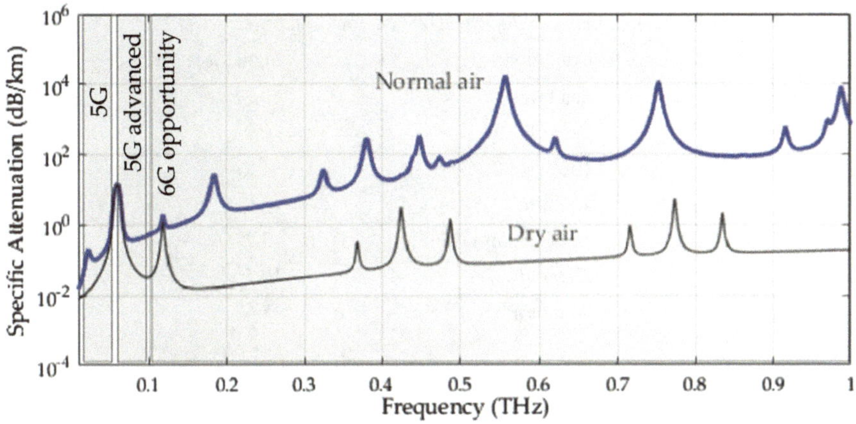

Fig. 3.5 Atmospheric attenuation of electromagnetic waves at sea level for normal (humid) air and dry air conditions

other objects appear as obstacles. This results in reflections and scattering of waves. The propagation of millimeter and terahertz waves is also influenced by atmospheric loss, polarization mismatch loss and pointing loss. Furthermore, these signals are heavily attenuated by walls, windows and other indoor and outdoor partitions, made from different materials.

Atmospheric Loss

Attenuation due to atmospheric loss is the major loss mechanism and tends to limit the range of millimeter-wave and terahertz communications in both indoor and outdoor settings. Two main contributors are the absorption of oxygen (O_2) and water vapor. Attenuation profiles for dry and normal (humid) air are shown in Fig. 3.5 for frequencies up to 1 THz. According to this figure, the oxygen absorption is most prominent at 60 GHz, where attenuation above 10 dB/km is present, as is evident from the graph for dry air. In most other frequency bands, the attenuation is below 10 dB/km. Water vapor (H_2O) absorption also influences the propagation, but it becomes prominent above 100 GHz, as is evident from the normal air graph.

Figure 3.5 does not take precipitation into account, which needs to be considered in outdoor settings. Attenuation due to precipitation arises because of the fact that even raindrops are comparable in size to the wavelength, which leads to wave scattering. As a result, outdoor links are seriously disrupted by precipitation. The attenuation figures for some common atmospheric conditions for frequencies between 70 and 80 GHz are shown in Table 3.3 [23].

Partition Loss

Another major type of loss is partition loss. Attenuation of waves due to various building objects can be so high that high-frequency signals are practically attenuated to the point at which they cannot be recovered anymore. Partition loss depends

Table 3.3 Atmospheric loss between 70–80 GHz in various conditions

Effect	Conditions	Signal loss (dB/km)
No precipitation	Sea level	0.22
Light rain	1 mm/h	0.9
Humidity	100% at 30 °C	1.8
Moderate rain	4 mm/h	2.6
Heavy fog	10 °C, 50 m visibility	3.2
Heavy rain	22 mm/h	10.7
Intense rain	50 mm/h	18.4

on the type of partition material and the thickness of the partition [16]. The attenuation figures of different materials are important for channel modeling and are usually determined experimentally. Attenuation values for some of the most common building materials are shown in Table 3.4. According to the data in this table, signals already have difficulty propagating through concrete walls at frequencies below 10 GHz, but can typically travel through glass, wood and drywall partitions with ease. As frequency increases towards 100 GHz (5G scenarios), signals will no longer propagate through wood but would still pass through glass and drywall. However, above 100 GHz (6G scenarios), even the thinnest of partitions will practically completely stop signal propagation. This is the main reason why deployment scenarios of 5G and 6G networks will have to shift from deploying microcells to deploying pico—and femtocells, especially in indoor environments.

Millimeter and terahertz waves are not blocked by only partitions (walls). Even smaller objects, such as furniture and people, have the potential to block the signal path. If the transmitted signal is strong, however, the waves will not be blocked completely; there will rather be a small area (proportional to the size of the object), where an attenuated version of the signal will exist. This effect is called shadowing [24], and is shown in Fig. 3.6. The shadowing concept gets its name from the similar effect related to light (e.g. a cloud passing "in front" of the sun on a clear day).

Scattering

Scattering is usually disregarded in the frequencies below 6 GHz, but becomes important at millimeter-wave and terahertz frequencies [16]. Scattering arises because the surfaces of buildings, walls and other objects cease to be electrically smooth at these frequencies, since the surface variations of these objects become comparable to the wavelength of incident waves, which was not the case for waves with wavelengths in centimeters. In the millimeter-wave regime, it is not uncommon that the scattered wave will have a power that is stronger than the reflected wave. Scattering is illustrated in Fig. 3.7.

The amount of power in the scattered wave can be calculated from the direct scattering model. Consider the incident wave, which impinges upon the surface at an angle θ_i, and the scattered wave at an angle θ_s, as shown in Fig. 3.7. The scattered electric field will be

Table 3.4 Partition loss of common materials Adapted from [16]

Material	Thickness (cm)	Frequency (GHz)	Attenuation (dB)
Drywall	2.5	2.5	5.4
	2.5	60	6.0
	38.1	28	6.8
	13.3	73	10.6
	13.3	73	11.7
	14.5	140	15.0
Clear glass	0.3	2.5	6.4
	1.2	28	3.9
	1.2	28	3.6
	0.3	60	3.6
	0.6	140	8.6
	1.3	140	16.2
	2.5	300	86.7
Heavily reinforced uniform concrete wall	35	4	22.0
	35	6	35.0
	35	9	64.0
Slightly reinforced uniform concrete wall	12.0	3	8.0
	12.0	7	13.0
	12.0	5	17.0
	12.0	10	27.0
	12.0	12	32.0
Concrete slab	3.0	45	13.9
Solid wood	2.0	45	4.8
	2.0	45	19.0
	2.0	100	20.4
	4.0	100	41.6
	3.5	300	65.5

$$|\mathbf{E}_s| = \left(\frac{SK}{d_t d_r}\right)^2 \frac{l\cos\theta_i}{F_{\alpha_R}} \cdot \left(\frac{1+\cos\psi}{2}\right)^{\alpha_R}, \tag{3.12}$$

where S^2 is the amount of power impinging on the wall element, which is spread in all directions, K is the antenna constant given by $\sqrt{60P_tG_t}$, d_t is the distance between the transmitter and the scattering surface, d_r is the distance between the scattering surface and the receiver, l is the length of the scattering object, ψ is the difference between the reflected angle θ_r and θ_s, α_R is the constant that determines the width of the scattering lobe and F_{α_R} is a result of an algebraic equation, as described in [25]

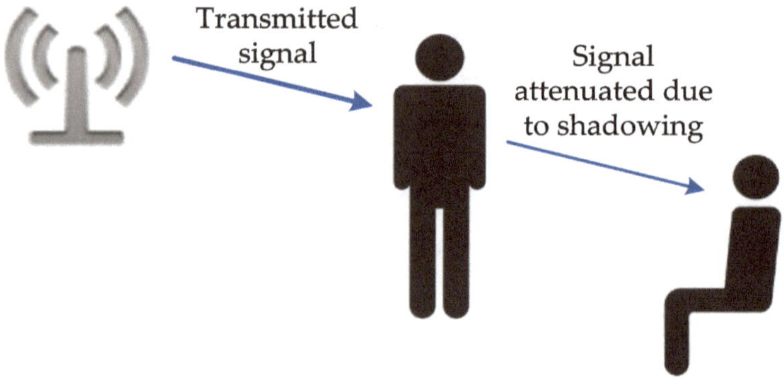

Fig. 3.6 Illustration of shadowing

Fig. 3.7 Scattering

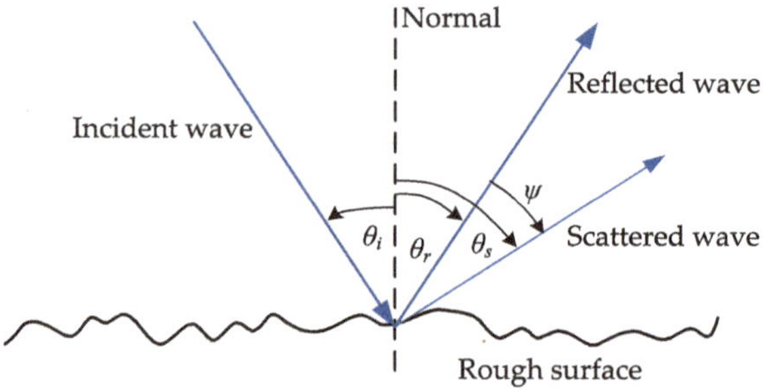

(derivation of this equation is beyond the scope of this book). The received power from the scattered wave can be worked out from the scattering electric field as

$$P_r = \frac{|\mathbf{E}_s|}{120\pi} \cdot \frac{G_r\lambda^2}{4\pi} = \frac{|\mathbf{E}_s| G_r\lambda^2}{480\pi}. \tag{3.13}$$

It can be shown that the amount of power in the scattered wave increases with frequency and can even surpass the reflected power. Thus, scattering becomes an important propagation mechanism above 30 GHz.

Antenna Losses

Other losses are associated with antennas used in the system. The two most prominent types of losses associated with antennas are pointing loss and polarization mismatch. In addition, impedance mismatch can result in attenuation of the signal to and from an antenna, but the concept of impedance mismatch loss is common in radio systems engineering, as it can occur when connecting any two RF blocks.

The pointing loss, or the pointing error, is the loss due to the offset between the receiver and the center of the LOS antenna beam. This offset can be due to incorrect aiming (e.g. during installation in fixed systems) or to various dynamic components if the transmitter and/or receiver are not stationary [26]. The significance of the pointing error increases with an increase in frequency, which is a consequence of the high directivity of millimeter-wave and terahertz beams, as will be discussed in the next section. The pointing loss is also an important consideration in optical communication systems, which will be discussed in Chap. 5.

The polarization loss is the loss that occurs when a receiver is not matched to the polarization of an incident electromagnetic field [27]. This is a consequence of the electromagnetic waves having a property called polarization—a direction of oscillation of the electric field portion of the electromagnetic wave. Usually, the terms horizontal and vertical polarization are used and antennas can be designed to have similar polarizations to minimize the loss [28]. Note that the wave polarization also tends to affect the partition loss discussed earlier. However, this fact was disregarded when compiling Table 3.4 and the best-case figures between horizonal and vertical polarizations were used.

Other Propagation Characteristics

Millimeter and terahertz wave loss mechanisms described up to now are those that are typically more prominent as the frequency increases. Naturally, there are other propagation characteristics that are common to all types of transmissions. Diffraction and multipath propagation are two propagation characteristics that will be briefly discussed here [24].

Diffraction, or bending of waves, occurs owing to obstructions in the path of a radio wave. Along its path, a signal may experience numerous obstructions, such as hills, vegetation, or man-made objects. Whether diffraction, shadowing, or complete signal blockage will occur depends on the relationship between the wavelength of the transmitted signal and the geometric size of the obstruction.

Reflected and scattered waves from walls and other objects can create simultaneous multiple propagation paths and lead to frequency-selective signal variability. Moving objects can also cause variability in the time domain. Thus, multipath-propagation characteristics are usually a combination of frequency and time domain shifts. Another multipath propagation effect is multipath fading, which is a result of amplitude and phase variation.

Lastly, it should be mentioned that the Doppler shift will also be more severe at terahertz frequencies. In reality, this means that the channel coherence time decreases, and that the channel status needs to be updated more frequently [16].

3.3.4 Overcoming Millimeter and Terahertz-Wave Link Challenges

The discussion on millimeter and terahertz waves in the previous section does not leave much room for optimism when it comes to establishing reliable communications. Many attenuation mechanisms have the tendency to render signals at these frequencies irrecoverable, which poses the question whether something can be done to overcome the challenges associated with transmission in these bands. Fortunately, there are several mechanisms.

First, it is a good idea to reconsider the path loss equation—Eq. (3.11). At this point, the following two assumptions can be made: (1) losses in the transmitting and the receiving antennas (L_T and L_R respectively) can be compensated for by adequate transmitting and receiving antenna gains (G_T and G_R), thus L_T and L_R can be omitted from this equation without loss of generality; and (2) the signal that is analyzed is transmitted via a LOS direct wave, so losses through blockage, which can be mitigated using relaying [29], are not applicable. Equation (3.11) then simplifies to

$$P_R = P_T + G_T + G_R - L_0 - L_A, \tag{3.14}$$

with the two main loss mechanisms attributed to the spreading loss L_0 and the atmospheric loss L_A.

It should firstly be noted that the combination of atmospheric attenuation and spreading loss should not be seen as an immediate problem. It was mentioned earlier that most experimentation with the millimeter-wave band was done in the free 60 GHz band, where there is a high amount of oxygen attenuation, which can be taken as a benchmark for establishing successful millimeter-wave links. Above 60 GHz, there are several bands, at least at frequencies up to 300 GHz, where the atmospheric attenuation is low enough (perhaps lower than 10 dB/km), so that the total attenuation is still lower than that seen in the 60-GHz band, even though the spreading loss increases as frequency increases [16, 30]. These frequency ranges are consequently suitable for transmission over several kilometers. Their rough location in the frequency spectrum is indicated in Fig. 3.8. This figure also shows the feasible transmission windows above 300 GHz. The widest bandwidth with steady attenuation can be found between 600 and 700 GHz where the attenuation is in the range of 100–200 dB. Although at these frequencies the attenuation prohibits transmission beyond several hundreds of meters, these frequencies will definitely be useful when designing femtocell links,

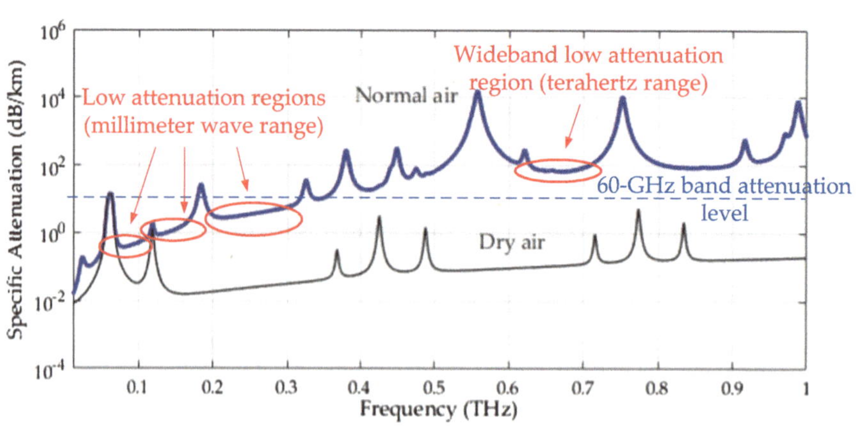

Fig. 3.8 Atmospheric attenuation of electromagnetic waves, with frequency bands with low attenuation indicated

proposed for 6G. Note, however, that the high attenuation bands can also be used to a certain advantage, albeit not in commercial wireless networks (5G or 6G). A high level of atmospheric attenuation is useful for the so-called whisper radio, i.e. short-range communication where privacy is a concern, for example between aircraft in military missions.

The attenuation problem is further alleviated by the high-level directivity that can be achieved at millimeter-wave and terahertz frequencies. Small antennas can be made extremely directional [6]. Recall that the required size of the antenna is proportional to the wavelength. For an antenna placed in air, the length of the antenna is exactly equal to the wavelength, while, for example, printed antennas can be made even smaller according to the relation [31]:

$$l = \frac{\lambda}{\sqrt{\mu_r \varepsilon_r}}, \tag{3.15}$$

where μ_r and ε_r are respectively the relative permeability and permittivity of the substrate on which the antenna is printed. The antenna radiation characteristics, in addition to the length, are also determined by the way in which it is excited.

The principle of antenna operation is based on the Ampere-Maxwell's law:

$$\Delta \times \boldsymbol{H}(t) = \boldsymbol{J}(t) + \frac{\partial \boldsymbol{D}}{\partial t} \tag{3.16}$$

D is the maximum dimension of the antenna, $\frac{\partial \boldsymbol{D}}{\partial t}$ is the displacement current, $\mathbf{J}(t)$ is the time varying current density and $\mathbf{H}(t)$ is the time varying magnetic field around the antenna. The radiation pattern is normally plotted separately for the horizontal and vertical dimensions. The radiation pattern of a directional antenna (in either direction) can be illustrated in Fig. 3.9 [32], with the main radiation lobe being distinctly larger than the side lobes. Furthermore, the directivity of highly directional antennas can be approximated by

$$D \approx \frac{32400}{\theta_H \theta_V}, \tag{3.17}$$

if θ_H and $\theta\text{-}_V$ are the widths of the beam in the horizontal and the vertical direction respectively, both expressed in radians. Highly directional antennas are often referred to as pencil beam antennas. Figure 3.10a–c illustrate how the directivity of beams compares between microwaves, millimeter-waves and terahertz waves [23].

For directional antennas, the gain of both transmitting and receiving antennas can be expressed in terms of antenna efficiency η_A and directivity:

$$G = \eta_A D. \tag{3.18}$$

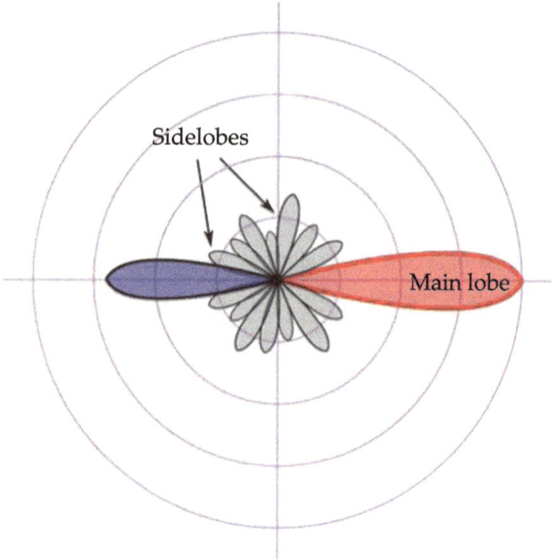

Fig. 3.9 Plot of the radiation pattern of a directional antenna [32]

As a result, directivity becomes a link budget parameter, which can be controlled:

$$P_R = P_T + \eta_{A,T} D_T + \eta_{A,R} D_R - L_0 - L_A. \tag{3.19}$$

According to Rappaport et al. [16], higher frequency links (e.g. above 100 GHz) are viable with less loss than lower frequency links (e.g at 30 or 70 GHz), because for the same transmitting power, antennas can be designed such that their directivity can overcome and surpass the additional spreading loss that occurs when moving the system from the lower frequency regime to the higher frequency regime. According to the same group of authors, pencil beam antennas, utilized in low atmospheric attenuation windows, will enable communication over the whole millimeter-wave range and well into the terahertz range. The current predictions specify "well into" in the previous sentence as the frequency of 800 GHz.

The directivity of millimeter and terahertz wave antennas can be used as an advantage in many other respects, including an increase in data throughput, which is the consequence of being able to pack multiple antennas in a small space (even on an IC or a package) and consequently being able to use multiple channels in a small geographical area. Note that this is different from the channel capacity discussed in Sect. 3.2. Furthermore, antennas can be made electronically steerable [22] in order to facilitate connection between mobile devices (not just stationary links). In this scenario, it is important to locate narrow beams, leading to the concept of beam finding [10]. Alternatively, multiple beams carrying the same information could be broadcast simultaneously, as shown in Fig. 3.10d. This leads to the MIMO, mMIMO

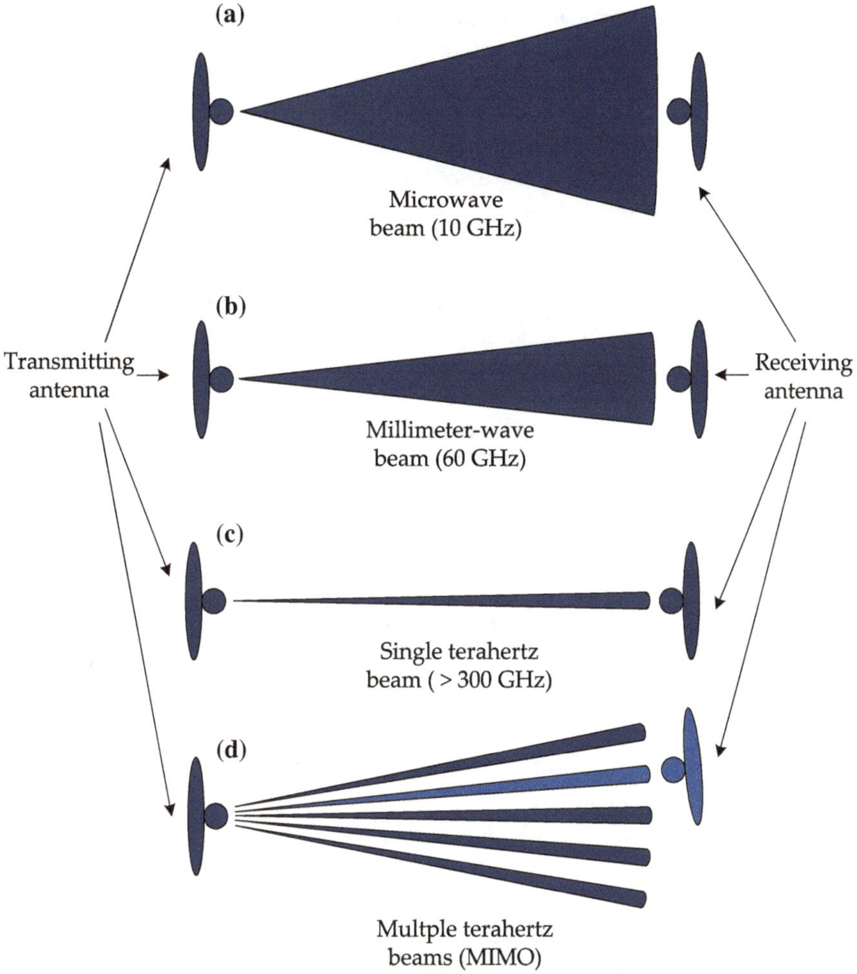

Fig. 3.10 Directivity of microwaves (**a**), millimeter waves (**b**) and terahertz waves (**c**); multiple terahertz waves used in the MIMO concept where one of multiple highly directional waves finds the receiver (**d**)

and umMIMO deployment scenarios. Yet another option is to use lenses to manipulate the magnetic field emitted by antennas [33]. However, by far the most promising approach is the manipulation of the electromagnetic wave propagation with the assistance of intelligent reflective surfaces (IRS), also called programmable metasurfaces. An IRS promotes the transmission from a source to a destination by adapting the propagation environment, which is achieved by beamforming its received signal towards the destination using passive metasurface elements [34]. The fact that metasurfaces do not require amplification in the form of power amplifiers, like traditional relaying, is one of the main advantages of this attractive concept.

Lastly, the transmitter power and the receiver sensitivity also play a major role in the link budget. With the emergence of new technologies, it becomes easier to build more efficient and more powerful power amplifiers and more sensitive receivers. In other words, P_T and P_R in Eqs. (3.14) and (3.19) can be made larger and smaller now more easily than in the past, and this trend is going to continue.

Discussions in this section just touch the surface of the multitude of technical aspects that are envisioned as making reliable communication in the millimeter and terahertz bands possible, and some of the discussions will continue in upcoming chapters. First, Chap. 4 will discuss challenges with building millimeter-wave and terahertz receivers and their main blocks, including power amplifiers and LNAs. Chapter 7 will, on the other hand, discuss innovative concepts such as umMIMO, holographic communication and IRS.

3.4 Multiplexing and Channel Access Methods for 5G and 6G

In order to efficiently utilize all the bandwidth of channels that are or will be made available for 5G, and later 6G, wireless protocols need to be designed such that they are able to utilize each channel as close to capacity as possible, when the need for that arises (i.e. as multiple users join the channel). Methods that can be used to achieve this are studied by channel coding theory [4, 5]. Two related components determine the use of the channel—multiplexing and channel access:

- Multiplexing determines how multiple signals can utilize a single resource, and includes concepts such as sampling, modulation and waveform design; while
- Multiple access determines which signal can utilize whatever resource is available at a specific moment and includes concepts such as frequency or time slot allocation.

Figure 3.1 at the beginning of this chapter showed the block diagrams of the transmitter (a) and the receiver (b). One aspect common to both diagrams included a modulation (or multiplexing) step in the signal processing (the prefix *de*-indicates the reverse process). Modulation implies that the properties of carrier signals, i.e. the signal that is suitable for amplification and transmission, are varied such that the information of interest is superimposed [4, 35]. Thus, before transmission, a certain modulation scheme needs to be deployed. The reverse process occurs in the receiver, where the received signal is stripped of the properties of the carrier signal; the carrier recovery step (not shown) is also required for this process. Modern telecommunication systems employ digital modulation, where the carrier signal is modulated by a discrete signal.

Some of the most common modulation schemes will be discussed in Sect. 3.4.1. However, it must be stressed that for upcoming wireless network generations, modulation, coding and synchronization of transmitted signals have to be reconsidered

[19, 36]. Traditional digital modulation schemes can be deployed, but the full benefits of millimeter-wave and terahertz channels will not be exploited in this manner. For example, when reaching terahertz frequencies, the distance plays a major role in how a signal can be transmitted, and it was proposed that distance-aware modulation schemes should be developed. When it comes to coding, new, ultra-low complexity coding schemes are required for terahertz transmission. Lastly, synchronization on the receiver side (or even in transmitters, in the case of MIMO, where multiple synchronized transmitters are deployed) becomes much more difficult when a signal is transmitted at rates approaching or even surpassing the 1-Tb/s range. This is, among other reasons, because it becomes much more complex to sample the signals and perform carrier recovery at these data rates. This leaves a lot of room for discussion of currently deployed and proposed channel access methods, which will be done in Sect. 3.4.2.

3.4.1 Common Modulation Schemes

Some of the modulation schemes commonly used at present include [1, 4]:

- Phase-shift keying (PSK);
- Frequency-shift keying (FSK);
- Pulse-amplitude modulation (PAM); and
- QAM.

Each of these modulation schemes has a different bandwidth utilization efficiency.

PSK, PAM and QAM are referred to as single-carrier modulation schemes, whereas FSK is a multi-carrier modulation scheme. All these modulation schemes are memoryless, because the mapping of the digital sequence is performed without insight into the information of previously transmitted signals.

The effectiveness of each modulation scheme can be evaluated by the bandwidth utilization efficiency (channel efficiency).

Phase Shift-Keying

PSK is a digital signal modulated onto the carrier by changing its phase. A finite number of phases (M) is used. Using two phases (0 and 180°) for bits 0 and 1 results in binary PSK (BPSK), using four (0, 90°, 180°, 270°) for bit combinations 00, 01, 10 and 11 results in quadrature PSK (QPSK) and using eight phases for eight three-bit combinations leads to octal PSK. If the signal pulse has the shape defined by $g(t)$, then M signal waveforms (symbols) are represented by

$$s_m(t) = g(t)\cos\frac{2\pi}{M}(m-1)\cos 2\pi f_c t - g(t)\sin\frac{2\pi}{M}(m-1)\sin 2\pi f_c t \quad (3.20)$$

for $m = 1, 2\ldots, M$, $0 \le t \le T$, where f_c is the carrier and T is the period of the signal. A typical waveform of the QPSK-modulated signal is represented in Fig. 3.11.

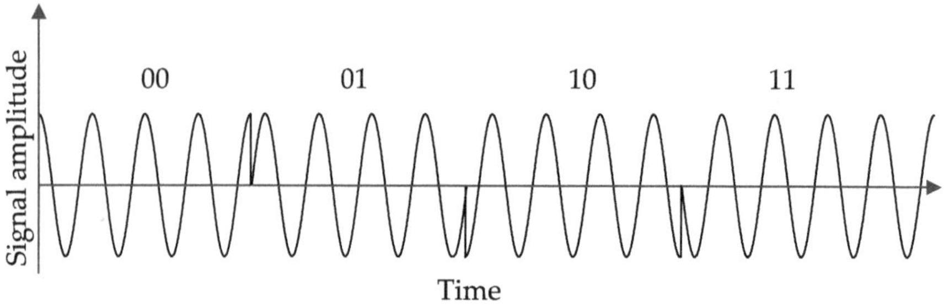

Fig. 3.11 The QPSK-modulated signal waveform

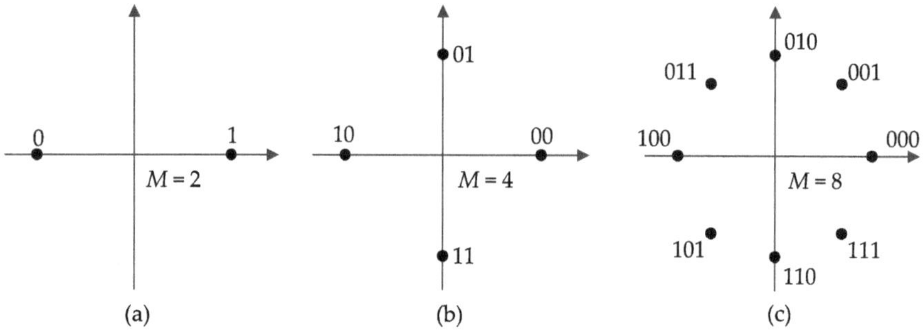

Fig. 3.12 Signal constellations of (**a**) BPSK, (**b**) QPSK and (**c**) octal PSK

Constellations, or space diagrams, of BPSK, QPSK and octal PSK are shown in Fig. 3.12.

The channel efficiency of BPSK is 1 bps/Hz, but it increases as the complexity of modulation is increased. For example, a data rate of 3.5 Gb/s can be achieved by QPSK in a 2.16-GHz channel in the 60-GHz band [37].

Frequency Shift-Keying

FSK is a result of modulation of the data onto the carrier with two local oscillators (LOs) running at different frequencies. Using two frequencies to represent 0 and 1 results in a binary FSK, as shown in Fig. 3.13.

Pulse-Amplitude Modulation

PAM is achieved by encoding of the signal as the amplitude level in a series of pulses. A finite number of amplitudes (M) is used. Each of the M signal waveforms can be represented as

$$s_m(t) = A_m g(t) \cos 2\pi f_c t \tag{3.21}$$

for $m = 1, 2\ldots, M$, $0 \le t \le T$, with A_m as the signal amplitude that can take discrete levels

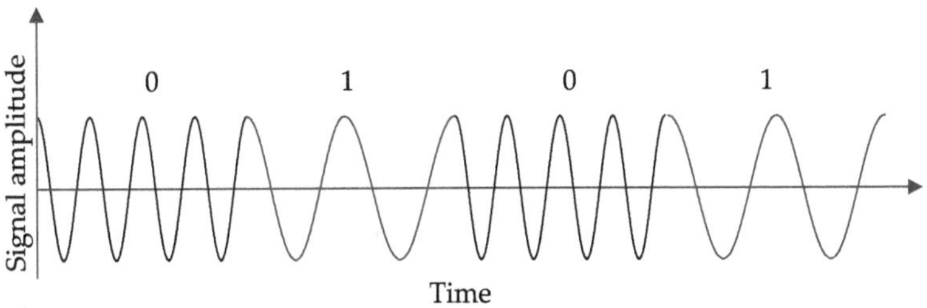

Fig. 3.13 Signal waveform resulting from binary FSK

$$A_m = (2m - 1 - M)d. \quad (3.22)$$

In Eq (3.22), the value of $2d$ is defined as the distance on the amplitude axis between two signal amplitudes, as illustrated in Fig. 3.14.

As with PSK, one-bit ($M = 2$), two-bit ($M = 4$) or three-bit ($M = 8$) symbol combinations are suitable for practical implementation, resulting in 2-PAM, 4-PAM and 8-PAM respectively. Constellations for three different cases are shown in Fig. 3.15.

Quadrature Amplitude Modulation

In QAM, two or more separately modulated signals are combined on the carriers that are out of phase. This results in a combination of a PSK scheme with another scheme, usually PAM. For example, a combination of PAM containing M_1 amplitude levels and PSK with M_2 phases will result in a signal constellation with $M = M_1 M_2$ waveforms. Each of the M waveforms M can be represented by

$$s_m(t) = A_{mc} g(t) \cos 2\pi f_c t - A_{ms} g(t) \sin 2\pi f_c t \quad (3.23)$$

for $m = 1, 2\ldots, M$, $0 \leq t \leq T$, where A_{mc} and A_{ms} are information-bearing signal amplitudes of the quadrature carriers and $g(t)$ is the signal pulse, as before. Several signal constellations ($M = 4$, 8, 16 and 32) are shown in Fig. 3.16.

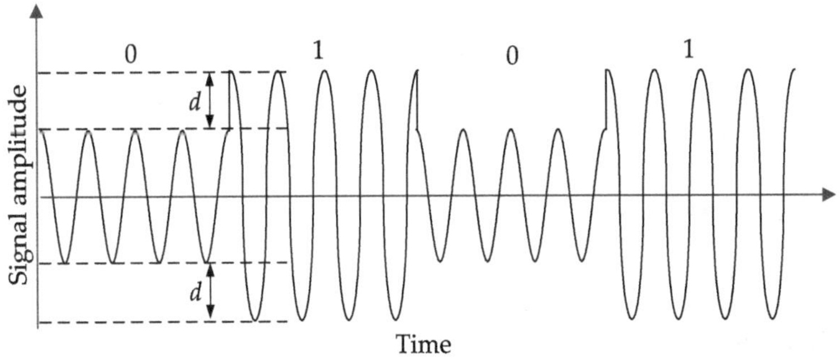

Fig. 3.14 The pulse-amplitude-modulated signal waveform ($M = 2$)

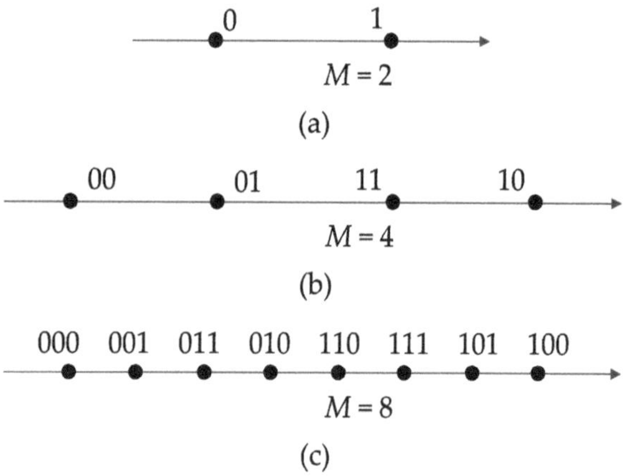

Fig. 3.15 PAM constellations for $M = 2$ (**a**), $M = 4$ (**b**) and $M = 8$ (**c**)

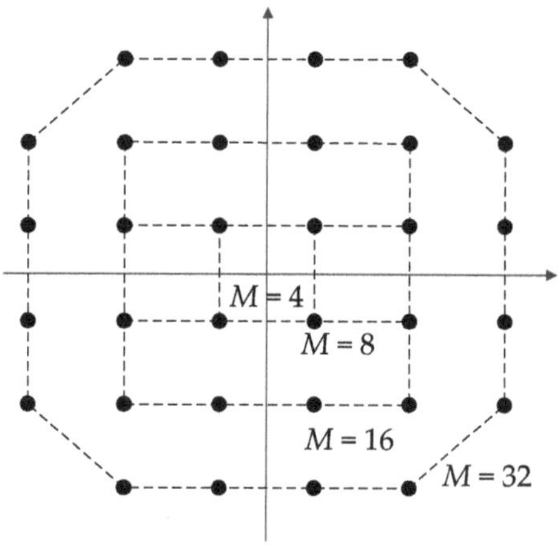

Fig. 3.16 QAM constellations for different values of M

QAM is one of the more bandwidth-efficient modulation schemes. 16-QAM can achieve 7 Gb/s data transfer rate in the 2.16-GHz wide channel in the 60-GHz band, which is double what can be achieved by QPSK [37]. However, realizing direct conversion QAM transceivers with dimensions of $M = 16$ and higher is challenging because the SNR requirement is much larger than that of QPSK and other simpler schemes. A more detailed analysis of the data rates that are realistically achievable in 5G networks is done in the next section.

Comparison of Modulation Schemes

Different modulation schemes can be compared in terms of the maximum data rate (capacity) that can be achieved in the fixed-bandwidth channels. Pärssinen [38] analyzed the data rates that can be achieved with different modulation schemes in 5G NR channels with bandwidths of 400 and 900 MHz. The results are shown in Table 3.5. These data rates would be achieved with OFDM, which will be discussed

Table 3.5 Maximum achievable data rates in 400-and 900-MHz bandwidth channels in the 5G NR standard (from [38])

Channel bandwidth (MHz)	BPSK R_{max} (Gb/s)	QPSK R_{max} (Gb/s)	16-QAM R_{max} (Gb/s)	64-QAM R_{max} (Gb/s)	256-QAM R_{max} (Gb/s)
400	0.36	0.71	1.42	2.13	2.85
900	0.76	1.53	3.06	4.59	6.11

Table 3.6 SNR required for transmission with different modulation schemes, assuming BER = 10^{-3} (from [38])

Modulation scheme	BPSK	QPSK	16-QAM	64-QAM	256-QAM
Link-level SNR	7	10	17.5	24	30

in Sect. 3.4.2, and depend on various factors, such as the coding rate, number of subcarriers in OFDM and subcarrier spacing.

From this table, it can be concluded that even a 10 Gb/s maximum rate is an incredibly challenging target for 5G with 400-MHz worth of bandwidth. However, about half of that requirement is reached in a 900-MHz channel, but only with 256-QAM. Even this achievement comes at an enormous price in the SNR, with at least 30 dB required even if the channel alone is considered and for BER of 10^{-3} (and other losses, such those in the receiver and transmitter are ignored). Lower-dimension modulation schemes, on the other hand, require a significantly lower SNR, as can be seen in Table 3.6.

3.4.2 *Advanced Channel Access Methods*

The spectrum is best utilized using advanced channel access methods, which combine advanced modulation schemes and sharing of radio resources between multiple users [39]. There are several classes of channel access methods, each using a different approach to reaching the channel capacity:

- Orthogonal multiple access systems, which use orthogonal waveforms in the time or frequency domain, or both.
- Spread spectrum systems or CDMA systems, which do not use time or frequency separation, but rather use spreading codes, which are themselves orthogonal, to encode the information.
- Multi-carrier with filtering schemes, which are based on orthogonal schemes, but add the filtering component for good spectral containment of transmitted signals.
- Non-orthogonal multiple access systems, which use multiplexing in the time, frequency, power and code domain.

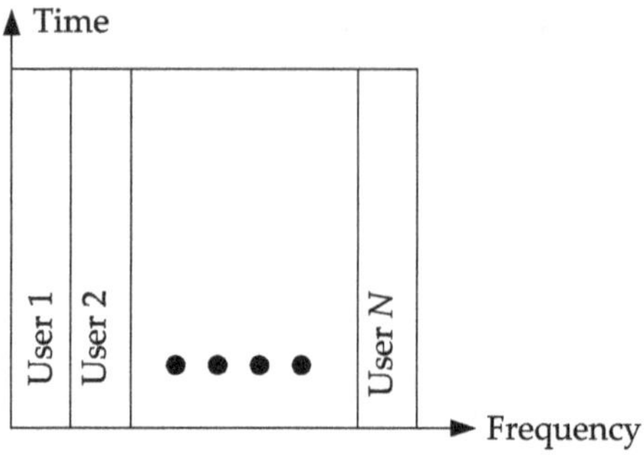

Fig. 3.17 Frequency resource sharing in FDMA

Wireless network generations from 1G to 5G used orthogonal multiple access schemes, apart from 3G that used CDMA. Early networks (1G and 2G) used FDMA and TDMA, which are the simpler orthogonal schemes. On the other hand, 4G and 5G deploy a more complex OFDMA system. However, future network generations, including B5G and 6G, will probably have to use non-orthogonal schemes, such as the non-orthogonal multiple access (NOMA) strategy, which might be able to support many more users than OFDMA [40].

Simple Orthogonal Multiple Access Systems

In a single-cell environment, all orthogonal multiple-access schemes can achieve similar capacity figures, but each has certain advantages over the others [39]. Orthogonal multiple access schemes include FDMA, TDMA and OFDMA. FDMA and OFDMA are similar, except that OFDMA allows overlapping frequency sub-bands, whereas that is not allowed for FDMA. OFDMA, as a more complex scheme, will be discussed in a separate section.

In FDMA, the total allowed frequency is divided into a set number of frequency sub-bands, as shown in Fig. 3.17, using a technique called frequency division multiplexing. FDMA is considered a simple, yet effective scheme, because the signals can be detected by band-pass filtering. This scheme needs a frequency guard band to separate the neighboring sub-bands.

TDMA is conceptually similar to FDMA, except that the signals are multiplexed in the time domain, while being allowed to share the complete bandwidth of the frequency band. This concept is illustrated in Fig. 3.18. Each time slot that is obtained in this manner forms a channel in the TDMA system. The TDMA system requires a time guard to separate the channels. In addition, time equalization between channels is needed for reliable transmission, which increases in complexity as the number of users increases. This is why the TDMA was used in GSM (as part of the hybrid TDMA/FDMA scheme), but was then abandoned in later generations of networks, which, with the exception of 3G, use only frequency domain equalization (FDE) as a part of the more advanced OFDMA.

Fig. 3.18 Time resource sharing in TDMA

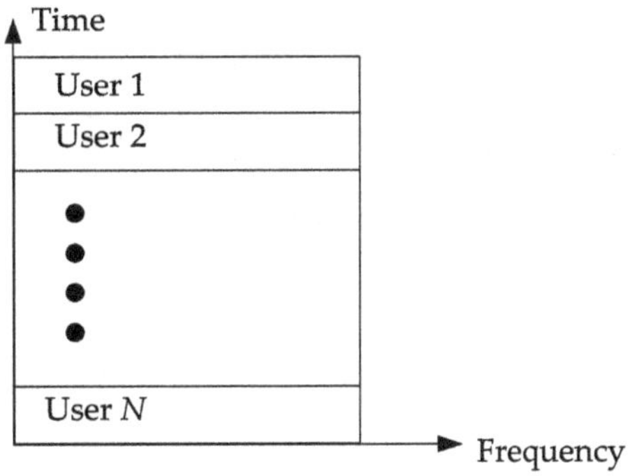

Orthogonal Frequency-Division Multiplexing

OFDMA is based on the modulation scheme called orthogonal frequency division multiplexing (OFDM). OFDMA is a technique that is used to achieve high data transmission rates in 4G and 5G. This is accomplished by transmitting data over a large number of carriers simultaneously rather than using a single carrier with a high data rate.

OFDM modulates the signals such that same-amplitude carriers are separated in frequency only just enough so that intermodulation products arising from one frequency are negligible at the frequencies of the other carriers, thus avoiding inter-symbol interference. The cyclic prefix (CP) insertion is used for this purpose. Carriers are generated with the aid of the inverse fast Fourier transform (IFFT) operations on blocks of M symbols at the transmitter. At the receiver, signals are extracted by performing fast Fourier transform (FFTs) on blocks comprised of M discrete samples [41–43]. The process of OFDM in the transmitter is shown in Fig. 3.19 [44] and addi-

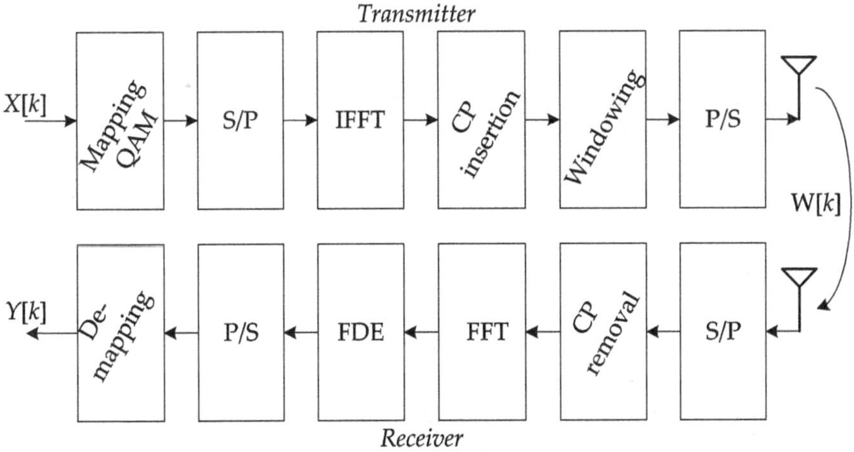

Fig. 3.19 Block diagram of the OFDM transmitter and the receiver. Adapted from [44]

tionally includes QAM-type mapping, serial-to-parallel (S/P) conversion, windowing and parallel-to-serial (P/S) conversion. The inverse of this process, which occurs in the receiver, is also shown in this figure, and it also includes FDE. Note that OFDM is particularly susceptible to nonlinear distortion caused in the transmitter.

Spread Spectrum Access Methods

Spread spectrum access or CDMA methods refer to techniques in which the signals with a specific frequency are intentionally spread in the frequency domain by using a spreading sequence. This results in signals with a wider bandwidth; however, spreading is done because it offers certain advantages, including making signals less prone to noise or preventing eavesdropping [45, 46]. Most popular spread spectrum schemes are frequency hopping systems and direct sequence spreading systems (DSSS) [39]. In frequency hopping schemes, signals are sent as a combination of TDMA and FDMA, but the frequency of the transmission by each user is changed according to a hopping sequence. In DSSS, there is no frequency or time division and users can transmit simultaneously, as long as they have unique spreading sequences for signal encoding. For DSSS to work, near perfect synchronization is required.

Multi-carrier with Filtering Schemes

OFDM can be extended by adding an integrated filtering component, which adds to spectral containment properties of the transmitted signals. The added waveform property enables partitioning of the spectrum into independent, individually controllable sub-bands. As a result, each sub-band, and the related carrier, can optimally adapt to signal conditions of individual user links [39, 44]. The most popular multicarrier with filtering scheme, called the filter bank multicarrier (FBMC), which allows subcarrier-based filtering with various pulse-shaping functions, was proposed for use in wireless networks beyond 5G.

A block diagram of the FBMC system is shown in Fig. 3.20. Two differences between the OFDM block diagram in Fig. 3.19 and the FBMC diagram in Fig. 3.20 can be noted. First, QAM signal mapping is replaced by offset QAM (OQAM). In OQAM, the orthogonality is constrained to the real-valued signal space. Secondly, CP insertion and CP removal are replaced by the appropriate synthesis and analysis filtering in the transmitter and the receiver, respectively.

Although OFDM and FBMC are conceptually the same and FBMC introduces several advantages, the implementation of FBMC systems is not as simple as that of OFDM systems. This is considered the primary drawback of FBMC schemes.

Non-orthogonal Channel Access Systems

Channel access systems that use non-orthogonal waveforms have recently received increased research attention, as these access schemes have the potential to outperform those with orthogonal waveforms, specifically in interference management [47]. Compared to non-orthogonal access schemes, orthogonal access schemes are limited in flexibility of the waveform design [39, 40, 44]. This advantage of NOMA schemes comes to light when a massive number of devices try to connect to the same network,

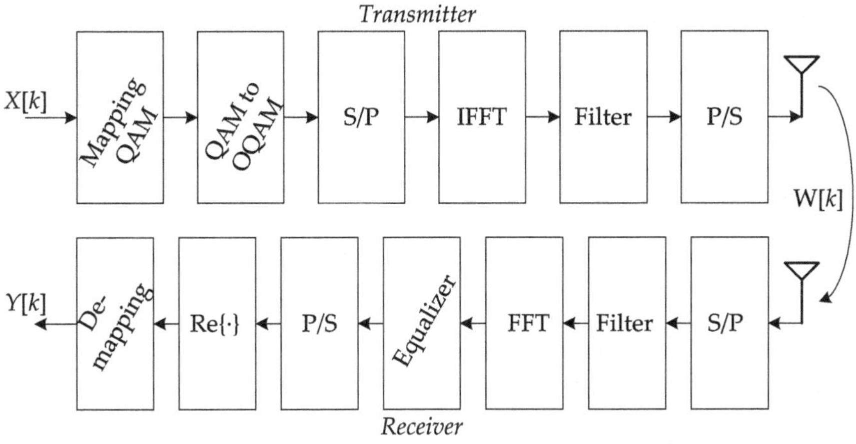

Fig. 3.20 Block diagram of the FMBC transmitter and receiver Adapted from [39, 44]

but to send or receive only small packets. In this case, the orthogonal schemes will [40]

- limit the maximum number of connections;
- be unable to achieve the sum rate capacity in the DL if knowledge of the channel is unavailable at the base station; and
- require dynamic scheduling in the UL (request and grant), which significantly increases the latency and communication overhead, especially when it comes to small packets.

On the other hand, NOMA schemes in this case would be able to assign multiple users on the same resource element. If compressive random access is used, grant-free UL is possible, thus significantly decreasing the scheduling overhead as the number of connections increases.

Two major directions in NOMA research are power domain NOMA (PD-NOMA) and code domain NOMA (CD-NOMA) systems. PD-NOMA systems use superposition coding in the power domain, whereas CD-NOMA systems use codebooks with different overloading factors.

Non-orthogonal channel access can also be achieved with sparse-code multiple access (SDMA). SDMA is a combination of NOMA and CDMA, thus combining power domain multiplexing with spectrum spreading. The codewords are multi-dimensional and have higher coding gain as well as shaping gain than those of single-dimensional codes. SMDA is more complex to implement than NOMA or orthogonal access systems on their own.

Other hybrid schemes that combine orthogonal and non-orthogonal access methods are also being researched at present [48]. Going into the detail of these scenarios goes beyond the scope of this book, but references cited in this section can serve to initiate further probing on this topic.

3.4.3 Full-Duplex Communication

There is another interesting way of increasing the data rate of wireless communications that has not been discussed in this chapter yet. It is called full-duplex communication [47, 49].

Traditionally, the spectrum is accessed in the so-called half-duplex mode, meaning that transmission and reception cannot occur at the same time. This is true for all wireless network generations prior to 5G, and for all spectrum access methods discussed in the previous section. In the full-duplex mode, however, the transceiver devices are designed in such a way that they can simultaneously transmit and receive signals, thus utilizing the spectrum twice as efficiently. The biggest obstacle towards realizing full-duplex communication is self-interference (echo), which arises from the fact that the transmitted signal is much stronger than the signal that needs to be received. Cancellation of self-interference is thus the main focus of full-duplex researchers. Some of the techniques proposed for self-interference cancellation within the receiver include:

- Passive signal suppression (achieving high isolation between transmit and receive antennas);
- Analog cancellation (crude self-interference cancellation); and
- Digital cancelation (fine self-interference cancellation).

To date (2020) the complexity of full-duplex transceiver implementation has not allowed its integration in 5G, although this was planned. However, there are scenarios, typically associated with B5G and 6G, in which full-duplex implementation may be more realistic. One such scenario is V2V communications [49]. Usually, vehicles communicate over shorter distances, allowing for relatively strong received signals, and the antennas can be isolated by a highly separated transmit and receive antenna arrangement. In addition, the attenuation can be offset by increasing the number of receive antennas. Other envisaged full-duplex applications in 6G include backhaul, relaying, vehicular radar, small-cell base stations and VR/AR/XR headsets [50].

Regrettably, most of the currently demonstrated full-duplex implementations have been at lower frequency bands (<6 GHz). Demonstrations of transceivers operating at millimeter-wave frequencies are still scarce. It should also be noted that full-duplex communication is even more difficult to achieve in mMIMO scenarios where multiple transmit and receive antennas are arranged in dense arrays. It is thus clear that more inter-disciplinary research is needed before commercialization of the full-duplex concept is possible [47].

3.5 Power Limitation

The last factor that determines the channel capacity—or the maximum achievable data rate—is the power of transmitted signals.

According to the discussion in Sect. 3.2, increasing the power while simultaneously increasing the bandwidth increases channel capacity. Secondly, from the discussion in Sect. 3.3, one can see that the increase in transmitted power relaxes the constraints in the link budget. Hence, the increase in the transmit power can sound like a "shortcut" to achieving high data rate communication. However, increasing power indefinitely is not feasible for various reasons. Some of these reasons are listed below.

- *Increased energy consumption.* Increasing maximum power above a certain level leads to unreasonable power consumption. This is specifically problematic with mobile devices or stationary devices such as smart sensors with no access to grid power. Secondly, considering that most of the energy used by humanity at present does not come from renewable resources, even deploying extremely high-power base stations could become problematic. Chap. 6 elaborates further on this issue.
- *Increased heat generation.* Cooling of high-power devices becomes increasingly challenging as the amount of transmit power increases. Most real-life power amplifiers have low efficiency figures, resulting in high power dissipation in the form of heat. Excessive heat dissipation can be a problem both in base stations and in mobile or stationary terminals, whenever these operate in the UL.
- *Physical power generation limits.* Power amplifiers use solid state devices to generate transmission power. The physics of solid-state devices dictates the maximum achievable power levels. There is also a tradeoff between the amount of power that can be generated and the frequency at which this power can be emitted. This means that extremely high-power transmissions are more readily achievable at lower frequencies (e.g. below 6 GHz) than at higher frequencies (e.g. above 30 GHz). Solid state devices (transistors) will be discussed later in this book, in Chap. 4.
- *Health issues.* Several health concerns associated with high power transmission have been raised in recent years, especially when it comes to 5G networks. These include the effect of electromagnetic radiation emitted at high power levels, as well as the health effect of heating of human (and animal) tissue as a result of the radiated heat. Although electromagnetic radiation is non-ionizing, electromagnetic power levels are subject to strict limits, as will be discussed in Chap. 9. These limits are imposed on the maximum power that can be emitted both by base stations and terminals.

Thus, the focus of future generations of wireless networks should rather be on the effective use of the available spectrum for moderate power levels, while researching other, non-electromagnetic types of communication.

3.6 Concluding Remarks

In this chapter, the unlocking of millimeter-wave and terahertz frequencies was discussed as the currently most feasible way of achieving the high-data figures required by generations of wireless networks beyond 5G. The capacity of any channel is governed by the capacity equation that identifies the bandwidth, which can be abundantly found in millimeter-wave and terahertz ranges, as one of only three main contributors to achieving high data rates. The transmit power and sharing of the channel resources are other contributing factors in the high data rate quest. Based on this analysis, various challenges and opportunities relating to all three of these aspects were discussed throughout this chapter, and some more advanced, in-depth discussions are left for later chapters.

In this chapter, it was assumed that device-level technologies that would allow access to all frequencies over the entire frequency spectrum of interest already exist or should become available in the near future. This is (still) not the case, thus the discussion on the present state of required technologies, as well as aspects relating to circuit design at millimeter-wave and terahertz frequencies, warrants a separate discussion. This will be conducted in the chapter that follows.

References

1. Božanić M, Sinha S (2017) Millimeter-wave low noise amplifiers. Springer, Cham
2. V.431 : Nomenclature of the frequency and wavelength bands used in telecommunications [Internet]. [cited 2020 Apr 27]. https://www.itu.int/rec/R-REC-V.431-8-201508-I/en
3. Božanić M, Sinha S (2015) Power amplifiers for the S-, C-, X- and Ku-Bands: An EDA perspective. Springer
4. Proakis JG, Salehi M (2002) Communication systems engineering. Prentice Hall
5. Proakis JG (2001) Digital communications. McGraw-Hill
6. du Preez J, Sinha S (2016) Millimeter-wave antennas: configurations and applications. Springer, Cham
7. Božanić M, Sinha S (2019) Emerging transistor technologies capable of terahertz amplification: a way to re-engineer terahertz radar sensors. Sensors 19:2454
8. Matinmikko M, Latva-aho M, Ahokangas P, Seppänen V (2018) On regulations for 5G: micro licensing for locally operated networks. Telecommun Policy 42:622–635
9. Alonge A, Afullo T (2016) 60 GHz millimeter-wave radio in South Africa: link design feasibility and prospects. 2016 progress in electromagnetic research Symposium (PIERS), pp 3686–3691
10. Osseiran A, Monserrat JF, Marsch P (2016) 5G mobile and wireless communications technology. Cambridge University Press
11. Hsiao Y, Chang Y, Tsai C, Huang T, Aloui S, Huang D et al (2016) A 77-GHz 2T6R transceiver with injection-lock frequency sextupler using 65-nm CMOS for automotive radar system application. IEEE Trans Microw Theory Tech 64:3031–3048
12. Hasch J, Topak E, Schnabel R, Zwick T, Weigel R, Waldschmidt C (2012) Millimeter-wave technology for automotive radar sensors in the 77 GHz frequency band. IEEE Trans Microw Theory Tech 60:845–860
13. Božanić M, Sinha S (2019) Systems-level packaging for millimeter-wave transceivers. Springer, Cham

14. Hagelen M, Briese G, Essen H, Bertuch T, Knott P, Tessmann A (2008) A millimetrewave landing aid approach for helicopters under brown-out conditions. 2008 IEEE Radar Conference, pp 1–4
15. Božanić M, Sinha S (2020) Millimeter-wave integrated circuits: methodologies for research, design and innovation. Springer Nature
16. Rappaport TS, Xing Y, Kanhere O, Ju S, Madanayake A, Mandal S et al (2019) Wireless communications and applications above 100 GHz: opportunities and challenges for 6G and beyond. IEEE Access 7:78729–78757
17. Kürner T, Priebe S (2014) Towards THz communications-status in research, standardization and regulation. J Infrared Milli Terahz Waves 35:53–62
18. Calvanese Strinati E, Barbarossa S, Gonzalez-Jimenez JL, Ktenas D, Cassiau N, Maret L et al (2019) 6G: the next frontier: from holographic messaging to artificial intelligence using subterahertz and visible light communication. IEEE Veh Technol Mag 14:42–50
19. Akyildiz IF, Jornet JM, Han C (2014) Terahertz band: next frontier for wireless communications. Phys Commun 12:16–32
20. Pozar DM (2012) Microwave engineering. John Wiley & Sons
21. Zhang L, Liang Y-C, Niyato D (2019) 6G visions: Mobile ultra-broadband, super internet-of-things, and artificial intelligence. China Commun 16:1–14
22. Niu Y, Li Y, Jin D, Su L, Vasilakos AV (2015) A survey of millimeter wave communications (mmWave) for 5G: opportunities and challenges. Wirel Netw 21:2657–2676
23. Adhikari P (2008) Understanding millimeter wave wireless communication. Loea Corporation, pp 1–6
24. Lambrechts W, Sinha S (2019) Last mile internet access for emerging economies. Springer
25. Degli-Esposti V, Fuschini F, Vitucci EM, Falciasecca G (2007) Measurement and modelling of scattering from buildings. IEEE Trans Antennas Propag 55:143–153
26. Brooker R, Vorderbrueggen D (2005) Antenna pointing accuracy impact on geostationary satellite link quality and interference. In: Proceedings of the 23rd AIAA International Communications Satellite Systems and 11th Ka and Broadband Communications Joint Conference, Rome, Italy
27. Polarization loss-MATLAB polloss [Internet]. [cited 2020 May 4]. https://www.mathworks.com/help/phased/ref/polloss.html#btrlmm_-2
28. Nayeri P, Lee K-F, Elsherbeni AZ, Yang F (2011) Dual-band circularly polarized antennas using stacked patches with asymmetric U-slots. IEEE Antennas Wirel Propag Lett 10:492–495
29. Stratidakis G, Papasotiriou EN, Konstantinis H, Boulogeorgos A-AA, Alexiou A (2020) Relay-based blockage and antenna misalignment mitigation in THz wireless communications. 2020 2nd 6G Wireless Summit (6G SUMMIT), pp 1–4
30. Xing Y, Rappaport TS (2018) Propagation measurement system and approach at 140 GHz-moving to 6G and above 100 GHz. In: 2018 IEEE Global Communications Conference (GLOBECOM), pp 1–6
31. Tummala R (2007) System on package: miniaturization of the entire system. McGraw Hill Professional
32. Preez J du, Sinha S (2017) Millimeter-wave power amplifiers. Springer
33. Kokkonen M, Myllymäki S, Jantunen H (2020) 3×3 Dipole lens antenna at 300 GHz with different permittivity lenses. 2020 2nd 6G Wireless Summit (6G SUMMIT), pp 1–4
34. Björnson E, Özdogan Ö, Larsson EG (2020) Intelligent reflecting surface versus decode-and-forward: how large surfaces are needed to beat relaying? IEEE Wirel Commun Lett 9:244–248
35. Raab FH, Asbeck P, Cripps S, Kenington PB, Popovic ZB, Pothecary N, et al (2003) RF and microwave power amplifier and transmitter technologies-Part 1. High Frequency Electron 2:22–36
36. Akyildiz IF, Jornet JM, Han C (2014) TeraNets: ultra-broadband communication networks in the terahertz band. IEEE Wirel Commun 21:130–135
37. Okada K, Li N, Matsushita K, Bunsen K, Murakami R, Musa A, et al (2011) A 60-GHz 16QAM/8PSK/QPSK/BPSK direct-conversion transceiver for IEEE802.15.3c. IEEE J Solid-State Circuits 46:2988–3004

38. Pärssinen A (2019) 5G transceivers from requirements to system models and analysis. In: Hueber G, Niknejad AM (eds) Millimeter-wave circuits for 5G and radar. Cambridge University Press, p 18
39. Schellmann M, Weitkemper P, Lähetkangas E, Ström E, Bockelmann C, Slimane B (2016) The 5G radio-access technologies. In: Osseiran A, Monserrat JF, Marsch P (eds) 5G mobile and wireless communications technology. Cambridge University Press, Cambridge
40. Zhu L, Xiao Z, Xia X-G, Oliver WuD (2019) Millimeter-wave communications with non-orthogonal multiple access for B5G/6G. IEEE Access 7:116123–116132
41. Cimini L (1985) Analysis and simulation of a digital mobile channel using orthogonal frequency division multiplexing. IEEE Trans Commun 33:665–675
42. Falconer D, Ariyavisitakul SL, Benyamin-Seeyar A, Eidson B (2002) Frequency domain equalization for single-carrier broadband wireless systems. IEEE Commun Mag 40:58–66
43. Thompson SC, Ahmed AU, Proakis JG, Zeidler JR, Geile MJ (2008) Constant envelope OFDM. IEEE Trans Commun 56:1300–1312
44. Ankarali ZE, Peköz B, Arslan H (2017) Flexible radio access beyond 5G: A future projection on waveform, numerology, and frame design principles. IEEE Access 5:18295–18309
45. Naudé N, Bozanic M, Sinha S (2006) Analogue CMOS direct sequence spread spectrum transceiver with carrier recovery employing complex spreading sequences. MELECON 2006-2006 IEEE Mediterranean Electrotechnical Conference. IEEE, pp 1227–1231
46. Sinha S, Bozanic M, Schoeman J, du Plessis M, Linde L (2009) A CMOS based multiple-access DSSS transceiver. In: 2009 South african conference on semi and superconductor technology, pp 19–24
47. Rajatheva N, Atzeni I, Bjornson E, Bourdoux A, Buzzi S, Dore J-B, et al (2020) White paper on broadband connectivity in 6G. arXiv:200414247 [eess] [Internet]. 2020 [cited 2020 Sep 16]. http://arXiv.org/abs/2004.14247
48. Liu X, Xu T, Darwazeh I (2020) Coexistence of orthogonal and non-orthogonal multicarrier signals in beyond 5G scenarios. 2020 2nd 6G Wireless Summit (6G SUMMIT), pp 1–5
49. Yuan Z, Ma Y, Hu Y, Li W (2020) High-efficiency full-duplex V2V communication. 2020 2nd 6G Wireless Summit (6G SUMMIT), pp 1–5
50. Krishnaswamy H, Dinc T (2019) RF and millimeter-wave full-duplex wireless for 5G and beyond. Millimeter-wave circuits for 5G and radar. Cambridge Univ. Press, pp 84–106

Chapter 4
Device Technologies and Circuits for 5G and 6G

Abstract This chapter will investigate the detail of available device technologies, as applicable to 5G, as well the research required in the near future that would enable these technologies to promote the emergence of 6G as envisioned in Chap. 1. Alternative technologies, including nanotubes manufactured from graphene and carbon, which may become the technologies of the future, will also be considered in this chapter. Secondly, this chapter will briefly look at the specifics of electronic design at millimeter and terahertz frequencies, which differs from regular electronics design in many respects. Specifically, any electronic circuit that is involved in signal processing in either a transmitter or a receiver requires a specific design methodology, that typically differs from the design of low-frequency equivalents. These specialized circuits include modulators, oscillators, filters, power amplifiers and LNAs, to name a few. Lastly, it should be noted that packaging has a much greater influence on circuits operating at high frequencies than on low-frequency circuitry, which opens up an opportunity for a short discussion on millimeter-wave and terahertz packaging approaches.

In the previous chapter, it was shown how electromagnetic communication at millimeter-wave and terahertz frequencies will assist in achieving data rates in Gb/s required by 6G. Meanwhile, 5G is already starting to utilize millimeter-wave frequencies. Thus, a great amount of effort in this century has been concentrated on overcoming the hurdles in establishing millimeter-wave, and more recently, terahertz communication links. A large portion of research contributions, specifically in last ten years or so, had the aim of increasing the speed of transistors without a large increase in the fabrication technology cost, which would enable signal processing at these frequencies.

This chapter will investigate the detail of available technologies, as applicable to 5G, as well the research required in the near future that would enable these technologies to promote the emergence of 6G as envisioned in Chap. 1. Alternative technologies, including nanotubes manufactured from graphene and carbon, which may become the technologies of the future, will also be considered in this chapter. Secondly, this chapter will briefly look at the specifics of electronic design at

M. Božanić and S. Sinha, *Mobile Communication Networks: 5G and a Vision of 6G*, Lecture Notes in Electrical Engineering 751,
https://doi.org/10.1007/978-3-030-69273-5_4

millimeter and terahertz frequencies, which differs from regular electronics design in many respects. Specifically, any electronic circuit that is involved in signal processing in either a transmitter or a receiver requires a specific design methodology, that typically differs from the design of low-frequency equivalents. These specialized circuits include modulators, oscillators, filters, power amplifiers and LNAs, to name a few. Lastly, it should be noted that packaging has a much greater influence on circuits operating at high frequencies than on low-frequency circuitry, which opens up an opportunity for a short discussion on millimeter-wave and terahertz packaging approaches.

4.1 Research into Solid-State Technologies Capable of Millimeter-Wave and Terahertz Amplification

A transistor is a solid-state semiconductor device used to amplify or switch electronic signals. As such, the transistor is the main active device in most electronic circuits, including those that are supposed to operate at millimeter-wave and terahertz frequencies (collectively called "high frequencies" in the rest of this chapter to remain concise). For a transistor to operate at high frequencies, it needs to have sufficiently high values of f_T and f_{max} [1]. High values of both frequencies are obtained with transistor scaling, that is, the decrease of transistor features. The transitional frequency f_T of a transistor is the frequency at which the transistor losses the capability of current amplification. In HBTs, f_T is mostly influenced by the vertical profile of the transistor [2]. The maximum oscillation frequency f_{max} is the frequency at which the transistor loses the ability of power amplification. In HBTs, this frequency is influenced by the transistor's lateral features. The parameter f_{max} is higher than f_T, which works in favor of high-frequency research and design, because at these frequencies, it is the power gain that is critical rather than the current gain and specialized circuits can be operated above f_T, as long as f_{max} is between 50 and 100% higher than the design frequency [3]. Note that balanced technology has f_{max} about 1.5 times that of f_T [4, 5].

CMOS technologies that were initially developed for digital applications have moderately high values of f_T and f_{max} [6]. Meanwhile, typical bipolar junction transistor (BJT) technologies are unfit for high frequency. The exception is the HBT, a bipolar transistor with a wide-gap emitter with a *pn*-junction built from differing (hetero-) materials, built from materials from groups III and V of the periodic table (e.g. InP or GaAs) or from different group-IV materials (e.g. SiGe). Group-IV materials, furthermore, are compatible with inexpensive CMOS technologies. HEMTs, field-effect transistors (FETs) with heterojunctions, also have satisfying high frequency. The highest performance of any technology to date is the 25 nm InP HEMT with f_{max} = 1.5 THz and f_T = 0.61 THz and amplification verified at 1.03 THz [7], although the use of this technology comes at a high price tag, which is another reason for active research into group-IV technologies. The DOTFIVE and

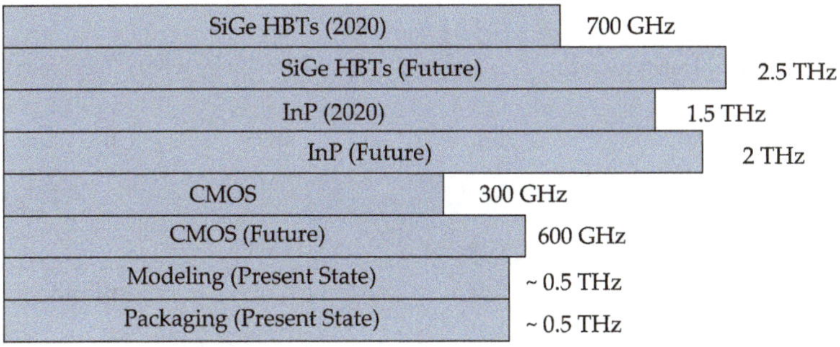

Fig. 4.1 Current state and predicted future of transistor technologies

DOTSEVEN projects, funded by the European Commission, resulted in a SiGe HBT with $f_T = 505$ GHz, $f_{max} = 720$ GHz, breakdown voltage of 1.6 V and gate delay of only 1.34 ps [8]. It was also predicted that the f_{max} values of at least the InP and SiGe technologies may reach or surpass 2 THz with further transistor scaling and use of new materials and techniques [9, 10]. The current and forecast state of active device technologies is illustrated in Fig. 4.1, with the current state of transistor modeling and circuit packaging also shown. The forecast of the future achievable frequencies is based on the predictions of possible decreases in transistor feature sizes [4, 11], which will become achievable as new lithography techniques emerge and increased precision of manufacturing becomes possible.

Considering that the high-frequency operation for 5G and 6G necessitates operation of miniature devices at low voltage levels (at least in handheld devices), at least basic understanding of the physics and modeling of miniature transistors is needed, which is discussed in this section. FETs and HBTs are fundamentally different devices, but many parallels between the two families can be drawn, especially when looking at small-signal modeling and high-frequency behavior. Many differences only come into play when it comes to sub-micron operation.

The following transistor and technology combinations can be discussed separately:

- *Metal oxide semiconductor (MOS) FETs (MOSFETs)*. These devices are used in Si CMOS, silicon on insulator (SOI) CMOS and BiCMOS technologies.
- *HEMTs*. These devices are fabricated in various materials on a variety of substrates. They are similar to MOSFETs in terms of operation
- *HBTs*. These devices are fundamentally different, as they are bipolar in nature (two *pn*-junctions; FETs are "unipolar").

The classification of common transistor types, together with semiconductor materials used in their fabrication, is shown in Table 4.1. Some of their most common properties relating to high-frequency operation are listed in Table 4.2.

Table. 4.1 Classification of transistors and technologies

Transistor group	Transistor type	Material	Technology type (1)	Technology type (2)
FET (Unipolar)	MOS	Silicon	CMOS	BiCMOS
	HEMT, pHEMT, mHEMT	GaAs, InP, GaN	HEMT	–
	MOS HEMT	GaN, SiC	HEMT	–
Bipolar	HBT	SiGe	HBT	BiCMOS
	HBT	InP	HBT	–

Table. 4.2 Comparison of parameters achieved by different technologies [16–18]

Parameter	InP HEMT	GaAs HEMT	GaN HEMT	CMOS	InP HBT	SiGe HBT
f_T, f_{max}	Very high	High	Low	Medium	Very high	High
Cost	Very high	High	Low	Very low	Very high	Low
Power	Average	Average	Very high	Very low	Average	Low
Noise	Very low	Very low	Very good	Very low	Low	Low
Linearity	Very good	Very good	Good	Average	Very good	Very good
Integration capability	Good	Good	Very poor	Very good	Poor	Very good
Manufacturing yield	Worse	Worse	Worse	Better	Worse	Better

4.1.1 Metal-Oxide Semiconductor Field-Effect Transistors

CMOS and BiCMOS technologies, both with the ability to deploy MOSFETs, are the most widely used semiconductor technologies. This is because of their low fabrication costs, simple fabrication steps, high integration, and high scaling capacity. In CMOS technology, the substrate is pure silicon (Si), but doping with *n*-type and *p*-type carriers is very controllable. The main physical properties of Si are listed in Table 4.3. In addition, Si has excellent thermal qualities and high resistance to mechanical stress. CMOS and BiCMOS technologies support low supply voltage operation and low power consumption. In high-frequency applications, pure CMOS technology is susceptible to higher circuit losses because of the low substrate resistance of Si. This problem is mitigated in CMOS SOI technology, where devices can be isolated from each other by laying out single-device islands [12–14]. High-resistance (HR) SOI technologies, with substrate resistivity in the $k\Omega \cdot cm$ range, can also be fabricated and perform even better at high frequencies [15].

Table. 4.3 Properties of silicon (from [19])

Property	Value
Bandgap energy (eV)	1.12
Electron mobility (cm^2/V-s)	1360
Hole mobility (cm^2/V-s)	480
Breakdown electric field (V/cm)	2×10^5
Saturation electron drift velocity (cm/s)	10^7
Relative dielectric constant (ε_r)	11.7

A MOSFET consists of highly doped drain and source regions separated by a channel region, as illustrated in Fig. 4.2a. Complementary positive (n) and negative (p) variations of this transistor are possible but the nMOS variant has better high frequency performance owing to higher mobility of carriers (electrons vs. holes, see Table 4.3 again). The transistor illustrated in Fig. 4.2a is the nMOS variant, where the drain (D) and source (S) are built from n-type material in a p-substrate. Figure 4.2 also shows one variant of a MOSFET in CMOS SOI technology—the fully depleted SOI nMOS (Fig. 4.2d). In a regular nMOS, control gate material is grown over a thin insulating oxide layer (typically silicon-dioxide) on top of the channel. The gate (G), placed above the oxide, is usually built from polysilicon (poly) material. A MOSFET is defined by the length of the channel (L) and the width of the doping region (W). The minimum technology dimension, used as a point of reference in technology scaling, refers to the minimum allowed MOSFET length. The p-substrate itself is a fourth terminal of the MOSFET, denoted as the body (B).

The principle of operation of the MOSFET will be described from the DC, low frequency and high frequency perspective.

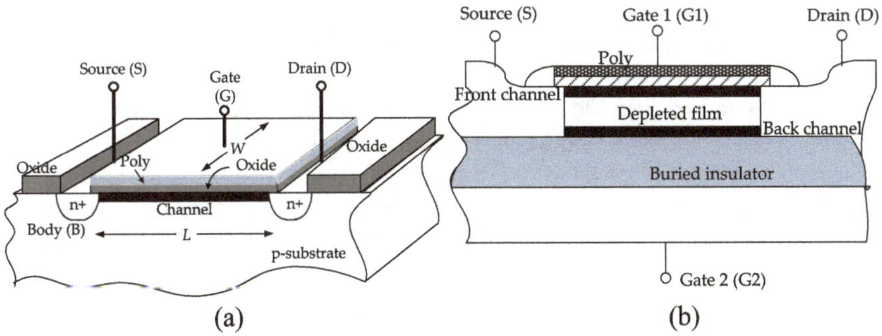

Fig. 4.2 Structure of a regular nMOS showing channel L and W (**a**) and fully depleted SOI nMOS (**b**) (adapted from [17, 20, 21, 22])

MOSFETs at DC and Low Frequencies

For high-frequency applications, MOSFETs are normally designed as short-channel devices. Note that in short-channel devices, various short-channel (or deep-submicron) effects start to emerge; however, different approaches are used in modern CMOS technologies to mitigate the short-channel effects as far as possible [23]. Thus, the emergence of short-channel effects is predominantly dependent on specific technology. In the following discussion on the principle of operation of the MOSFET, it will be assumed that short-channel effects can be either mitigated or understood.

Analysis of the MOSFET can start by first looking at the performance at DC and at low frequency. The large-signal operation of FETs at DC is fully described by the drain-source current I_D [24–29]. For I_D in an *n*MOS to start flowing, the transistor must be excited by varying the gate and drain voltages with respect to the source, which is grounded (V_{GS} and V_{DS} respectively). There will also be a voltage difference between the gate and the drain, V_{GD}. The fourth terminal, the body, is also connected to ground because the substrate itself is grounded. With the V_{DS} and V_{GS} increasing, horizontal and vertical electric fields are generated. Assuming that $-V_{GD} = V_{DS}$—V_{GS}, differing values of V_{DS} and V_{GS} will result in three regions of operation:

1. Cutoff, where V_{GS}, $V_{GD} < V_t$;
2. Saturation, pinch-off or active region, where $V_{GS} \geq V_t$ and $V_{GD} < V_t$ or $V_{GS} < V_t$ and $V_{GD} \geq V_t$; and
3. Triode, where V_{GS}, $V_{GD} \geq V_t$.

Above, the parameter V_t is the transistor threshold voltage, which is dependent on transistor doping. Si in MOSFETs is typically doped such that V_t is between 0.6 and 0.7 V. In both the triode and saturation regions the transistor is in inversion, which indicates that the current conduction channel has formed and current I_D between the drain and the source has started flowing. The inversion actually already starts appearing at V_{GS} voltages slightly lower than $V_{t.}$. This is termed weak inversion, whereas the transistor operating in the active region will be in strong inversion. In the cutoff region, however, the voltage V_{GS} is not enough for the inversion to happen and the conduction channel between the drain and source to form. In the triode or linear region, on the other hand, the transistor behaves as a resistor, which is not suitable for many practical applications. This leaves the active region as the region where the transistor is typically operated. In the active region, pinch-off occurs and a conduction channel forms, as illustrated in Fig. 4.3. The three regions of operation are illustrated in Fig. 4.4, where the change in I_D with different values of V_{DS} and V_{GS} is shown.

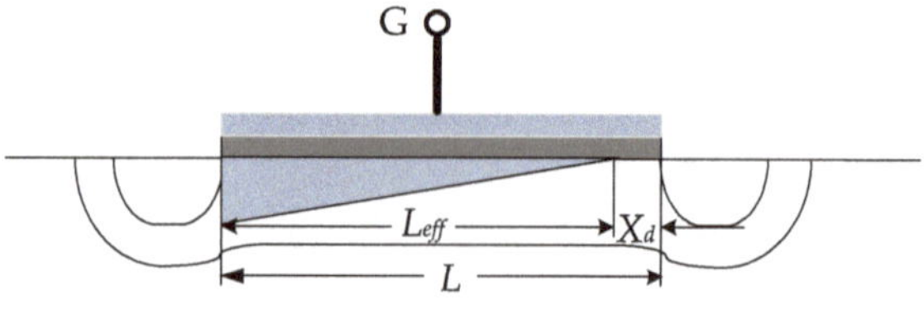

Fig. 4.3 Illustration of the pinch-off

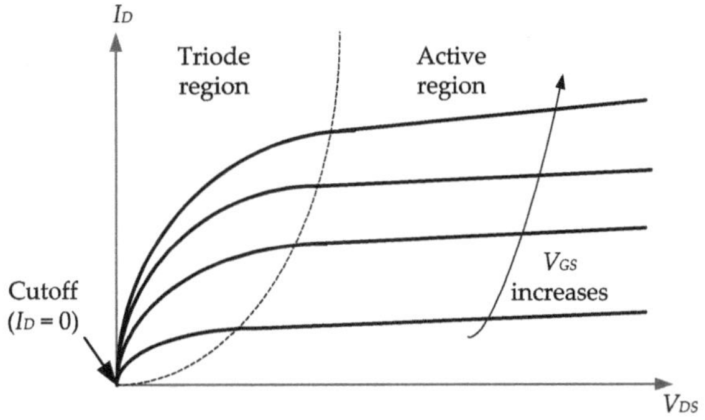

Fig. 4.4 I_D in different regions of nMOS operation

In the active region of operation, I_D behaves according to the following equation:

$$I_D = \frac{k'}{2}\frac{W}{L}(V_{GS} - V_t)^2(1 + \lambda V_{DS}), \tag{4.1}$$

where, k' is the parameter defined as

$$k' = \mu_n C_{ox}, \tag{4.2}$$

C_{ox} is the oxide capacitance and λ is the channel-length modulation parameter, defined as $\lambda = \frac{1}{L_{eff}}\frac{dX_d}{dV_{DS}}$, with L_{eff} and X_d shown in Fig. 4.3. Equation (4.1) is valid if carrier velocity saturation does not occur, which is true for long-channel devices. However, as the transistor length decreases, at a certain point the electric field becomes sufficient to ensure that the electron drift velocity starts approaching a constant scattering-limited velocity (v_{scl}). As a result, the length of the transistor and the carrier mobility do not play a role in the amount of current that flows anymore and I_D current starts following a linear relationship:

$$I_D = WC_{ox}(V_{GS} - V_t)v_{scl}(1 + \lambda v_{DS}). \tag{4.3}$$

Velocity saturation is one of the most important short-channel effects in submicron MOSFETs. The relationship in Eq. (4.3) is true for fully saturated devices only, however, and partially saturated devices have currents in between those predicted by Eqs. (4.2) and (4.3). High-power devices, such as those in power amplifiers, reach velocity saturation sooner than devices in low-power circuits.

Additionally, when channel length decreases to sizes comparable to the mean free path, which is the average distance traveled by a moving particle between successive impacts, electrons can acquire a velocity much higher than the scattering-limited velocity [23], leading to a phenomenon called ballistic transport or velocity overshoot. The probability of ballistic transport increases as transistor features become

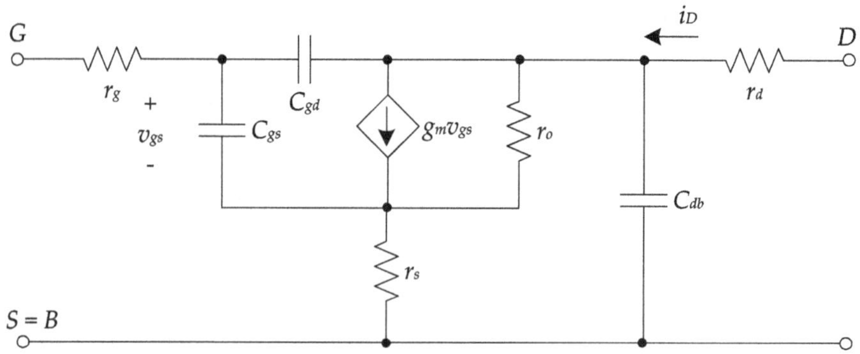

Fig. 4.5 The high-frequency small-signal model of the *n*MOS

smaller with technology scaling, which offers an additional incentive in transistor scaling research.

In analog and RF design, the transistor is usually used as an amplifier (some exceptions include oscillators). In most of these designs, the signals are in the order of several millivolts. If there are small changes in the gate-source voltage denoted by v_{gs}, the small-signal model of the transistor can be introduced. This model is illustrated in Fig. 4.5, where the transistor is connected in the common-source configuration. The capacitors in this model are the parasitic capacitors between different device terminals and are applicable for high-frequency operation, as will be described later. The same holds true for terminal resistances r_g, r_d and r_s. In the small-signal model, the small-signal change in v_{gs} with will also result in small changes in drain current (i_d) and this allows for a transconductance g_m to be introduced to relate v_{gs} and i_d. g_m is computed as:

$$g_m = k'\frac{W}{L}(V_{GS} - V_t) = \sqrt{2k'\frac{W}{L}I_D} \tag{4.4}$$

for long-channel devices and

$$g_m = WC_{ox}v_{scl} \tag{4.5}$$

for short-channel devices. The output resistance r_o appears owing to the channel length modulation:

$$r_o = \frac{1}{\lambda I_D}. \tag{4.6}$$

MOSFET at Millimeter-Wave and Terahertz Frequencies.

The high-frequency response of the transistor can be analyzed by looking at all components in the high-frequency small-signal model in Fig. 4.5. Additional capacitances and resistances appear owing to the transistor fabrication technology and as such, they are technology-dependent. Transistor scaling, typically, results in the decrease of these quantities. The most important parasitic quantity is the gate-to-source capacitance, C_{gs}. The value of the C_{gs} in the saturation is

$$C_{gs} = \frac{2}{3} C_{ox} W L. \tag{4.7}$$

At the same time, pinch-off causes $C_{gd} \cong 0$. In the triode region,

$$C_{gs} = C_{gd} = \frac{1}{2} C_{ox} W L. \tag{4.8}$$

Resistances r_g, r_d and r_s are resistances appearing because of the physical resistivity of material used in the contacts and are bias-independent [30–32]. They are normally small and start influencing the performance of the transistor close to its frequency limits.

At this point, mathematical definitions of f_T and f_{max} can be given. It can be shown by doing the high-frequency circuit analysis on the small-signal model that in the active region of operation, where C_{gd} becomes $0, f_T$ is

$$f_T = \frac{g_m}{2\pi(C_{gs} + C_{gd})} \cong \frac{g_m}{2\pi C_{gs}}. \tag{4.9}$$

Submicron devices result in smaller parasitics, including capacitances. This means that the devices with the highest f_T will inevitably be velocity-saturated. If g_m is replaced with the value in Eq. (4.4) and C_{gs} with the value of Eq. (4.7), it will be shown that f_T is only dependent on v_{scl} and L:

$$f_T \propto \frac{v_{scl}}{L}. \tag{4.10}$$

This explains why the speed of transistors is influenced by the technology node (defined by the length of the transistor).

Maximum operation frequency f_{max} may be a more important metric for high-frequency research. Up to the frequency at which the power gain becomes 1, it is still possible to build oscillators and amplifiers, irrespective of whether the circuit is operated above f_T. It can be shown that

$$f_{max} = \sqrt{\frac{f_T}{8\pi C_{gd} r_g}}. \tag{4.11}$$

In the f_{max} definition, physical gate resistance now also plays a major role in transistor response.

There are many effects that are not modeled by the small-signal model. These include distributed effects, which are considered the most important high-frequency effects [33]. There are lateral and longitudinal distributed effects. Lateral distributed effects are also known as non-quasi-static (NQS) effects; these are the consequence of the limited transport time of charges along the drift region of the MOSFET channel. The extent of NQS depends on the velocity of carriers and the effective channel length. Longitudinal distributed effects (LDE) arise when the wavelength of the current wave in the transistor finger is comparable to the finger length, resulting in the transistor finger itself behaving like a transmission line. NQS effects, LDE and other effects are too complex to be modeled in hand circuit analysis but are included as a part of RF simulation models.

4.1.2 High-Electron-Mobility Transistors

HEMTs or heterojunction FETs (HFETs) are FETs that have a heterostructure that is formed through two semiconductor materials with differing bandgaps [23]. In these devices, the channel is formed at the heterointerface between wide-energy-gap material and the undoped narrow bandgap layer and consists of two-dimensional electron gas (2DEG). This arrangement results in no impurity scattering and subsequently in higher electron mobility. In a typical HEMT, multiple layers of III-V semiconductor materials are stacked vertically. HEMTs can be fabricated either as part of process technologies or as discrete transistors [34].

Properties of semiconductor material combinations commonly used in the fabrication of HEMTs are listed in Table 4.4 [19, 34–36]. HEMT materials have much higher mobility than does Si, while the listed saturation velocities are mostly comparable to those of Si. This results in superior performance of HEMTs when compared to Si MOSFETs, because higher mobility allows higher drain currents and small-signal gains to be produced with the same width-to-length ratio, for both long and short devices.

The most common heterointerfaces in HEMTs combine Al, Ga As and In. Typical interfaces are AlGaAs/GaAs, AlGaAs/InGaAs and InAlAs/InGaAs. An AlGaAs/GaAs HEMT is illustrated in Fig. 4.6 [37]. The transistor conducts in the

Table. 4.4 Properties of common HEMT material combinations (from [19, 34, 35, 36])

Property	SiC	GaN	AlN	GaAs	InP
Bandgap energy (eV) (room temperature)	3.26	3.42	6.2	1.42	1.34
Electron mobility (cm^2/V-s)	900	2000	300	8500	5400
Breakdown electric field (V/cm)	2.2×10^6	3.5×10^6	1.1×10^7	4×10^5	5×10^5
Saturation electron drift velocity (cm/s)	2.7×10^7	2.5×10^7	–	1.2×10^7	2×10^7
Relative dielectric constant (ε_r)	9.7	9	8.5	12.9	12.5

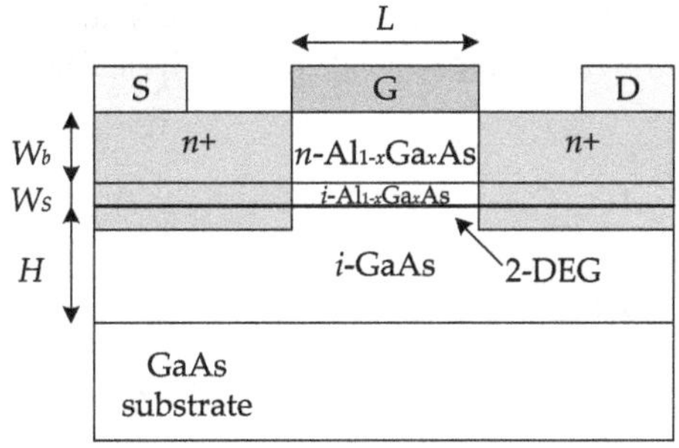

Fig. 4.6 Structure of the AlGaAs/GaAs HEMT. From [37]

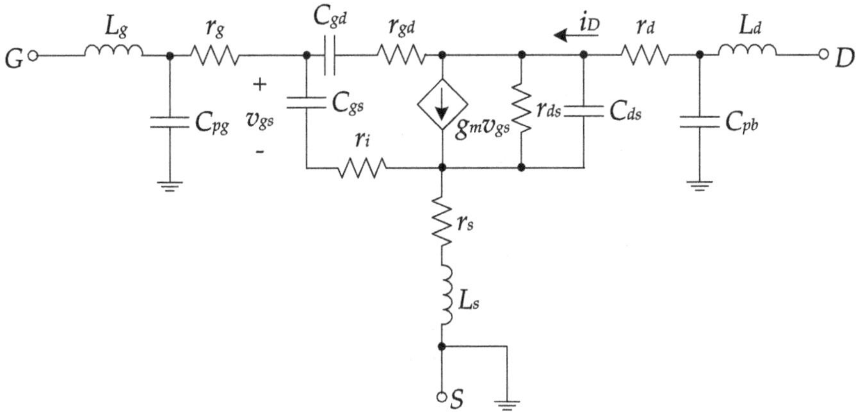

Fig. 4.7 Small-signal high-frequency model of a HEMT. From [38]

joining layer of AlGaAs/GaAs owing to the polarization difference that causes a positive net charge that attracts the electrons, which become confined in a triangular-shaped potential well [34]. The whole conduction band surface is dragged upwards, and confined electrons form the 2DEG at the GaAs channel.

When 2DEG appears, it is not necessary to have the insulating oxide under the gate. This is the most important structural difference between HEMTs and MOSFETs. This means that an additional resistance r_{gd} appears in the small-signal high-frequency model of the HEMT, as illustrated in Fig. 4.7 [38]. As a result, HEMTs, for the most part, can be described by equations that are similar to those used to describe MOSFETs.

In the active region, the drain current of long devices is

$$I_D = \frac{k'}{2}\frac{W}{L}(V_{GS} - V_t)^2, \tag{4.12}$$

where k' is now defined by

$$k' = \mu_n C_o \tag{4.13}$$

and C_o represents the capacitance in the 2DEG channel. The transconductance g_m in this case is the same as in Eq. (4.4). The drain current in velocity-saturated devices is

$$I_D = WC_o(V_{GS} - V_t)v_{scl} \tag{4.14}$$

and g_m is

$$g_m = WC_o v_{scl}. \tag{4.15}$$

The transitional frequency f_T is approximately

$$f_T = \frac{g_m}{2\pi(WLC_o)} \tag{4.16}$$

and f_{max} is described in the same way as that of the MOSFET (Eq. 4.11). With higher g_m values than those in MOSFETs, as well as with lower capacitances, both f_T and f_{max} values are much higher than those of MOSFETs with the same minimum dimension.

Different materials that can be used in the composition of HEMTs allow for a range of differing applications of these transistors. On the one hand, GaAs-based HEMTs are an excellent choice for high power, high efficiency, low noise and good linearity, even at millimeter-wave frequencies, but they have low integration capacity [39]. On the other hand, InP devices can reach into the terahertz domain. Gallium nitride (GaN) and silicon carbide (SiC) are the best devices for power switching [40]. SiC substrates are also capable of operating in extremely high temperatures. Furthermore, additional transitional (metamorphic) layers can be used to connect substrate of one material type to the active device layers of another type. These devices are known as mHEMTs. The combination of GaAs substrate with active device layers built from InP composites results in an mHEMT with excellent high-frequency properties.

4.1.3 Bipolar Transistors

In a bipolar transistor, two *pn*-junctions are formed. One junction appears between the base (B) and emitter (E), and the other one appears between the base and the collector (C). The structure of the *n*-type bipolar device (*npn*) is shown in Fig. 4.8 [23, 25, 26, 28]. Two junctions result in two depletion regions being formed.

Homojunction Si bipolar technologies suffer from low gains, and their performance was surpassed by that of CMOS almost 30 years ago. However, if differing semiconductor materials are used on each side of the emitter–base or both junctions, HBTs or double HBTs (DHBTs) are built. HBTs have significantly higher current gains than homojunction bipolar transistors. HBTs can be composed of III-V materials in Table 4.4, or alternatively of two group-IV materials, as in the case of a SiGe HBT. Group-III-V HBTs typically have low integration capacity, which is also true

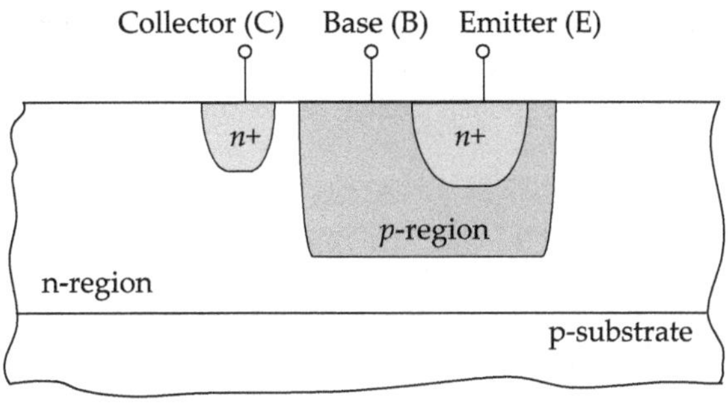

Fig. 4.8 The structure of an *npn* bipolar transistor

for InP transistors, which have the best high-frequency performance of all HBTs. On the other hand, only a handful of extra processing steps turns a standard Si CMOS process into a SiGe BiCMOS process (process with MOSFETs and HBTs available simultaneously) with only a slight increase in fabrication costs. HBTs that appear as a result of the merger of the two processes can achieve higher transconductance gains and suffer less from noise than MOSFETs in the same process (Table 4.5).

HBTs at DC and Low Frequencies.

The operation of bipolar transistors (HBTs) can be analyzed by looking at *npn* configured in the common-emitter configuration. The geometric structure of a bipolar transistor results in a small amount of current that flows into the base of the transistor, whereas in the case of the MOSFET, there is no current flowing into the gate (at least not at DC). As a result, bipolar transistors have finite current gains, denoted by β. The two *pn*-junctions that are formed by the transistor geometry each has built-in potentials determined by the junction doping characteristics: $V_{BE(\text{on})}$ and $V_{BC(\text{on})}$ for the base-emitter and base–collector junction respectively. If the *npn* is excited by the base-emitter voltage V_{BE} and the base–collector voltage V_{BC}, four regions of operation can be identified:

1. Cutoff, when $V_{BE} < V_{BE(\text{on})}$ and $V_{BC} < V_{BC(\text{on})}$;
2. Forward active region, when $V_{BE} \geq V_{BE(\text{on})}$ and $V_{BC} < V_{BC(\text{on})}$;
3. Reverse active region, when $V_{BE} < V_{BE(\text{on})}$ and $V_{BC} \geq V_{BC(\text{on})}$; and
4. Saturation region, when $V_{BE} \geq V_{BE(\text{on})}$ and $V_{BC} \geq V_{BC(\text{on})}$.

Note that when $V_x \geq V_{x(\text{on})}$, junction x is forward-biased and when $V_x < V_{x(\text{on})}$, the same junction is reverse-biased. The bipolar transistor is doped to perform better in the forward active region. As a result, the reverse operation results in inferior

Table. 4.5 Properties of SiGe (after [36])

Property	Value
Bandgap energy (eV)	0.945
Electron mobility (cm^2/V-s)	7700
Breakdown electric field (V/cm)	4×10^5
Relative dielectric constant (ε_r)	13.5

speed of the transistor. The current gain is dependent on the operating conditions, frequency, temperature and collector current, and the transistor will operate as long as the voltage applied to each of the junctions is below the breakdown voltage of that junction.

Large-signal (DC) behavior of the bipolar transistor can be described by three current equations. The first equation describes the collector current. This current is dependent on the voltage applied over the base-emitter junction, V_{BE}, as well as the voltage between the collector and emitter, V_{CE}:

$$I_C = I_S\left(1 + \frac{V_{CE}}{V_A}\right)\exp\frac{V_{BE}}{V_T}, \tag{4.17}$$

where I_S is the transistor saturation current, V_A is the Early voltage and V_T is the thermal voltage, defined as

$$V_T = \frac{kT}{q} \tag{4.18}$$

with k as the Boltzmann's constant (1.3807×10^{-23} J/K), q as the electron charge (1.6022×10^{-19} C) and T as the absolute temperature. V_T is equal to about 26 mV at 300 K. I_S is a property of transistor doping and the layout and it typically ranges from 10^{-14} to 10^{-16}. V_A in bipolar transistors refers to effective base-width modulation with the applied voltage V_{CE}. The second equation relates to the base current I_B and I_C:

$$I_C = \beta_F I_B, \tag{4.19}$$

where β_F is the forward gain. The third and last equation shows that the emitter current is the sum of the collector and base currents:

$$I_E = I_B + I_C = (\beta_F + 1)I_B. \tag{4.20}$$

The small-signal low-frequency operation of an HBT can be described with the aid of a small-signal diagram in Fig. 4.9. g_m is the change of the collector current, i_c, due to small changes in the base-emitter voltage, v_{be}:

$$g_m = \frac{i_c}{v_{be}} = \frac{I_C}{V_T}. \tag{4.21}$$

The finite current gain results in the presence of a finite resistor r_π between the base and the emitter. This input resistance is the ratio of the input voltage to the base current, and can be computed as follows:

$$r_\pi = \frac{v_{be}}{i_b} = \frac{v_{be}}{i_c/\beta_0} = \frac{\beta_0}{g_m}. \tag{4.22}$$

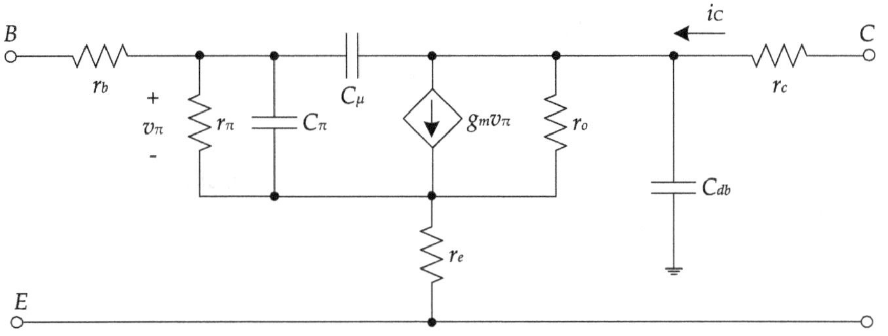

Fig. 4.9 High-frequency small-signal model of a bipolar transistor

Above, the small-signal relationship $i_c = \beta_0 \ i_b$ was used, which introduces the small-signal forward current gain β_0. The output resistance is

$$r_o = \frac{V_A}{I_C}. \tag{4.23}$$

The Early voltage effect also influences the base current, resulting in a resistor that appears between the base and the collector with a value greater than

$$r_\mu = \beta_0 r_o. \tag{4.24}$$

HBTs at Millimeter-Wave and Terahertz Frequencies

To investigate the high-frequency response of the HBT, capacitors in the high-frequency small-signal model need to be described.

A capacitor C_π is the sum of the base-charging capacitance C_b and the emitter–base depletion layer capacitance C_{je}:

$$C_\pi = C_b + C_{je}. \tag{4.25}$$

The base-charging capacitance at the specific bias point is the product of the base transit.time in the forward direction and g_m:

$$C_b = \tau_F g_m \tag{4.26}$$

and is typically between 1 and 3 ps in the state-of-the-art SiGe HBTs [8] and less than 1 ps in modern InP transistors [41]. The capacitance of the base-emitter junction (C_{je}), which is forward-biased, is approximately equal to twice the junction capacitance with no potential applied. The remaining two capacitances, the capacitance between the base and collector, C_μ, and the collector-source capacitance, C_{cs}, are due to junctions that are reverse-biased.

The transitional frequency f_T is

$$f_T = \frac{g_m}{2\pi\left(C_\pi + C_\mu\right)}. \tag{4.27}$$

The sum of C_π and C_μ is usually dominated by C_b, which allows for Eq. (4.27) to simplify to

$$f_T = \frac{1}{2\pi\tau_F}. \tag{4.28}$$

This means that f_T is strongly dependent on the base-charging time τ_F, which in turn depends on the vertical profile (base width) of the transistor [2]. The maximum oscillation frequency f_{max} is

$$f_{max} = \sqrt{\frac{f_T}{8\pi C_\mu r_b}}, \tag{4.29}$$

where r_b is the physical resistance of the base, which is decreased by lateral scaling of transistors as fabrication technologies scale to lower process nodes. It should be noted that the f_{max} of an HBT is much larger than the f_{max} of a MOSFET that has a similar f_T value [42].

4.1.4 *Future of Solid-State Devices*

Values of f_T and f_{max} that can be reached by present state-of-the art technologies were already listed at the beginning of Sect. 4.1. At this point in time, this is sufficient for the planned deployment of 5G networks. In this section, however, this discussion is extended to include the prediction of frequencies that could be achieved in the near future, perhaps before commercial development of 6G begins.

Outlook for SiGe BiCMOS Technology

The core reason for promoting BiCMOS technologies is twofold: they have lower fabrication costs compared to the costs of other high-end technologies, and much better integration complexity than those technologies.

BiCMOS technologies with transistors having f_T and f_{max} simultaneously exceeding 300 GHz were only developed in the last decade. Two important projects truly aimed at extending transistor capacity into the terahertz range were the back-to-back DOTFIVE and DOTSEVEN projects. The second of the two projects, the DOTSEVEN project [43], had the main objective of extending the f_{max} of the HBT beyond 0.7 THz. This goal was achieved for a standalone transistor and the integration of the HBT into a BiCMOS process was left for a follow-up TARANTO[1] project [44], which reported 130 nm BiCMOS technology with an HBT reaching 600 GHz in 2019 as an interim result. It is anticipated that by the time this project is completed, transistors with f_{max} surpassing 720 GHz achieved in the DOTSEVEN project will be reported.

This will, however, not be the end of the road of the BiCMOS technology. According to Schröter et al. [4, 10, 11], further scaling of transistors in the vertical and lateral direction should still, at least theoretically, be possible, probably up to f_{max} in the range of 1.5–2.5 THz.

Outlook for InP Technologies

The wide gap of InP (1.34 eV) and the high electron drift velocity (more than 2×10^7 cm/s) results in the unprecedented performance of InP-based technologies. An f_{max} value of over 1 THz in HEMTs was achieved more than ten years ago and in 2013, a 25-nm HEMT was reported, with $f_{max} = 1.5$ THz and $f_T = 0.61$ THz. High-frequency operation is a result of scaling of the gate by using electron beam lithography, which resulted in reduced gate capacitance and resistance. The channel was created by inserting an InAs layer between two InGaAs layers. Predictions say that even faster transistors will be possible if the indium composition inside the channel is increased, which is not possible with currently available fabrication technologies [45].

InP DHBTs also have particularly good frequency response. They have a higher breakdown voltage and higher current drive capability at a given f_T than InP HEMTs [41, 46, 47]. In 2016, a DHBT employing a 130 nm process was demonstrated [41], with f_T of 0.52 THz, f_{max} of 1.15 THz and a common-emitter breakdown voltage of 3.5 V. InGaAs and InAs semiconductors were used again, this time in the emitter to minimize the contact resistance. The base was composed of carbon-doped InGaAs and the collector was composed of InGaAs/InAl/As. A combination of InP and GaAsSb [48], similarly, resulted in $f_{max} = 1.2$ THz and $f_T = 0.5$ THz [49].

When it comes to the outlook of InP technologies, the technology roadmap [9] and some follow-up findings [46] indicate that a future 32-nm process node would allow for fabricating transistors with f_{max} much higher than 2 THz and f_T in excess of 1 THz.

Outlook for GaAs Technologies

GaAs substrates are less expensive and more robust than InP substrates. A GaAs transistor is actually an active device with metamorphic layers built out of InP, InGaAs or InAlAs [45]. For this transistor type, f_T values above 500 GHz and f_{max} values above 1 THz are readily achievable [50]. GaAs technologies have an excellent noise response that extends into the terahertz range.

---

[1] According to the official project website (https://tima.univ-grenoble-alpes.fr/taranto/), TARANTO stands for "Towards Advanced BICMOS nanotechnology platforms for RF to the applications".

Outlook for GaN Technologies

GaN technology continues to be researched in the context of 5G and 6G because of its potential in high-power applications. GaN transistors are known to have possibly the best power conversion efficiency, outperformed only by that of SiC transistors, which makes them suitable for use in high-power 5G base stations [51]. At present, transistors with $f_T = 454$ GHz and $f_{max} = 444$ GHz are available [52]. Current technology performance was achieved by reducing fringing gate capacitances and it is expected that a further increase in speed should be achievable by decreasing mesa-edge capacitances that form during etching [53].

Although the power density that can be achieved with GaN devices is high, devices on SiC substrates still retain the highest level of performance when high power and high temperature operations are concerned. However, while SiC technology has been used extensively in a multitude of high-power applications, their low carrier mobility does not allow these transistors to operate beyond several GHz. Thus, GaN technology research is focused on modifying existing GaN technologies to include SiC technology capabilities in the high power/temperature domain. Recently, a group led by Romanczyk et al. [54, 55] reported two GaN HEMT devices fabricated on two different substrates, both yielding significantly better performance in terms of power density than the figure of 2 W/mm that is typically reported for these devices at frequencies above 90 GHz and power-added efficiency (PAE) of about 30%. In fact, N-polar GaN-on-sapphire deep recess HEMTs reported in [54] could produce 5.5 W/mm of output power at 94 GHz, with an associated PAE of 20.6%. Meanwhile, SiN-passivated N-polar GaN deep recess HEMTs, reported in [55], produced 8.8 W/mm of power at the same frequency, with PAE of above 30%. The reported operating frequency makes both devices more than suitable for high-power base stations that will be required in 6G networks, which are expected to operate at frequencies above the 90-GHz mark. However, the f_T and f_{max} of these devices are only around 100 GHz and 200 GHz respectively at present, meaning that more research is required if these results are to be replicated at frequencies closer to those of the terahertz regime.

Outlook for CMOS Technology

At present, CMOS transistors can readily support 5G technologies [56]. Many researchers, fortunately, predict that these technologies may be found suitable for 6G or terahertz applications in the near future. The theoretical first-order approximation value [57] of the maximum f_T of any given CMOS technology is

$$f_T = \frac{10}{L} \text{THz} \cdot \text{nm}, \tag{4.30}$$

where L is the gate length in nanometer. According to this prediction, the maximum transitional frequency of a 5-nm technology would be 2 THz. However, some process modifications from standard CMOS will be needed to achieve such high-frequency operation. One of the options is to replace regular FETs with FinFETs when the minimum gate length scales below 22 nm. To facilitate scaling even further, lower than 14 nm, it may be necessary to start considering different materials in FinFET composition, including SiGe or even III–V materials, in order to improve channel mobility and the threshold voltage [58].

The FinFET structure was originally introduced as a way of reducing short-channel effects [20]. In FinFETs, the channel is a vertical fin that is surrounded by the gate, as illustrated in Fig. 4.10. FinFET is a multi-gate transistor, since it can be controlled electrostatically from more than one side. Depending on whether the gate control is independent or not, the FinFET can be classified as either an independent gate FinFET (IG-FinFET) or shorted gate FinFET (SG-FinFET) [59].

In comparison with regular MOSFETs, FinFETs are less prone to short-channel effects (as already stated), have higher gains when used as amplifiers, have smaller footprints that allow higher circuit density to be reached and have excellent RF performance. Both *n*-type and *p*-type FinFETs can achieve similar values of f_T and f_{max}. According to Singh et al. [60], state-of-the-art 14-nm *n*- and *p*-FinFETs can reach f_T values of 314 GHz and 285 GHz receptively, and f_{max} of 180 GHz and 140 GHz respectively. However, most importantly, the new structure of the transistor will allow for further scaling of CMOS technologies and even higher frequencies to be reached in future.

When it comes to the future of CMOS technologies, the nanoelectronics research community is anticipating the results of the T-Music project funded by DARPA. One of the goals of this project is to develop a CMOS transistor operating up to 1 THz [61].

Cooling of Nanoelectronic Circuits for Better Frequency Performance

An interesting property of semiconductors is that cooling of these devices to cryogenic temperatures improves various aspects of their performance. SiGe HBTs, WBG HBTs, HEMTs and Si CMOS transistors can all benefit from this phenomenon. The most important benefits of cooling are shown in Table 4.6 [62, 63].

For 5G and 6G researchers, the fact that f_T, f_{max} and thus the bandwidth increase with cooling is the most important benefit of cooling. The highest operating frequency to date in SiGe HBTs was reported for a transistor cooled to cryogenic temperatures. In 2014, Chakraborty et al. [64] reported that f_{max} of 798 GHz had been reached at a temperature of 4.3 K for a transistor that only has $f_{max} = 417$ GHz at 300 K (room temperature). In theory, $f_{max} > 1$ THz should already be reachable by state-of-the art SiGe transistors, which can at present reach $f_{max} = 700$ GHz at room temperature. Regrettably, cryogenic cooling is expensive and requires heavy equipment, but it is useful in researching the future scaling of devices. The key idea of this approach is that

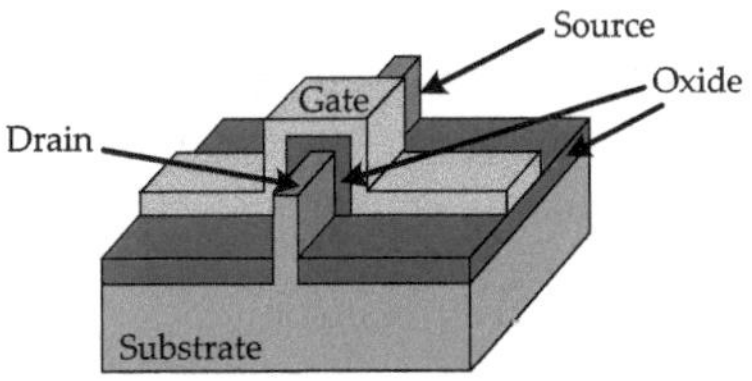

Fig. 4.10 The FinFET structure. Adapted from [20]

Table. 4.6 Performance improvement of high-frequency circuits as a result of operating at cryogenic temperatures

Performance improvement	Reason
Higher transconductance gain (g_m)	Increased carrier mobility
Higher f_T, f_{max}, bandwidth	Increased g_m
Lower power dissipation	Reduced voltage, improved switching characteristics and lower leakage currents
Lower interconnect parasitics	Resistance decreases with decrease in temperature
Higher integration density	Transistor and even ICs can be densely packed or stacked with no heat buildup

the operation of cooled, already available, devices can be observed in experimentation to quantify the key parameters that trigger the increase in their performance, including the influence of parasitics and noise, the role of junction widths as well as many other factors. These findings can then guide optimization of devices for the next technology node at room temperature [5].

4.2 Future Alternatives to Traditional Semiconductor Technologies

At present, there are ongoing research efforts into alternatives to traditional semiconductor technologies. Three promising options, each based on an allotrope of carbon, will be discussed in this section.

4.2.1 Graphene-Based Electronics

Graphene [65–68] is a nanomaterial that belongs to a group of 2D semiconductors, that is, semiconductors that are only one or several atom layers thick. It was discovered only in 2004. It is composed of a single layer of carbon atoms (mostly ^{12}C isotope), aligned in the pattern of a honeycomb. Like Si and Ge, carbon is located in group IV of the periodic table, which means that it has several properties that are similar to those of Si and Ge. Graphene in its pure form has high carrier mobility (in the region of 200 000 $cm^2/V \cdot s$), high thermal and excellent mechanical properties.

A cross-section of graphene FETs (GFETs) on silicon substrate is shown in Fig. 4.11 [69]. GFETs are essentially zero-band-gap transistors, because the ratio between the ON and OFF currents in such a transistor is not sufficiently high. The zero-bandgap energy is due to the symmetry of two carbon atoms in a unit cell of graphene. This means that the bandgap in GFETs needs to be engineered.

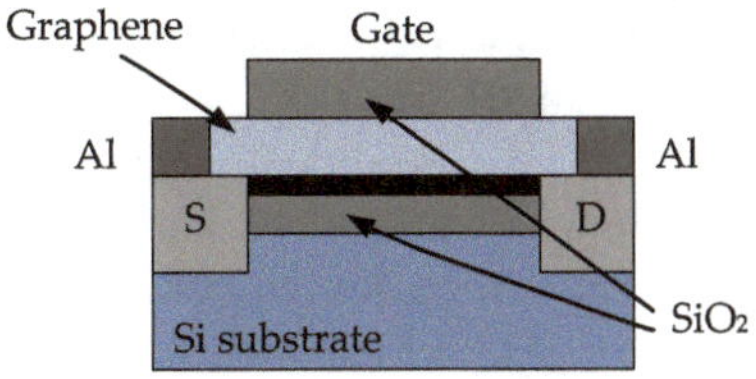

Fig. 4.11 A cross-section of the graphene/Si FET. Adapted from [69]

One way to do that is to use what is essentially a one-dimensional structure that has a narrow width and smooth edges in the transistor channel [65, 68, 70, 71]. Figure 4.12 illustrates how the honeycomb graphene structure can be cut along two lattice directions to produce this structure, called the graphene nanoribbon (GNR). Graphene can be cut such that either a zigzag or an armchair edge is produced. GNRs with a width in the range of 2–3 nm become semiconductors with bandgaps in the region of 0.5 eV (500 meV). Some problems with this approach include the fact that the available lithography equipment does not support mass production of GNRs, and that the thickness of the graphene sheet prohibits the use of traditional doping techniques.

Alternatively, bandgap can be engineered by placing two monolayers of graphene on top of each other [65, 72]. It has been shown that the bandgap of such a structure depends on the applied vertical electrical field. For the zero applied field, the bandgap is also zero, and it increases to 0.4 eV for the applied field in the region of several hundreds of meV. Some researchers also reported that chemical altering (functionalizing) of graphene could lead to bandgap appearance [65, 68, 72]. Some approaches include adding hydrogen, fluorine or lithium, or even a combination of these materials, to graphene to introduce asymmetry for bandgap opening. The carrier

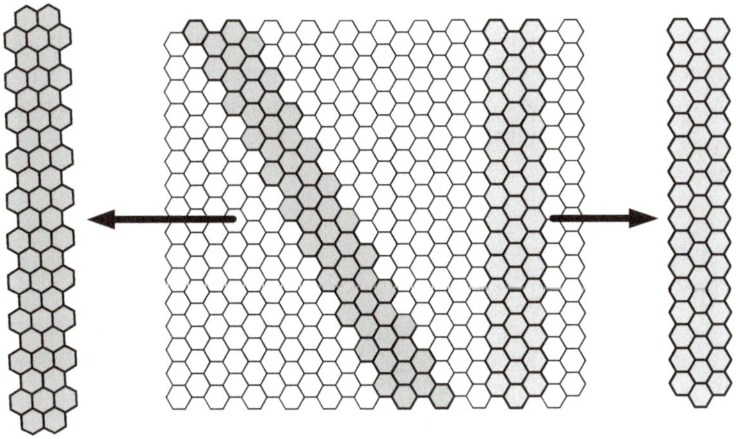

Fig. 4.12 Cutting of the graphene to produce either an armchair-edge (left) or a zigzag-edge (right) GNR

mobilities in the hydrogenated graphene, for example, can be made similar to those of the bulk Si.

Once GFET fabrication limitations are overcome, the performance of GFETs could surpass that of regular Si MOSFETs [65, 70, 72]. Top-gated GFETs with gate lengths in the region of 0.5–2 μm have been reported in [73], reaching extrinsic f_T and f_{max} of up to 34 and 37 GHz respectively. In the same paper, it was predicted that dropping the gate length to 50 nm should result in f_{max} surpassing 100 GHz. Since the graphene technology performance scales in similar fashion to that of CMOS MOSFETs, the theoretical performance of GFETs with a certain gate length is better than that of MOSFETs with the same length dimension.

4.2.2 Diamond-Based Electronics

Diamond is another form of carbon. Diamond has the highest known hardness and thermal conductivity (33 W/cm · K) [62]. This conductivity makes it the best option for high-temperature and high-power operation [74]. Electrically, it has properties of semiconductors. It has a bandgap of 5.5 eV, which is about five times higher than that of Si. Other valuable properties of diamond include an operating temperature of beyond 1000 °C, electric breakdown fields in the region of 10 MV/cm^3 and electron mobility of about 2000 cm^2/V · s.

Active diamond devices that have successfully been fabricated to date include *pn* junction diodes, Schottky diodes, BJTs, metal–semiconductor FETs (MESFETs), junction FETs and metal–insulator FETs (MISFETs). On the other hand, there is a multitude of practical problems with diamond electronics. For example, some diamond transistors are not activated at room temperature. The transistor structures that do work at the feasible range of temperatures still need to be researched and optimized before they will acquire any commercial value. Should these efforts become successful, there would be many potential uses of these devices in 5G and 6G networks, for example in high-power base stations, where GaN technology is currently the main alternative.

4.2.3 Carbon Nanotubes

CNTs are nanoscale devices with a cylindrical structure, constructed from a rolled monolayer graphene-like sheet [75–77]. CNTs were discovered over a decade before graphene in 1991. Sometimes, CNT cylinders are nested. The diameter of CNTs is about 1 nm for non-nested structures.

CNTs have several excellent properties. In addition to nanoscale dimensions, these include the fact that their structure is practically one-dimensional, that they can achieve ballistic electron conduction, that they have low susceptibility to electromigration and lastly, that they have high thermal conductivity. Furthermore, CNTs can

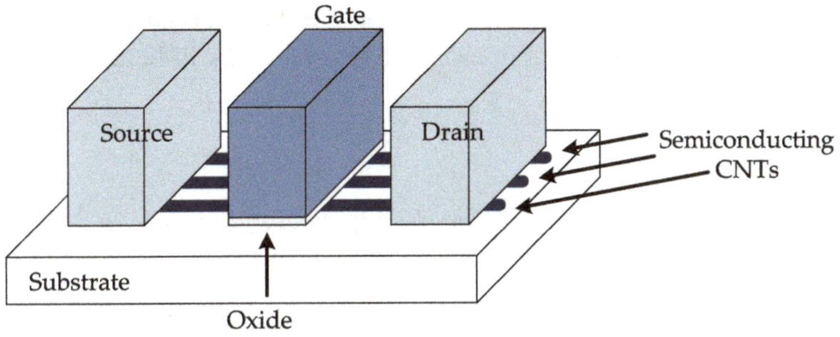

Fig. 4.13 The structure of a FET with the channel built from several parallel CNTs. Adapted from [79]

be either metallic or semiconducting. Metallic CNTs can carry current densities three orders of magnitude higher than copper and can be used in nanoscale interconnects. Semiconducting CNTs have high carrier velocities ($8 \cdot 10^7$ cm/s) and can be used in the fabrication of CNT-FETs with differing bandgap energy values, depending on the CNT diameter. Whether a CNT is metallic or semiconducting depends on the way in which the graphene sheet is rolled. There are three types of rolling, leading to zigzag CNTs, armchair CNTs or helical CNTs. Zigzag and helical CNTs can be either semiconducting or have metallic behavior, while armchair CNTs can only be semiconducting.

CNT-FETs are fabricated by using one or more CNTs to form a transistor channel between two electrodes [75–77], as shown in Fig. 4.13. In theory, extremely small transistors can be fabricated (10 nm and smaller) because of the tiny CNT dimension. CNT-FETs are compatible with CMOS processes in terms of architecture and operating voltage. It is expected that tiny carbon nanotube transistors will have excellent high-frequency properties. A recent example showed that the f_T and f_{max} of the 110-nm long CMOS-compatible CNT-FETs are both greater than 100 GHz [78], which indicates that theoretically possible much smaller technology nodes could reach operation even into the terahertz range. However, fabrication of CNTs, like that of graphene transistors, still needs process improvements.

4.3 Millimeter-Wave and Terahertz Circuits

Circuits required to build millimeter-wave and terahertz transmitters and receivers are fundamentally different from circuits that are operated at low frequencies. First, they are quite often built using transistors operated close to the frequency limit of their operation. Second, passive components, including capacitors and inductors, start performing differently at high frequencies. In many instances, lumped components have to be replaced with transmission-line counterparts, which fortunately tend to shrink in size because their length is proportional to the wavelength.

Power amplifiers, LNAs, oscillators, multiplexers and filters are some of the high-frequency circuits that are used in every transceiver and will be briefly reviewed here.

4.3.1 Millimeter-Wave and Terahertz Power Amplifiers

The role of the power amplifier is to increase the amount of power of modulated signals so that they can be transmitted over the channel. A power amplifier converts DC power from the supply voltage rails into a significant amount of RF power, which follows the shape of the modulated signal that is to be transmitted [19, 80–82]. Decent power amplifiers have high efficiency, high linearity, high power gain and output power, as well as a large dynamic range.

Key Characteristics of Power Amplifiers

A general high-frequency power amplifier is shown in Fig. 4.14. In this figure, V_{DD} is the supply voltage, R_L is the load, and RFC is the RF choke, and the driving transistor is any MOSFET, HEMT or HBT. In some power amplifier configurations, RFC can be replaced by a finite inductor L_D, and in that case, this inductor is part of the output filter, the role of which is in harmonic tuning waveform shaping. The filter is also shown in the general power amplifier diagram. Some power amplifiers use a complementary push–pull transistor configuration with one n-type and one p-type device; however, in high-frequency design, these configurations are avoided because of the inferior performance of p-type transistors, as described earlier.

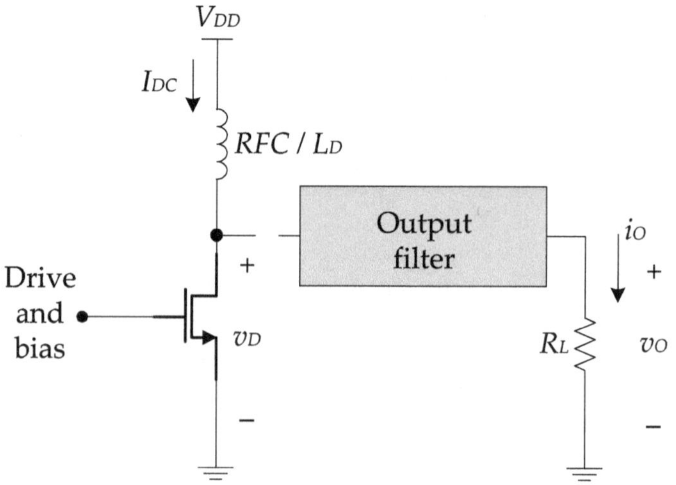

Fig. 4.14 General power amplifier

The power delivered into the load is determined by the supply voltage V_{DD} and the size of the load R_L. The maximum power in the load is

$$P_{out} = \frac{{V_{DD}}^2}{2R_L}, \tag{4.31}$$

for sinusoidal waveforms, but it can be higher if the waveforms are shaped differently. The power gain is the ratio of output power to input power:

$$G = \frac{P_{out}}{P_{in}} \tag{4.32}$$

and the total DC power is the current drawn from the voltage supply over a period of time, or

$$P_{DC} = V_{DD} I_{DC}, \tag{4.33}$$

where I_{DC} is the DC component of the drain current waveform.

Although the gain of the power amplifier is important, the efficiency is the parameter that is regarded as the most important measure of the performance of a power amplifier. Usually, the efficiency is specified as drain or collector efficiency or as the PAE. The drain or collector efficiency is defined as

$$\eta = \frac{P_{out}}{P_{DC}}. \tag{4.34}$$

The PAE, on the other hand, also takes the input power into account:

$$PAE = \frac{P_{out} - P_{in}}{P_{DC}} = \eta\left(1 - \frac{1}{G}\right). \tag{4.35}$$

The PAE becomes a better indication of the performance of a power amplifier as amplifier gains increase. The PAE shows that power amplifiers achieve their greatest efficiency when the gain of the amplifier is at its maximum, i.e. when the amplifier is saturated. If this is not the case, the power is wasted (turned into heat); however, the modulation schemes described in Chap. 3 are usually designed to keep the power amplifier in saturation. Maximum drain (collector) efficiency is achieved if no power can be dissipated over the drain (collector), meaning that either i_D or v_D must be equal to zero at any given time. As a result, the best power amplifiers, with theoretical efficiency of 100%, are those with non-overlapping current and voltage waveforms. However, many factors, including parasitics, non-ideal passives in amplifier design, tuning errors and transistor saturation generally result in efficiency figures much lower than ideal values predicted by equations.

Output power capability is the ratio of the maximum power delivered to the load and the product of the maximum values of i_D and v_D:

$$c_p = \frac{P_{out(max)}}{I_{DM(max)} V_{DS(max)}}. \tag{4.36}$$

The maximum operating frequency for a predetermined power and supply voltage can also be defined and is determined by the power amplifier circuit, the transistor layout and various technology-related parameters.

Finally, the dynamic range of the power amplifier can be defined. The dynamic range is the range of power levels between the lowest and highest level that the power amplifier can deliver reliably (usually specified in dBm).

Classification of Power Amplifiers

Power amplifiers are typically classified according to the nature (shape) of the voltage and current waveforms of the driving transistor. Two large groups of power amplifiers exist:

- Continuous mode power amplifiers: Classes A, B, AB and C (the driving transistor conducts continuously), and
- Switch-mode power amplifiers: Classes D, E, F and their variations—Classes D^{-1}, E^{-1}, F^{-1} EF, E/F, etc. (the driving transistor operates as a switch).

Typically, only switchmode classes are suitable for frequencies in the millimeter-wave and terahertz range.

Many power amplifiers operate with current and voltage waveforms that can be classified somewhere between these different classes. It is also possible to combine power amplifiers of different classes to optimize power and efficiency, as is the case in Doherty power amplifiers [83] and outphasing power amplifiers [84].

High-Frequency Power Amplifier Example: Class-E Zero-Voltage-Switching Power Amplifier.

The Class-E power zero-voltage-switching (ZVS) amplifier requires one transistor, a shunt capacitor, a series resonant filter and an RFC or a finite DC feed inductor, as illustrated in Fig. 4.15. In the Class-E amplifier, the output filter shapes the waveforms such that they are never on simultaneously, resulting in theoretical 100% efficiency. Its output power is just over half of $V_{DD}{}^2/R$ and the output power capability is only 0.098.

L_0 and C_0 are part of a resonant circuit with

$$f_0 = \frac{1}{2\pi\sqrt{L_0 C_0}}. \tag{4.37}$$

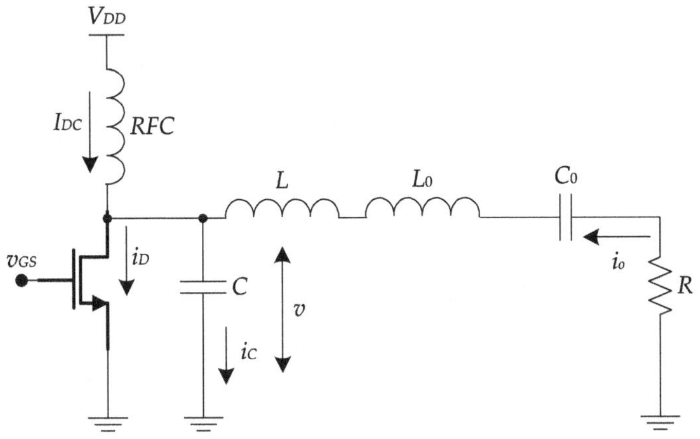

Fig. 4.15 Class-E power amplifier circuit

The loaded Q-factor of the resonant tank is dependent on total inductance $L + L_0$:

$$Q_L = \frac{2\pi f_0(L + L_0)}{R}. \tag{4.38}$$

If the driving transistor is modeled as an ideal switch, the drain voltage resulting from the transistor gate being driven by a signal with switching frequency f_0 will be determined by the switch itself when it is turned on and by the response of the resonant circuit when it is turned off. If the duty cycle of the switching waveform is 50%, Class-E ZVS power amplifier waveforms will look like those shown in Fig. 4.16. Note that in the case of the drain current waveform, because switching happens abruptly, the current will take some time t_f to stop flowing completely when the switch starts conducting again, as illustrated, causing some overlap in waveforms and a decrease in efficiency. The resonant filter ensures that the output voltage reaching the load is mostly sinusoidal.

This Class-E amplifier is described below. First, the maximum drain voltage can be approximated by [80]

$$V_{max} = 3.562 V_{DD}. \tag{4.39}$$

The optimum load resistance for a specified output power P_{out} can be worked out from the following approximation:

$$R = 0.577 \frac{V_{DD}^2}{P_{out}}. \tag{4.40}$$

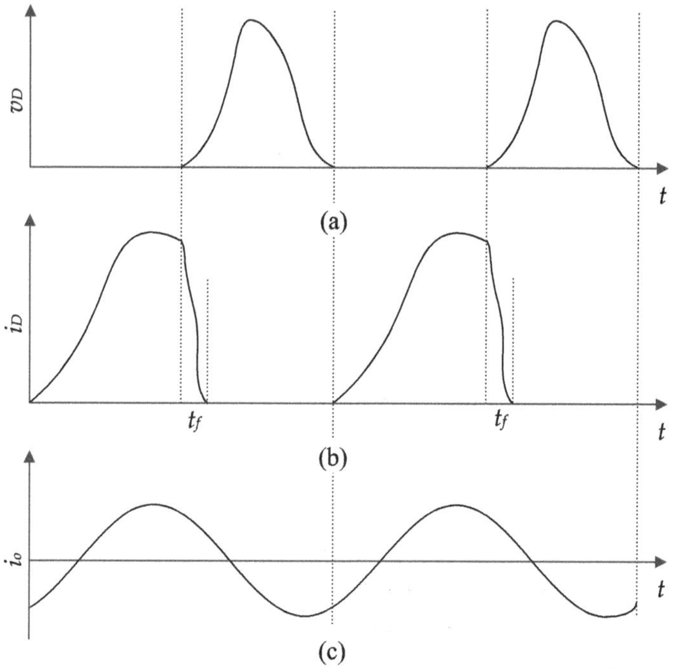

Fig. 4.16 Class-E power amplifier waveforms: **a** Drain voltage, **b** Drain current and **c** Load current

The filter components can now be worked out. The total series inductance ($L + L_0$) is, with $\omega_o = 2\pi f_o$:

$$L + L_0 = \frac{Q_L R}{\omega_o} \tag{4.41}$$

and the shunt capacitance is

$$C = \frac{1}{5.447\omega_0 R}. \tag{4.42}$$

The shunt capacitance value must be able to absorb the output capacitance of the transistor (C_{OUT}), therefore, the instanced physical capacitor component must have a value of C–C_{OUT}.

The series capacitance is approximately

$$C_0 = \frac{5.447C}{Q_L}\left(1 + \frac{1.42}{Q_L - 2.08}\right). \tag{4.43}$$

This results in the DC current

$$I_{DC} = \frac{V_{DD}}{1.734R} \tag{4.44}$$

drawn from the supply and in a maximum drain current of

$$I_{max} = 2.8621 I_{DC}. \tag{4.45}$$

The minimum inductance value of the RFC required to separate V_{DD} adequately from AC signals is approximately

$$L_{RFC} = 8.669\frac{R}{f_0}. \tag{4.46}$$

At high frequencies, losses in inductors and capacitors placed in the power amplifier circuit become quite prominent. There are many ways in which gain and efficiency are improved in modern amplifiers. A configuration that does not require a major redesign from what was just described is the one where a single driving transistor is replaced with multiple stacked driving transistors, as conceptually shown in Fig. 4.17. In this circuit, the gates of $M_2, M_3 \ldots M_n$ are connected to some finite impedance (and not to a fixed voltage, which is the RF ground), resulting, ideally, in the effect where the drain voltages of all transistors are added together. The output filter deploys the same waveform shaping as in the case of the single-transistor amplifier but the output swing is n times higher than the swing at the drain (V_{DM}) of the driving transistor (M_1).

4.3.2 Millimeter-Wave and Terahertz Low-Noise Amplifiers

Whereas the power amplifier is the integral part of the transmitter, the LNA is the first component that appears in the front end of a high-frequency receiver [85]. The LNA can connect either directly to an antenna or a low-loss bandpass signal selection filter [86, 87]. The role of the LNA is to amplify weak signals detected from the transmission channel. These signals are typically just a little stronger than the noise floor, so the LNA needs to add as little additional noise to the signal as possible.

A typical LNA is built around a transistor in the common-source (emitter) or common-gate (base) configuration. This is as shown in the conceptual LNA diagram in Fig. 4.18, where the driving transistor is, as in the case of power amplifiers, any MOSFET, HEMT or HBT. The single driving transistor is often replaced with a cascode amplifier, which allows for independent matching at the input and output, increased output impedance and higher system gains, owing to the independent matching at the input and the output. The input signal first passes through the input-matching network with the role to provide simultaneous noise and power match

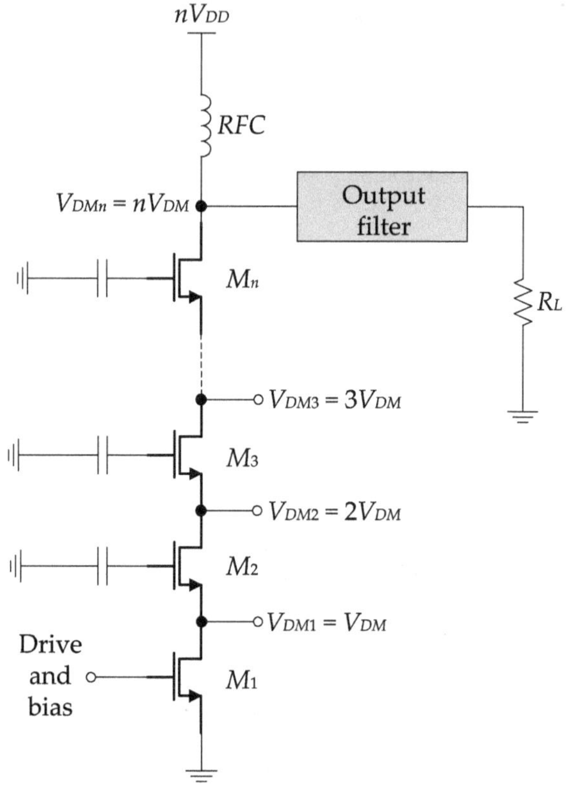

Fig. 4.17 Stacked Class-E Amplifier

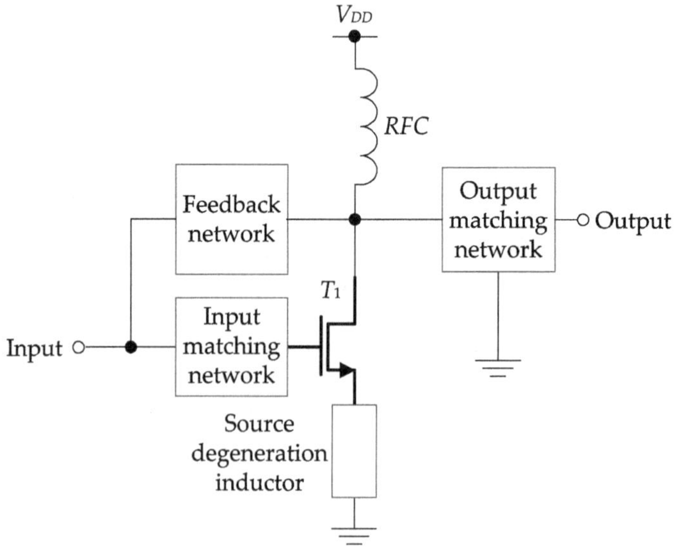

Fig. 4.18 General circuit diagram of an LNA

(in other RF amplifiers, matching is usually only performed for optimum power). In common-source (emitter) and cascode configurations, the source (emitter) of the transistor is sometimes inductively degenerated. The output signal passes through an output-matching network and in some instances, there is also a feedback network in an LNA.

LNAs can be either narrowband or wideband. While wideband designs allow for the greater reusability of a single LNA configuration over a wide frequency range, narrowband LNAs are able to reach much higher frequencies by manipulating the gain-bandwidth product.

Core Characteristics of Low-Noise Amplifiers

The core characteristics of an LNA are the gain, noise figure, power consumption, reverse isolation, bandwidth and gain flatness, sensitivity, selectivity and linearity.

The power gain of the LNA is defined as the ratio of power delivered to the load to input power:

$$G = \frac{P_L}{P_{in}}. \tag{4.47}$$

Reverse isolation can be defined in the gain if the power is applied at the output and measured at the input. While it is desirable to achieve large values of forward gain, reverse isolation should be as low as possible.

The noise performance is considered the most important parameter in the LNA. In a general electronic circuit, noise consists of small current and voltage fluctuations, which are generated in different components. The noise in an LNA is quantified with the noise figure. The noise figure is the noise factor converted into decibels. The noise factor is the ratio of the input SNR to the output SNR:

$$F = \frac{SNR_i}{SNR_o}. \tag{4.48}$$

and is always greater than 1. The noise figure is thus

$$NF = 10 \log F. \tag{4.49}$$

The noise figure of a multi-stage system is dominated by the noise of the first stage, if the gain of this stage is at least several dB. In that case, the noise contribution of subsequent stages can be ignored. Sensitivity is the lowest signal that the amplifier can detect with no errors.

The total DC power consumption is less of a consideration in LNAs than in power amplifiers, because LNAs operate on very small signal levels. However, it can be worked out in the same manner as for the power amplifier.

The bandwidth and gain flatness are characteristics of wideband LNAs. The 3-dB bandwidth refers to 3 dB below (half of) the maximum gain magnitude. Gain flatness is the metric that describes the variation or ripple in gain over the desired bandwidth.

The noise bandwidth can also be considered in LNAs; however, it is approximately equal to the 3-dB bandwidth of the system. For narrowband LNAs, selectivity is as important as the bandwidth in wideband amplifiers. Selectivity is the ability of a circuit to select the desired signal at a given frequency, while rejecting any other signals present in the channel.

Finally, linearity is also an important consideration of an LNA, since the nonlinear behavior on an LNA as the first stage can result in signal distortion in the complete receiver.

Narrowband Low-Noise Amplifiers

As in the case of power amplifiers, popular LNAs are single-ended configurations. Four basic types of single-ended LNA stages are shown in Fig. 4.19. Sometimes, the required gain, noise figure and linearity can simultaneously be achieved in a single

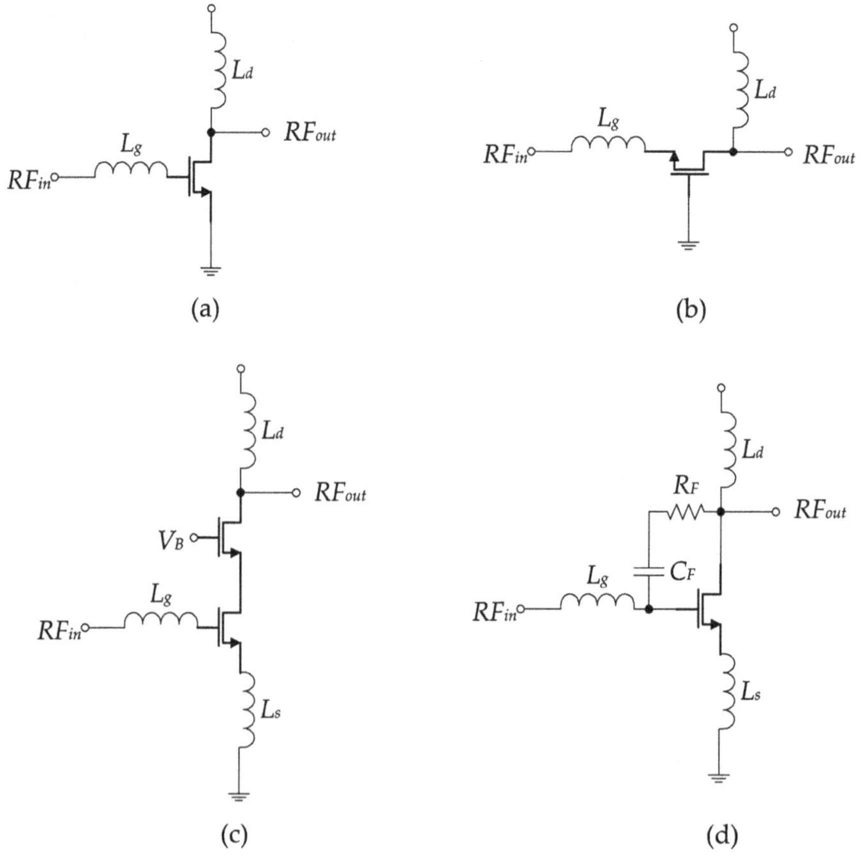

Fig. 4.19 **a** Common-source LNA, **b** Common-gate LNA, **c** Cascode LNA with inductor degeneration, **d** Feedback LNA

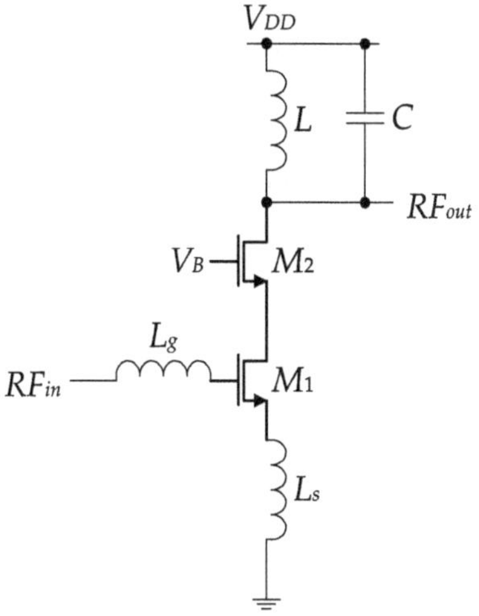

Fig. 4.20 A cascode configuration with a resonant tank and source degeneration

LNA stage, but it is often necessary to design multi-stage LNAs. When connecting multiple stages, interstage matching becomes as important as input matching.

One of the best topologies of narrowband LNAs is the cascode LNA. Cascode LNAs are regarded as simple, yet effective, as they exhibit high stability, good reverse isolation and good overall gain. Additionally, the cascode configuration minimizes the Miller effect in the LNA in much the same way it is achieved in low-frequency amplifiers. If a MOSFET is used as a driving transistor of the LNA, a lower noise figure value is obtained with the input transistor biased at around 0.8 V. The frequency of operation can be set by a resonant tank tuned at the frequency of interest added as part of the output filter, which is shown in in Fig. 4.20.

The design process for an LNA starts with the design of the input-matching network, which needs to satisfy conditions of simultaneous noise and impedance match. The gain of the LNA is a secondary consideration at this point. At frequency f (angular frequency $\omega = 2\pi f$), the input impedance of the cascode LNA will be the input impedance of M_1, which can be modeled with the small-signal model in Fig. 4.5 for MOSFETs and Fig. 4.9 if M_1 is replaced with an HBT. This value is

$$Z_i = \frac{1}{j\omega C_{gs1}} \tag{4.50}$$

for the MOSFET cascode or

$$Z_i = \frac{1}{j\omega C_{\pi 1}} || r_{\pi 1} \tag{4.51}$$

for the HBT cascode, provided that L_s and L_g are both zero. The computed impedance value in both cases is different from the value of the impedance where the minimum

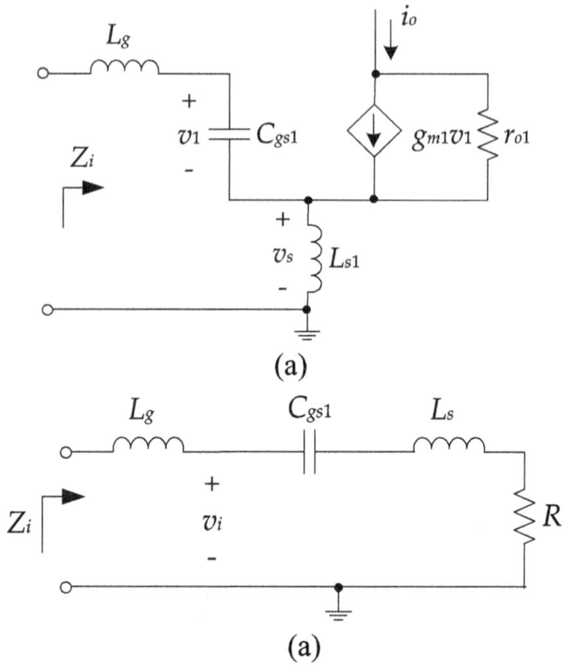

Fig. 4.21 Input impedance transformation applicable when the source of M_1 is degenerated with inductor L_s: **a** Before transformation and **b** after transformation

attainable noise figure is reached, which can be denoted by Z_{opt}. This means that an impedance transformation needs to be performed to make Z_i and Z_{opt} overlap. This is done by bringing both Z_{opt} and Z_i to 50 Ω, which is the standard antenna impedance value [85, 88].

With L_s and L_g as shown in Fig. 4.20, the small-signal diagram of M_1 would look as pictured in Fig. 4.21a. The input impedance of this circuit is

$$Z_i = j\omega L_g + \frac{1}{j\omega C_{gs1}} + j\omega L_s + \frac{g_{m1}}{C_{gs1}} L_s \tag{4.52}$$

which, by noting that $\omega_T = 2\pi f_T = g_m/C_{gs}$ and if C_{gd} is ignored, can be rewritten as

$$Z_i = j\omega L_g + \frac{1}{j\omega C_{gs1}} + j\omega L_s + R, \tag{4.53}$$

where $R = \omega_T L_s$. The result of this transformation is shown in Fig. 4.21b. This equation now has real as well as imaginary terms, meaning that it is possible to equate the input impedance Z_i to a real value ($Z_0 = 50$ Ω) by ensuring that the imaginary part is zero. The imaginary part becomes zero if the sum $L_g + L_s$ resonates C_{gs1} out at frequency f. This results in the following design equations. First, L_s is determined from

$$L_s = \frac{R}{\omega_T} = \frac{Z_0}{\omega_T}. \tag{4.54}$$

This means that L_g is then

$$L_g = \frac{1}{\omega^2 C_{gs1}} - L_s. \tag{4.55}$$

The resonant tank ensures the narrowband operation at the center frequency.

$$f_0 = \frac{1}{2\pi\sqrt{LC}}. \tag{4.56}$$

The simultaneous impedance and noise match comes at a loss of some gain. The new value of transconductance becomes

$$G_m = \frac{g_{m1}}{\omega C_{gs1}(Z_0 + \omega_T L_s)} = \frac{f_T}{2fZ_0} \tag{4.57}$$

showing that with f_T and Z_0 practically constant, gain only depends on the frequency of operation and the transitional frequency, which is a technology parameter. The noise factor achieved as a result of the matching described above (if MOSFETs are used) is

$$F = 1 + \frac{r_g}{Z_0} + \frac{\gamma}{\alpha} g_m Z_0 \left(\frac{\omega_0}{\omega_T}\right)^2, \tag{4.58}$$

where γ is the channel-noise coefficient (ranges from 2/3 in saturation mode and beyond 2 for short-channel devices) and α is another technology-specific coefficient, which decreases from unity when moving from long to short channels.

The analysis above assumed that the effect of feedback capacitors C_{gd1} and C_{gd2} was negligible. As frequency increases, their influence on the circuit performance increases, casing the Miller effect to start happening. If C_{gd1} and C_{gd2} are included in the small-signal model in Fig. 4.21a, it can be shown [85, 89] that a complex Miller impedance, of the value

$$Z_M = \frac{\omega^2 g_{m1} L_s C_{gd1}}{1 + (\omega g_{m1} L_s)^2} + j\omega C_{gd1} \frac{2 + (\omega g_{m1} L_s)^2}{1 + (\omega g_{m1} L_s)^2} \tag{4.59}$$

will appear in shunt in the circuit in Fig. 4.21b, with the positioning as illustrated in Fig. 4.22. Impedance matching in this situation becomes much more complicated and it is difficult to obtain an analytical solution. In this and similar cases, it becomes much simpler to use graphical or simulation-driven analysis. Some techniques usually used include simulating for the optimum reflection coefficient (for achieving a noise

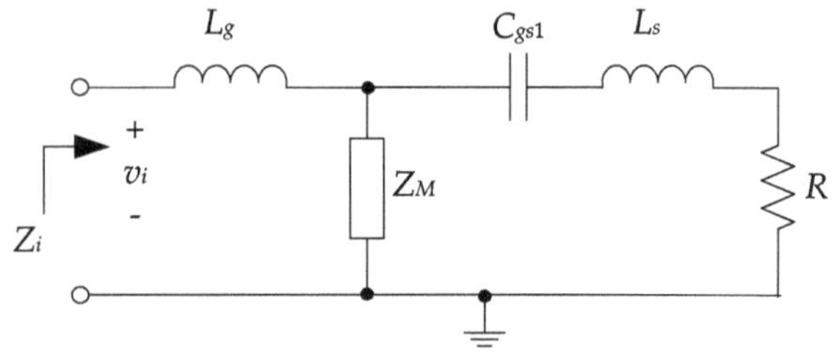

Fig. 4.22 Model of the inductively degenerated transistor M_1 with Miller impedance included

match) for different component values or establishing acceptable regions of operation specified in terms of various design LNA parameters.

Broadband Low-Noise Amplifiers

Broadband LNA configurations are often favored because a single configuration can operate over a range of frequencies, making them reusable in a range of applications. The wideband operation comes at the expense of the worsened LNA noise performance. Feedback LNAs are an example of wideband LNAs.

In feedback LNAs, a negative feedback network is included at the expense of the achievable gain. Additional LNA stages can be cascaded to compensate for the gain loss, while still retaining a good noise figure and small footprints. The latter happens because feedback LNAs do not generally require matching inductors. Figure 4.23 shows a possible feedback LNA configuration. The voltage gain of this circuit is [82, 90]

$$A = \frac{v_o}{v_i} = \frac{\frac{R_L}{R_F} - g_m R_L}{1 + \frac{R_L}{R_F}} \approx -\frac{g_m R_L}{1 + \frac{R_L}{R_F}}. \tag{4.60}$$

The input impedance of this amplifier has a real value:

$$Z_{in} = \frac{R_F + R_L}{1 + g_m R_L} \approx \frac{R_F + R_L}{g_m R_L}, \tag{4.61}$$

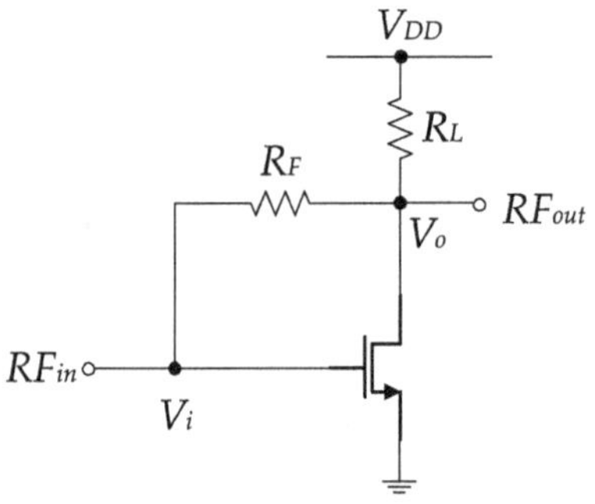

Fig. 4.23 Resistive-feedback wideband LNA

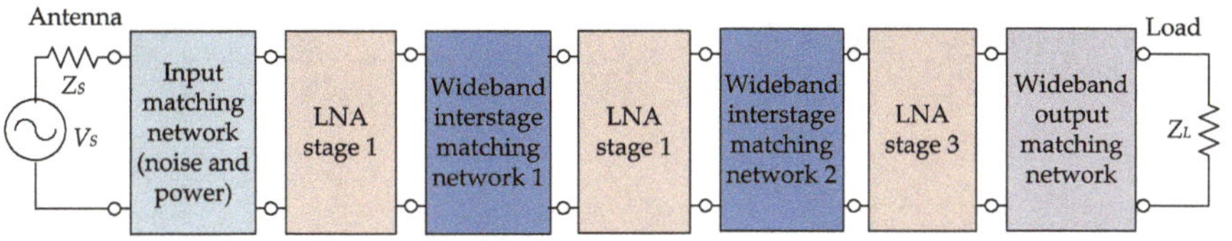

Fig. 4.24 Block diagram of a generic cascaded LNA

which allows for simultaneous noise and power matching. The noise figure of this circuit, in terms of small-signal parameters, is

$$F = 1 + \frac{R_F}{R_s}\left(\frac{1 + g_m R_s}{1 - g_m R_F}\right) + \frac{1}{R_s R_L}\left(\frac{R_F + R_s}{1 - g_m R_F}\right)^2 + \frac{\gamma}{\alpha}\frac{g_m}{R_s}\frac{(R_F + R_s)^2}{1 - g_{m1} R_F}. \quad (4.62)$$

Cascaded Low-Noise Amplifiers

To reach millimeter-wave and terahertz frequencies, multiple LNA stages usually need to be cascaded. Several cascode, common-source or common-emitter stages can be connected with wideband interstage matching networks, as shown in Fig. 4.24. In these LNAs, the input-matching network remains the most critical part of the LNA circuit and is designed in the same fashion as for the single stage (this holds true if the gain of the first stage is high). Furthermore, at high frequencies, it is not as easy to achieve high gain figures and good stability of any of the single stages, so it is not uncommon to have five or even as many as eight cascaded stages. This results in an increased noise figure, and in the case of cascaded LNAs, the design once again becomes an exercise in which a compromise among the most important design parameters must be found.

4.3.3 *Millimeter-Wave and Terahertz Oscillators*

Oscillators are circuits that provide a periodic output signal without any AC signal applied to the circuit [19]. In a transmitter, the LO generates a carrier frequency for transmission and in the receiver, the voltage-controlled oscillator (VCO) forms a part of the carrier detection circuit.

Local Oscillators.

The main characteristics of an LO are its oscillation frequency (f_o), frequency stability, spectral purity and output voltage, current or power amplitude and their stability [19]. The oscillations are established by means of a positive feedback loop [91]. The general oscillator topology can normally be the same as that of a feedback amplifier, except that the circuit is intentionally designed to be unstable, leading to oscillation if a startup condition is provided. The frequency of an oscillator is tuned by including a resonant tank in the feedback loop.

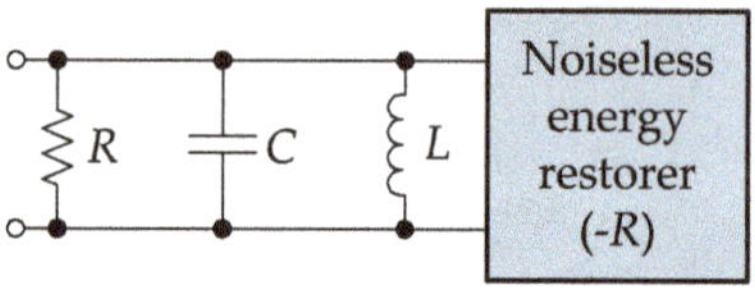

Fig. 4.25 Simplified oscillator model

In general, any resonant tank could in theory sustain oscillations, if all the components and connections in the circuit were lossless. Because this is not the case, the losses have to be compensated for with active components. Therefore, the losses in the resonant tank need to be compensated for by a circuit called a noiseless energy restorer. This concept is illustrated in Fig. 4.25, where the resonant tank is modeled by L and C and losses are represented by R [92]. Oscillations of this circuit are sustained if the noiseless energy restorer provides a negative resistance with magnitude that is equal or larger than R.

Whereas there are standard design procedures for low-frequency oscillators, designing oscillators at high frequency is not an easy task. The design steps include finding the suitable configuration, choosing the transistor type and evaluating the non-linear parameters of the transistor involved at both small and large signal conditions, while evaluating the performance in simulation that includes all the possible losses in the circuit and components [93]. Moreover, many millimeter-wave and terahertz oscillators are often superharmonic oscillators, that is, the required oscillation frequency is generated at a multiple of the base tone. Configurations such as push-push, triple-push and self-mixing topologies are some of the examples of superharmoic oscillators [94, 95].

On the other hand, the Colpitts oscillator is a topology that was successfully implemented at frequencies over 100 GHz in the past. One such oscillator can be seen in Fig. 4.26a, with its model shown in Fig. 4.26b. This circuit can be designed such that the frequency at which the real part of the impedance looking into the base of the transistor becomes zero is maximized. This frequency is then.

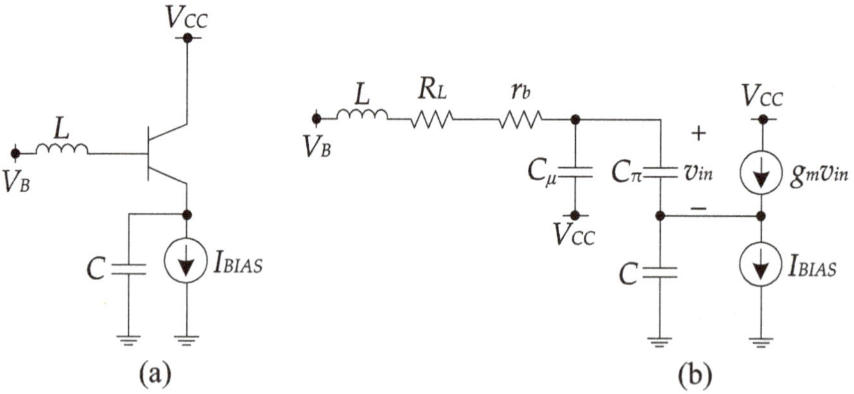

Fig. 4.26 Millimeter-wave Colpitts oscillator seen in [94] (**a**), its model (**b**)

$$f_{max,colp} = \frac{f_{max}}{\sqrt{1 + g_m r_b \frac{C_\mu}{C_\pi}}}. \tag{4.63}$$

where f_{max} is still the maximum oscillation frequency of the transistor and is fixed for a transistor geometry and bias current. The oscillator oscillates at

$$f_0 = \frac{1}{2\pi\sqrt{L\left(\frac{C_\pi C}{C_\pi + C} + C_\mu\right)}}, \tag{4.64}$$

and tuning is still accomplished with the aid of L and C.

Power amplifiers can also be turned into oscillators. These oscillators have a self-amplifying capability. In [93], a Class-E power amplifier was reconfigured as an oscillator by including its resonant tank in the feedback loop.

Oscillator imperfections usually show as phase noise [96], which can be observed as jitter if the carrier signal is plotted in the time domain, as done in Fig. 4.27. Causes of phase noise include the noise in the active devices and the finite Q-factor of the L-C resonator. High-frequency systems are much more prone to phase noise than lower-frequency systems. Fundamental oscillators at millimeter-wave and higher frequencies, interestingly, often have poorer phase-noise performance than superharmonic oscillators.

Voltage-Controlled Oscillators

In VCOs, the oscillation frequency is controlled electronically by a reference voltage. The voltage control is useful in carrier recovery, which is deployed as part of a phase-lock loop (PLL) in the receiver circuit. A simple PLL is shown Fig. 4.28.

Frequency control at lower frequencies is normally achieved by utilizing varactors (variable capacitors) or arrays of switched capacitors. Varactors or switched

Fig. 4.27 Illustration of jitter in time domain

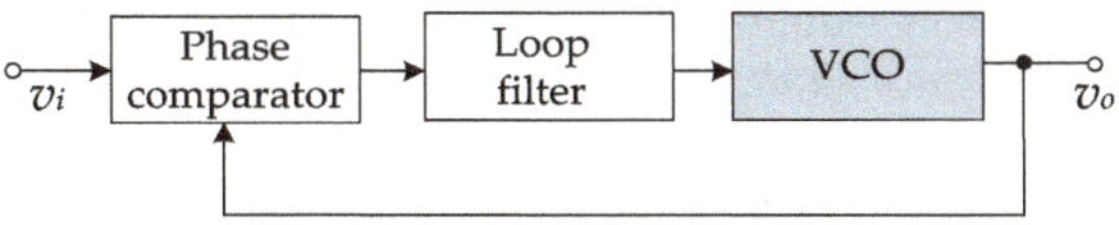

Fig. 4.28 The simplest form of PLL

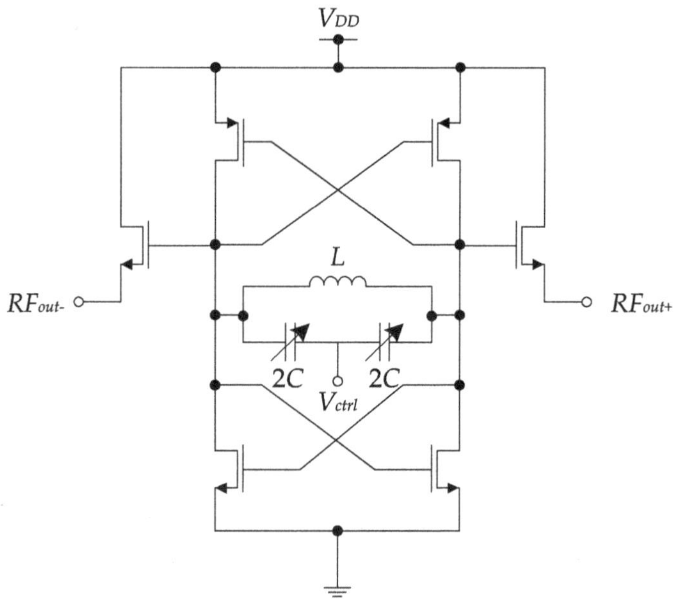

Fig. 4.29 Cross-coupled millimeter-wave VCO [93]

capacitors in the VCO replace the cap in the resonator circuit of the LO. Varactors, however, have parasitic quantities that are limiting at high frequencies, so for Millimeter-Wave and terahertz frequencies, alternative VCO configurations need to be considered. One example of an oscillator suitable for millimeter-wave frequencies is the cross-coupled topology shown in Fig. 4.29. In this figure, the tuning is performed with the aid of voltage V_{ctrl}.

4.3.4 Mixers

Mixers, which are used in both the transmitter and the receiver, translate the signal spectrum from one frequency to another [97]. On the transmitter side, the low-frequency signal is multiplied with the oscillator signal to generate the high-frequency signal that is afterwards amplified in power and transmitted. On the receiver side, the mixer multiplies the received signal with the carrier signal and restores the scaled version of the original, low-frequency signal. Important mixer parameters include conversion loss or gain, isolation, conversion compression, dynamic range and third-order intermodulation distortion.

The principle of modulation is based on multiplication of two sinusoidal signals of different frequencies. Multiplying a (local information signal) with b (the oscillator signal) results in

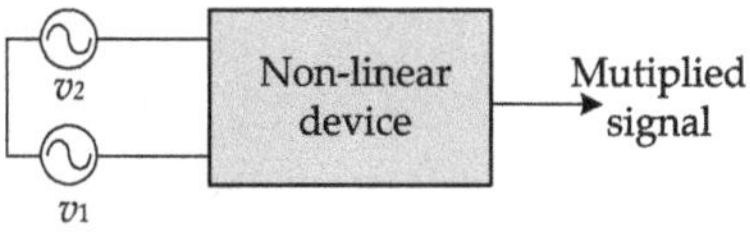

Fig. 4.30 Generic non-linear device as a multiplier/mixer/modulator

$$a \cdot b = Acos(\omega_a t) \cdot Bcos(\omega_b t) = \frac{1}{2}AB[\cos(\omega_a t + \omega_b t) + \cos(\omega_a t - \omega_b t)]. \tag{4.65}$$

If $\omega_b > > \omega_a$, multiplication will result in two signals centered close to ω_b with a small frequency variation ω_a. Usually, the sum of the two signals is the part of the modulated signal that is amplified and can be denoted by c. Thus, $c = Ccos(\omega_c t) = Ccos(\omega_a t + \omega_b t)$.

Demodulation works in exactly the same way. If the same mixer is applied to signals c and b, the result is,

$$c \cdot b = \frac{1}{2}AC[\cos(\omega_c t + \omega_b t) + \cos(\omega_c t - \omega_b t)]. \tag{4.66}$$

If the second (difference) term is passed through a band pass or a low-pass filter, the result of multiplication is actually a signal of the same shape as a, because $\omega_a = \omega_c - \omega_b$.

Any non-linear device can serve as a multiplier, because the product term will exist in the frequency spectrum of the amplified signal as long as the two signals are applied simultaneously to its input, as shown in Fig. 4.30. Even transistors are non-linear devices when large signals are applied. Single-transistor mixers, however, are impractical [98], because the requirement for DC biasing of such mixers will result in signal feedthrough from both signals that are applied to the mixer input. The DC offset of one of the two signals can be removed by using a differential mixer cell producing a balanced mixer, but to be able to apply both modulation signals differentially, a double-balanced mixer configuration is needed. This configuration is illustrated in Fig. 4.31. A practical narrowband configuration that can be used at millimeter-wave frequencies is shown in Fig. 4.32.

4.3.5 Filters

In the transmitter and receiver, filters can appear at any stage of signal processing. The role of the filter is to allow the wanted range of frequencies to pass through and reject unwanted out-of-band signals. Thus, the filter is characterized by the attenuation profile [86, 87].

One of the most complex filters to build at high frequencies is the band-pass filter (BPF). The attenuation profile of this filter is shown in Fig. 4.33 [91]. Good BPFs are small in size, have a high Q-factor, low insertion loss, high return loss, good

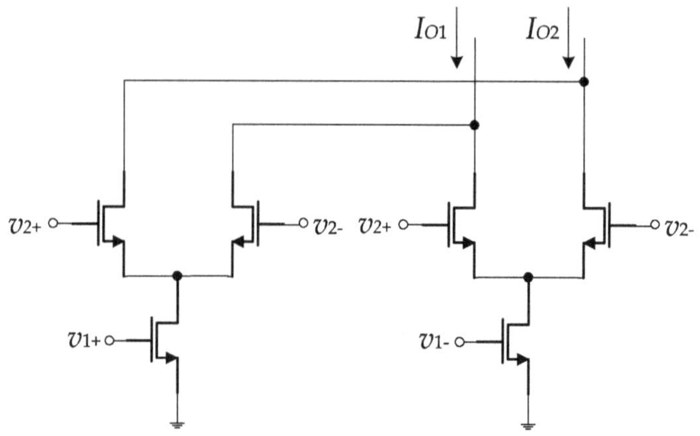

Fig. 4.31 Double-balanced mixer cell

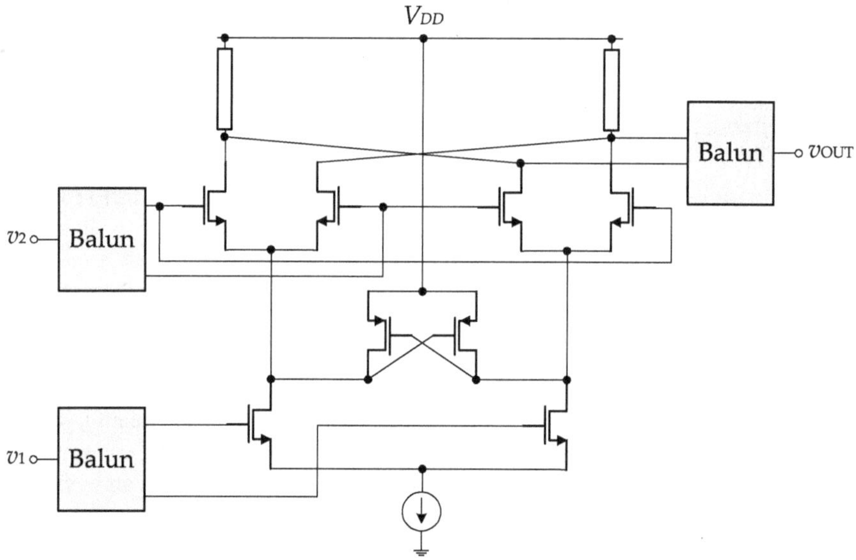

Fig. 4.32 Narrowband millimeter-wave mixer [98]

selectivity and high stopband rejection. BPFs placed at the receiver input between the antenna and the LNA also need to have a low noise figure. Active, passive and hybrid BPFs are possible.

Passive BPFs suffer from high loss and low Q-factors, which allows active BPFs to focus on loss compensation and Q-enhancement. The active variant also allows for good selectivity, a high level of integration and freedom of electronic tuning capability.

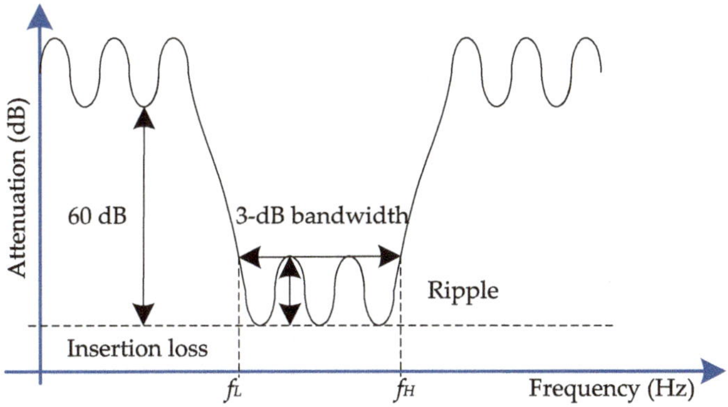

Fig. 4.33 The attenuation profile of the BPF

If a passive lossy filtering action is modeled by a transmission line, as shown in Fig. 4.34a, then the compensation can be achieved with negative resistance, which is implemented as an active circuit, as shown in Fig. 4.34b. This principle was used in the work by Chaturvedi et al. [99, 100], where the transmission-line resonator is deployed to provide equivalent values of inductance (L_{eq}) and capacitance (C_{eq}) for filtering action, while a *p*-type (RF *p*MOS) and an *n*-type (RF *n*MOS or *npn* HBTs) device

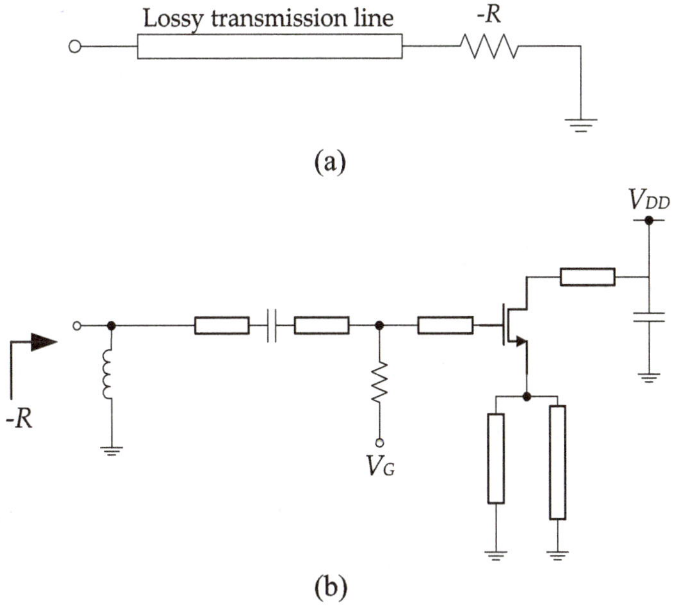

Fig. 4.34 **a** Negative resistance loss compensation with the aid of negative resistance, **b** One active implementation possibility for negative resistance (adapted from [87])

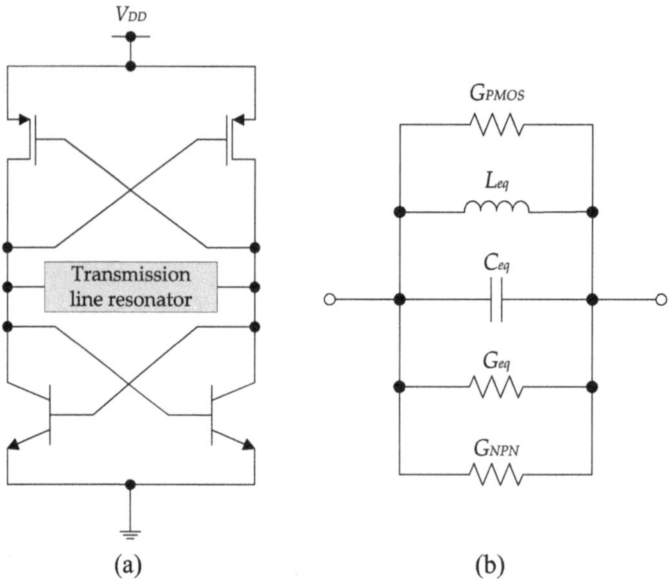

Fig. 4.35 Transmission-line resonator as reported in [100] (**a**), Model of the same filter (**b**)

each generates negative conductance ($-G_p$ and $-G_n$) in the arrangement of elements shown in Fig. 4.35a. The compensated resonant tank is shown in Fig. 4.35b, where the positive conductance of the resonant tank (G_{eq}) is supposed to be compensated for by the sum of $-G_p$ and $-G_n$. To achieve second-order bandpass filtering action, two compensated resonant tanks with different transmission-line lengths were cascaded in these works.

4.4 Packaging for Millimeter-Wave and Terahertz Frequencies

The proposed increase in the frequency of operation for the 5G and 6G circuits introduces another factor into the circuit design: the influence of the packaging on the circuit operation. There are two related aspects of packaging:

1. The influence of the physical packaging, which is used to protect electronic circuits from the influence of the environment (e.g. pressure, impact and moisture) on the high-frequency behavior of the circuit; and
2. The use of packaging as an advantage in circuit design, by deploying the so-called systems approach.

This section focuses on the second problem.

In technologies such as CMOS and BiCMOS, numerous system blocks can be implemented on the same die. In addition to baseband digital and analog circuits, it is now possible to integrate all blocks described in Sect. 4.3: high-frequency oscillators, power amplifiers, LNAs, multiplexers and filters. This approach is called system-on-chip (SoC) and has the advantages of low cost, low power dissipation, high integration capability and short design cycles for application-specific ICs [17]. In many applications, especially as frequency increase and design constraints (e.g. high power) become more stringent, one specific technology is not suitable to achieve all the specifications. In these cases, dies from different process technologies can be included in a single package, with each technology being more suitable for a specific function. As an example, a BiCMOS technology can be used for baseband and high-frequency processing and GaN technology can be used to deliver high power. This approach is called multi-chip module (MCM) packaging. The ultimate system flexibility is achieved in the silicon-on-package (SoP) approach [101], which also allows passives (e.g. resistors, inductors and capacitors) to be fabricated on different, more suitable substrates instead of integrated (IC) substrates, but still to be packaged together with active devices fabricated on a semiconductor die. Mixing of integrated and discrete substrates is encouraged in SoP, with passive components fabricated on discrete substrates, such as organics or laminates, optimized for greater Q-factors. Even components such as antennas and heat sinks, coplanar waveguides, microstrip lines, baluns, discrete filters and microelectromechanical system (MEMS) components can be placed under the same enclosure, with the overall effect being that the components are placed more closely together, which results in shorter interconnects, decreased parasitic effects and overall improvement of system performance [101, 102]. Different packaging strategies are shown in Fig. 4.36, and their relationship is best understood with the aid of the Venn diagram in Fig. 4.37.

This section will predominantly look at the two mainstream approaches in millimeter-wave and terahertz packaging: SoC and SoP.

4.4.1 System-On-Chip

SoC is an approach in which the complete system is placed on a single die and packaged without any external components, with the exception of voltage supplies and stabilization capacitors, integration of which is limited by size.

The integration of RF circuitry was never simple in SoC. Historically, problems arose when integrating blocks such as power amplifiers and LNAs on the same chip. It is even more complex to integrate antennas where the wavelength influences the antenna size. It was only possible because frequencies as high as those in the millimeter-wave regime allowed antennas to become sufficiently small to fit single or multiple antennas inside the package [103].

The core advantages of SoC packaging, compared to other packaging strategies, are low cost, the highest level of integration, miniature size and the lowest bill of materials. In SoC, sensitive RF signals always remain inside the package. Challenges

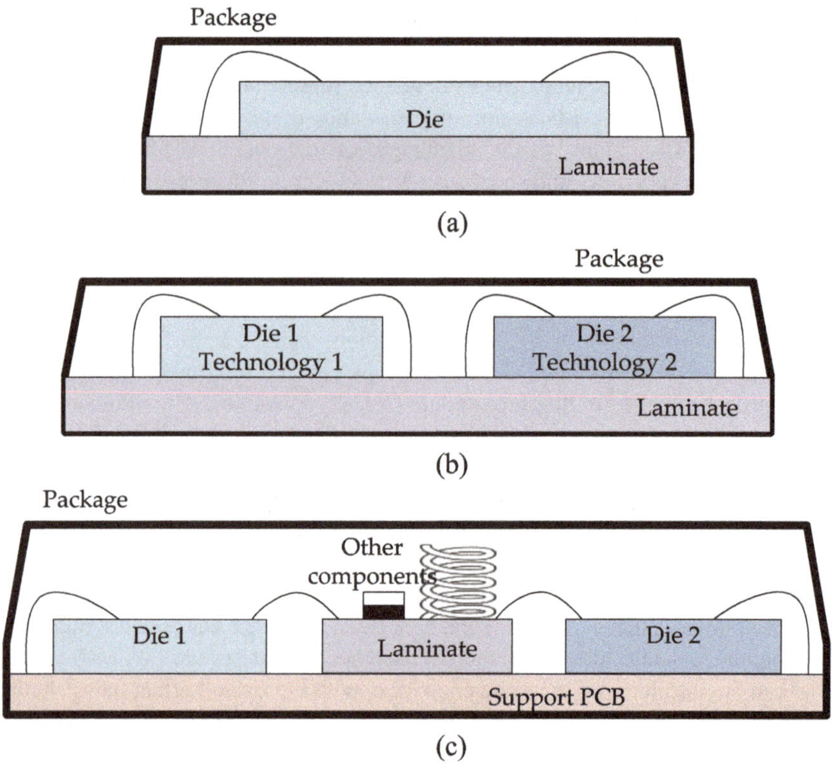

Fig. 4.36 Different packaging strategies: **a** SoC, **b** MCM and **c** SoP

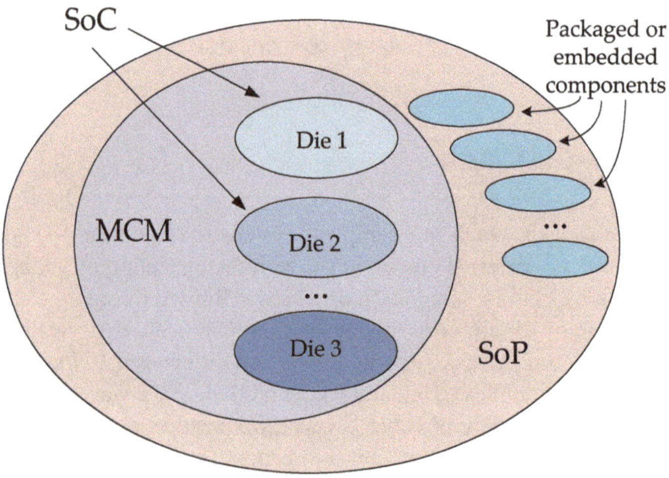

Fig. 4.37 Relationship between SoC, MCM/3D ICs and SoP

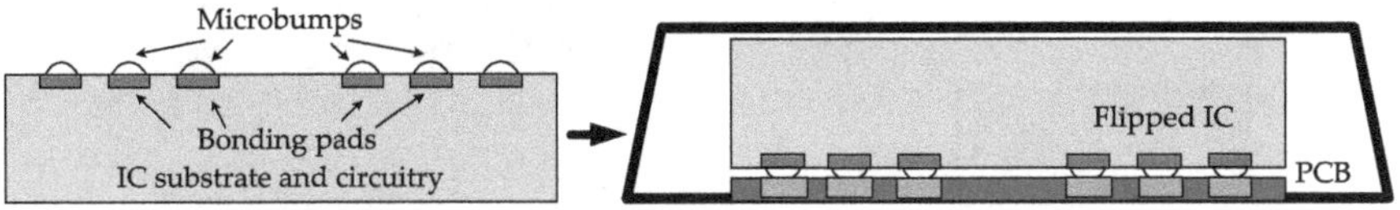

Fig. 4.38 Flipped IC making contact with support PCB

of SoC for millimeter-wave applications include finding suitable technology to satisfy the integration of all the required blocks, including those that require high on-chip passives. Available substrates usually have high losses, which influence the integrated antennas and transmission lines and high circuit density leads to cross-coupling [104]. Lastly, the amount of power that can be delivered from an SoC is typically limited, but despite this, high-power devices can suffer from heating issues.

When it comes to the physical package, it must be chosen to interfere minimally with sensitive signals when they cross the package boundary, while also taking thermal and mechanical package requirements into account [105]. Inside the package, various types of connections are used to connect the circuit pads to the package boundary. Bond wires and ribbon bonds used at low frequencies are usually not suitable, unless the bonding method is specifically modified for high frequencies (see [106, 107]). This is because they introduce high insertion loss and are typically longer than one-tenth of the operating wavelength. Flip-chip packaging, or microbumping [108, 109], is more suitable. Microbumps are metallic solder bumps (typically AuSn or PbSn) that are deposited on top of the circuit pads, which can make direct contact with the support PCB when flipped around, placed face-down on the supporting PCB. A flip chip package is illustrated in Fig. 4.38. Sometimes, a redistribution layer is needed, which allows for a more regular bump arrangement, independent of pad locations on the wafer. This way, flip-chip solder bump dimension parameters can be closely controlled, which is very beneficial for high frequencies [28]. Even though microbumps still have a small amount of parasitics, both inductive and capacitive (with capacitive effects dominating), they are less prominent and random in appearance than those of bond wires. Microbumps are also shorter and thicker and have better signal integrity and power capability.

Microbumped flip-chips can be packaged in two ways: if the size of the package matches that of the naked die, chip-scale packaging (CSP) is achieved and if the package is larger than the IC, an embedded wafer-level ball grid array (eWLB) is achieved. Whereas the CSP has the lowest possible footprint of any IC, in eWLB packaging, a fan-out area with a redistribution layer around the die is introduced, which offers space for placing additional off-chip components (see Sect. 4.4.2 on SoP) and interconnects [110]. The difference between the two is best understood with the aid of the diagrams in Fig. 4.39.

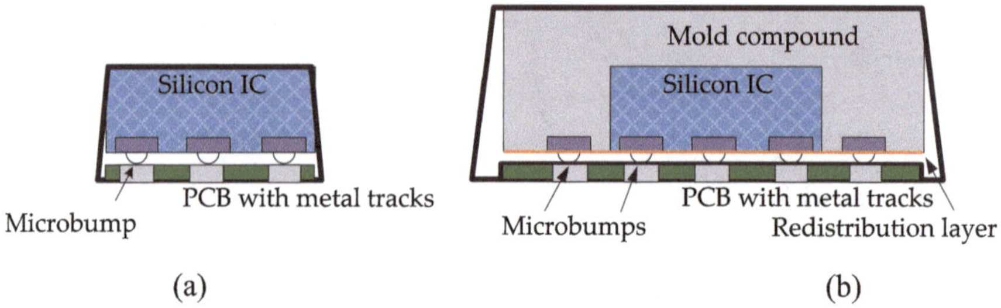

Fig. 4.39 CSP (**a**) and eWLB (**b**) packages

4.4.2 System-On-Package

SoP is the concept that allows for the miniaturization and optimization of the system. This approach is particularly appealing for millimeter-wave and terahertz frequencies, allowing RF circuits fabricated in specialized technologies to be included on the same package with a multitude of other components and dies, such as memory, processors and digital circuitry, analog circuitry, power control, energy harvesting (solar cells), sensors, actuators, MEMS, as well as a large number of different passives and arrays of antennas [17, 101, 111]. Sometimes even two or more complete transceiver circuits are integrated into the same package and the arrangement of components can be in two or three dimensions. Different components and dies are embedded into or placed on substrates that are suitable for high-frequency operation. Typically, ceramics such as low-temperature fired ceramics (LTCC) or polymers such as liquid–crystal polymer (LCP) exhibit the best performance, as they are able to support embedded passives.

The core advantage of SoP is that it is a highly flexible and low-cost integration solution [112], especially when it comes to smart electronics for 5G, 6G and related verticals, including IoT and Industry 4.0 devices, wearables, smartphones and medical devices. Most of the functionality of these devices can only be accomplished if multiple technologies and techniques are combined, while decreasing the overall power consumption and increasing the operating bandwidth. This is particularly important with the increase of power consumption, which is expected as the required data rates of future networks increase.

Another advantage of SoP for high frequencies is the ability to support embedded passive components. Resistors and capacitors can be fabricated in different, typically low-loss thin-film substrates, which yield much better performance than integrated on silicon. Inductors use the third dimension, while some technologies allow for the embedding of ferrites, thus increasing the range of inductance values that can be achieved [113]. Transmission lines and MEMS structures, as well as antennas, are also supported. Antennas on package have better radiation patterns and bandwidths than integrated antennas (antennas-on-chip) [114].

Challenges associated with SoP include issues with floor planning and power distribution [115] or those with making interconnects. A large number of integration choices also bring about different tradeoffs, for example when considering which components to fabricate onto a die and which components to place on the discrete substrate. Furthermore, the high component count with highly specialized functions also leads to a complex mix of various materials, each with its own material properties, including thermal properties such as heating and coefficient of thermal expansion (CTE) that influence the overall reliability of the system. CTE matching of different materials is thus one of the very important steps in SoP design.

An SoP is normally classified according to the type of circuitry that it supports. However, not every type is suitable for high-frequency applications. Some that are will be discussed below. The eWLB technique was mentioned earlier in the context of SoCs. The redistribution layer of the eWLB package can also be used to embed passives, such as inductors with high Q-factors, and multiple dies can be placed into the mold compound [116]. Multiple redistribution layers allow for building metal insulator metal capacitors, waveguide microstrip lines or antennas. If an interposer is added on top of an eWLB package, it can also serve as a carrier for discrete components [117]. Compared to other SoP strategies, eWLB is simple and has lower manufacturing costs [118]. An example of an eWLB used as SoP is shown in Fig. 4.40.

LTCC substrates are also suitable for high frequencies, as they allow for multilayered structures (nine to six dielectric layers) to be built with flexible fabrication techniques supporting passive devices and antennas-in-package [119]. In LTCC, antenna arrays are compact because they can be spread over different dielectric layers. Most benefits of LTCC packages also apply to an LCP package. The additional advantage of LCP is that it is a dual medium, acting both as a substrate and thae package because of its hermetic properties, while also supporting lamination and embedding of bare IC dies.

SoPs that integrate MEMS modules allow for the expansion of high-frequency systems to include a wide range of sensor modules. MEMS technologies are used in the fabrication of pressure and tilt sensors, accelerometers, gyros, chemical microsensors, microfluidic devices, resonators, RF switches, passives and other components [120]. In SoP, MEMS structures are placed inside a hermetically sealed cavity [121] and the transceiver IC, which interacts with the MEMS structure, can also serve as a means of support, as shown in Fig. 4.41. Alternatively, MEMS can be embedded into LTCC or LCP packages.

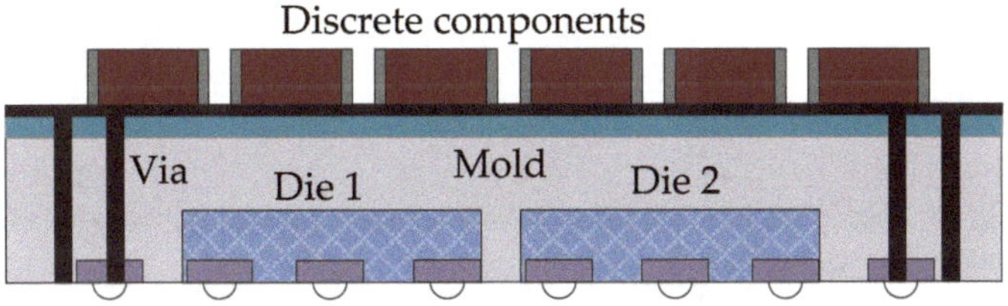

Fig. 4.40 An eWLB-based SoP: two dice and an interposer for the support of discrete components

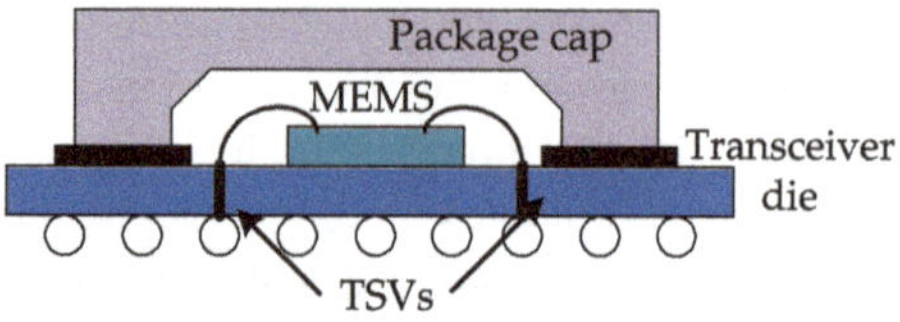

Fig. 4.41 MEMS on a transceiver die

Other suitable substrates for SoP at high frequencies include glass, alumina and organic substrates [122–124], while research into novel substrate technologies, including paper, textiles and plastic continues [125].

4.5 Concluding Remarks

This chapter expanded the discussion on electromagnetic communication at millimeter-wave and terahertz frequencies to include an analysis of how practical this type of communication is from the device, circuit and packaging perspective. At present, frequencies up to 100–200 GHz are readily reachable with current semiconductor device technologies. When it comes to communication at frequencies above 200 GHz, a number of technologies are capable of supporting circuits up to the lower parts of the terahertz band (just above 300 GHz), but this comes at great fabrication cost and causes integration issues. Thus, more research into high-frequency devices and circuits is needed to improve existing technologies such as SiGe BiCMOS to decrease costs to the point where mass production of high-frequency systems becomes feasible. It is hoped that a time span of one decade leading to the predicted commercialization of 6G will be sufficient for this. Otherwise it may be proven that 6G data rate targets discussed in previous chapters have been set to unrealistic figures. This decade may also be sufficient for some of the new technologies, such as nanotubes made of graphene or carbon, to rise to the point where they become competitive with semiconductor technologies, when achievable frequency of operation is concerned.

Fortunately, in 6G, VLC is another feasible option for achieving fast data links in several scenarios. Optical communications are, thus, the topic of Chap. 5.

References

1. Božanić M, Sinha S (2020) Millimeter-Wave integrated circuits: methodologies for research, design and innovation. Springer Nature
2. Rinaldi N, Schröter M (2018) Silicon-germanium heterojunction Bipolar transistors for mm-Wave systems: technology, modeling and circuit applications. River Publishers
3. Deal W, Mei XB, Leong KMKH, Radisic V, Sarkozy S, Lai R (2011) THz monolithic integrated circuits using InP high electron mobility transistors. IEEE Trans Terahertz Sci Technol 1:25–32

4. Schroter M, Krause J, Rinaldi N, Wedel G, Heinemann B, Chevalier P et al (2011) Physical and electrical performance limits of high-speed Si GeC HBTs—Part II: lateral scaling. IEEE Trans Electron Devices 58:3697–3706
5. Yuan J (2013) Using temperature to explore the scaling limits of SiGe HBTs. Extreme Environ Electron. Boca Raton: CRC Press
6. Antonopoulos A, Bucher M, Papathanasiou K, Mavredakis N, Makris N, Sharma RK et al (2013) CMOS small-signal and thermal noise modeling at high frequencies. IEEE Trans Electron Devices 60:3726–3733
7. Mei X, Yoshida W, Lange M, Lee J, Zhou J, Liu P et al (2015) First demonstration of amplification at 1 THz using 25-nm InP high electron mobility transistor process. IEEE Electron Device Lett 36:327–329
8. Heinemann B, Rücker H, Barth R, Bärwolf F, Drews J, Fischer GG, et al (2016) SiGe HBT with fT/fmax of 505 GHz/720 GHz. 2016 IEEE international electron devices meeting (IEDM), pp 3.1.1–3.1.4
9. Rodwell MJW, Le M, Brar B (2008) InP Bipolar ICs: scaling roadmaps, frequency limits, manufacturable technologies. Proc IEEE 96:271–286
10. Schröter M, Rosenbaum T, Chevalier P, Heinemann B, Voinigescu SP, Preisler E et al (2017) SiGe HBT technology: future trends and TCAD-based roadmap. Proc IEEE 105:1068–1086
11. Schroter M, Wedel G, Heinemann B, Jungemann C, Krause J, Chevalier P et al (2011) Physical and electrical performance limits of high-speed SiGeC HBTs—Part I: vertical scaling. IEEE Trans Electron Dev 58:3687–3696
12. Nishi Y, Doering R (2017) Handbook of semiconductor manufacturing technology. CRC Press, Boca Raton
13. Parlak M, Buckwalter JF (2013) A passive I/Q Millimeter-Wave mixer and switch in 45-nm CMOS SOI. IEEE Trans Microw Theory Tech 61:1131–1139
14. Kushwaha P, Khandelwal S, Duarte JP, Hu C, Chauhan YS (2016) RF modeling of FDSOI transistors using industry standard BSIM-IMG model. IEEE Trans Microw Theory Tech 64:1745–1751
15. Gianesello F, Gloria D, Montusclat S, Raynaud C, Boret S, Dambrine G et al (2007) 1.8 dB insertion loss 200 GHz CPW band pass Filter integrated in HR SOI CMOS technology. 2007 IEEE/MTT-S international microwave symposium, pp 453–6
16. Roodaki PM, Taghian F, Bashirzadeh S, Jalaali M (2011) A survey of Millimeter-Wave technologies. 2011 international conference on electrical and control engineering. Yichang, pp 5726–8
17. Božanić M, Sinha S (2019) Systems-level packaging for Millimeter-Wave transceivers. Springer, Cham
18. Juneja S, Pratap R, Sharma R (2020) Semiconductor technologies for 5G implementation at millimeter wave frequencies—Design challenges and current state of work. Engineering Science and Technology, an International Journal [Internet]. 2020 [cited 2020 Aug 6]; Available from: https://www.sciencedirect.com/science/article/pii/S2215098620300537
19. Kazimierczuk MK (2014) RF power amplifiers. Wiley
20. Ndjountche T (2018) Amplifiers, comparators, multipliers, filters, and oscillators. CRC Press, Boca Raton
21. Inac O, Cetinoneri B, Uzunkol M, Atesal YA, Rebeiz GM (2011) Millimeter-Wave and THz circuits in 45-nm SOI CMOS. 2011 IEEE compound semiconductor integrated circuit symposium (CSICS), pp 1–4
22. Inac O, Uzunkol M, Rebeiz GM (2014) 45-nm CMOS SOI technology characterization for Millimeter-Wave applications. IEEE Trans Microw Theory Tech 62:1301–1311
23. Sze SM, Ng KK (2006) Physics of semiconductor devices. Wiley
24. Hodges D, Jackson H, Saleh R (2003) Analysis and design of digital integrated circuits. McGraw-Hill Companies, Incorporated
25. Gray PR, Hurst P, Meyer RG, Lewis S (2009) Analysis and design of analog integrated circuits. Wiley

26. Lee TH (2003) The design of CMOS radio-frequency integrated circuits. Cambridge University Press
27. Hastings RA (2006) The art of analog layout. Pearson Prentice Hall
28. Sedra AS, Smith KC (2013) Microelectronic circuits. 6th ed. Oxford University Press
29. Neamen DA (2007) Microelectronics: circuit analysis and design. McGraw-Hill
30. Tinoco JC, Raskin J-P (2010) New RF Extrinsic resistances extraction procedure for deep-submicron MOS transistors. Int J Numer Model Electron Networks Devices Fields 23:107–126
31. Arutyunyan SS, Pavlov AYu, Pavlov BYu, Tomosh KN, Fedorov YuV (2016) On a two-layer Si3N4/SiO2 dielectric mask for low-resistance Ohmic contacts to AlGaN/GaN HEMTs. Semiconductors 50:1117–21
32. Dhakad S, Sharma N, Periasamy C, Chaturvedi N (2017) Optimization of Ohmic contacts on thick and thin AlGaN/GaN HEMTs structures. Superlattices Microstruct 111:922–926
33. Hasani JY (2016) Three-port model of a modern MOS transistor in millimeter wave band, considering distributed effects. IEEE Trans Comput Aided Des Integr Circuits Syst 35:1509–1518
34. Fletcher ASA, Nirmal D (2017) A survey of Gallium Nitride HEMT for RF and high power applications. Superlattices Microstruct 109:519–537
35. Weimann NG, Stoppel D, Schukfeh MI, Hossain M, Al-Sawaf T, Janke B, et al (2016) SciFab—a wafer-level heterointegrated InP DHBT/SiGe BiCMOS foundry process for mm-wave applications. Physica Status Solidi (a) 213:909–16
36. Lambrechts W, Sinha S (2016) SiGe-based re-engineering of electronic warfare subsystems. Springer
37. Ryndin EA, Al-Saman AA, Konoplev BG (2019) A quasi-two-dimensional physics-based model of HEMTs without smoothing functions for joining linear and saturation regions of I-V characteristics [Internet]. Active and Passive Electronic Components. 2019 [cited 2019 Apr 16]. Available from: https://www.hindawi.com/journals/apec/2019/5135637/abs/
38. Zhang H, Ma P, Lu Y, Zhao B, Zheng J, Ma X et al (2017) Extraction method for parasitic capacitances and inductances of HEMT models. Solid-State Electronics. 129:108–113
39. Alim MA, Rezazadeh AA (2018) Device behaviour and zero temperature coefficients analysis for microwave GaAs HEMT. Solid-State Electron 147:13–18
40. Alshahed M, Heuken L, Alomari M, Cora I, Tóth L, Pècz B et al (2018) Low-dispersion, high-voltage, low-leakage GaN HEMTs on native GaN substrates. IEEE Trans Electron Devices 65:2939–2947
41. Urteaga M, Hacker J, Griffith Z, Young A, Pierson R, Rowell P, et al (2016) A 130 nm InP HBT integrated circuit technology for THz electronics. 2016 IEEE Int Electron Dev Meeting (IEDM), pp 29.2.1–29.2.4
42. Joseph A, Jain V, Ong SN, Wolf R, Lim SF, Singh J (2018) Technology positioning for mm wave applications: 130/90nm SiGe BiCMOS vs. 28nm RFCMOS. 2018 IEEE BiCMOS and compound semiconductor integrated circuits and technology symposium (BCICTS), pp 18–21
43. Lachner R (2014) (Invited) Towards 0.7 Terahertz Silicon Germanium Heterojunction Bipolar Technology—the DOTSEVEN Project. ECS Trans 64:21–37
44. Rücker H, Heinemann B (2019) Device architectures for high-speed SiGe HBTs. 2019 IEEE BiCMOS and compound semiconductor integrated circuits and technology symposium (BCICTS), pp 1–7
45. Samoska LA (2011) An overview of solid-state integrated circuit amplifiers in the subMillimeter-Wave and THz regime. IEEE Trans Terahertz Sci Techno 1:9–24
46. Urteaga M, Griffith Z, Seo M, Hacker J, Rodwell MJW (2017) InP HBT technologies for THz integrated circuits. Proc IEEE 105:1051–1067
47. Seo M, Urteaga M, Hacker J, Young A, Griffith Z, Jain V, et al (2011) InP HBT IC technology for terahertz frequencies: fundamental oscillators up to 0.57 THz. IEEE J Solid-State Circuits 46:2203–14

48. Chevalier P, Schröter M, Bolognesi CR, d'Alessandro V, Alexandrova M, Böck J et al (2017) Si/SiGe: C and InP/GaAsSb heterojunction Bipolar transistors for THz applications. Proc IEEE 105:1035–1050
49. Bolognesi CR, Flückiger R, Alexandrova M, Quan W, Lövblom R, Ostinelli O (2016) InP/GaAsSb DHBTs for THz applications and improved extraction of their cutoff frequencies. 2016 IEEE international electron devices meeting (IEDM), pp 29.5.1–29.5.4
50. Tessmann A, Leuther A, Massler H, Hurm V, Kuri M, Zink M, et al (2014) A 600 GHz low-noise amplifier module. 2014 IEEE MTT-S international microwave symposium (IMS2014), pp 1–3
51. Kalnoskas A. Why GaN will be key to feeding power-hungry 5G networks [Internet]. 5G Technology World. 2019 [cited 2020 May 16]. Available from: https://www.5gtechnologyworld.com/why-gan-will-be-key-to-feeding-power-hungry-5g-networks/
52. Tang Y, Shinohara K, Regan D, Corrion A, Brown D, Wong J et al (2015) Ultrahigh-speed GaN high-electron-mobility transistors with fT/fmax of 454/444 GHz. IEEE Electron Device Lett 36:549–551
53. Du J, Wang K, Liu Y, Bai Z, Liu Y, Feng Z et al (2017) Influence of mesa edge capacitance on frequency behavior of Millimeter-Wave AlGaN/GaN HEMTs. Solid-State Electron 129:1–5
54. Romanczyk B, Li W, Guidry M, Hatui N, Krishna A, Wurm C et al (2020) N-polar GaN-on-sapphire deep recess HEMTs with high W-band power density. IEEE Electron Device Lett 41:1633–1636
55. Romanczyk B, Zheng X, Guidry M, Li H, Hatui N, Wurm C et al (2020) W-band power performance of SiN-passivated N-Polar GaN deep recess HEMTs. IEEE Electron Dev Lett 41:349–352
56. Feng G, Boon CC, Meng F, Yi X, Li C (2016) An 88.5–110 GHz CMOS low-noise amplifier for Millimeter-Wave imaging applications. IEEE Microwave and Wirel Components Lett 26:134–6
57. Lee TH (2014) Terahertz CMOS integrated circuits. 2014 IEEE international symposium on radio-frequency integration technology, pp 1–2
58. Lambrechts W, Sinha S, Abdallah JA, Prinsloo J (2018) Extending Moore's law through advanced semiconductor design and processing techniques. CRC Press, Boca Raton
59. Chen D, Jha NK (2019) Introduction to nanotechnology. In: Jha NK, Chen D (eds) Nanoelectronic circuit design [Internet]. New York, NY: Springer; 2011 [cited 2019 Nov 25]. p. 1–22. Available from: https://doi.org/10.1007/978-1-4419-7609-3_1
60. Singh J, Ciavatti J, Sundaram K, Wong JS, Bandyopadhyay A, Zhang X et al (2018) 14-nm FinFET technology for analog and RF applications. IEEE Trans Electron Dev 65:31–37
61. T-MUSIC Selects Performers to Develop Integrated Mixed-Mode RF Electronics in Onshore Foundries [Internet]. [cited 2020 May 16]. Available from: https://www.darpa.mil/news-events/2020-02-04
62. Khanna VK (2017) Extreme-temperature and Harsh-wnvironment electronics: physics, technology and applications. IOP Publishing, Bristol
63. Rajashekara K, Akin B (2013) A review of cryogenic power electronics—status and applications. 2013 international electric machines drives conference, pp 899–904
64. Chakraborty PS, Cardoso AS, Wier BR, Omprakash AP, Cressler JD, Kaynak M et al (2014) A 0.8 THz fMAX SiGe HBT operating at 4.3 K. IEEE Electron Dev Lett 35:151–3
65. Mohanram K, Yang X (2011) Graphene transistors and circuits. In: Jha NK, Chen D (eds) nanoelectronic circuit Design [Internet]. New York, NY: Springer; 2011 [cited 2019 Nov 25]. p. 349–76. Available from: https://doi.org/10.1007/978-1-4419-7609-3_10
66. Zhao J, Liu L, Li F (2014) Graphene oxide: physics and applications. Springer
67. Zheng Q, Kim J-K (2015) Graphene for transparent conductors: synthesis, properties and applications. Springer
68. Liao L, Peng H, Liu Z (2014) Chemistry makes graphene beyond graphene. J Am Chem Soc. 136:12194–12200
69. Ma X, Gu W, Shen J, Tang Y (2012) Investigation of electronic properties of graphene/Si field-effect transistor. Nanoscale Res Lett. 7:677

70. Singh R, Kumar, D, Tripathi CC (2015) Graphene: potential material for nanoelectronics applications. Indian J Pure App Phys (IJPAP). 53:501–513–513
71. Dröscher S, Molitor F, Ihn T, Ensslin K (2019) Graphene Constrictions. In Aoki HS, Dresselhaus M (eds) Physics of graphene [Internet]. Cham: Springer International Publishing; 2014 [cited 2019 Nov 25]. p. 141–69. Available from: https://doi.org/10.1007/978-3-319-026 33-6_5
72. Fiori G, Iannaccone G (2013) Multiscale modeling for graphene-based nanoscale transistors. Proc IEEE 101:1653–1669
73. Bonmann M, Asad M, Yang X, Generalov A, Vorobiev A, Banszerus L et al (2019) Graphene field-effect transistors with high extrinsic fT and fmax. IEEE Electron Device Lett 40:131–134
74. Rouger N, Pham T-T, Perez G, Cédric M, LEFRANC P, Jeannin P-O et al (2017) Diamond devices for power electronics. 11th conference on new diamond and nano carbons [Internet]. Cairns, Australia; 2017 [cited 2019 Dec 4]. Available from: https://hal.archives-ouvertes.fr/hal-01590457
75. Dong C, Chilstedt S, Chen D (2019) FPCNA: a carbon nanotube-based programmable architecture. In: Jha NK, Chen D (eds)nanoelectronic circuit design [Internet]. New York, NY: Springer; 2011 [cited 2019 Nov 25]. p. 307–48. Available from: https://doi.org/10.1007/978-1-4419-7609-3_9
76. Kulkarni SK (2014) Nanotechnology: principles and practices. Springer, Cham
77. Appenzeller J (2008) Carbon nanotubes for high-performance electronics—progress and prospect. Proc IEEE 96:201–211
78. Rutherglen C, Kane AA, Marsh PF, Cain TA, Hassan BI, AlShareef MR et al (2019) Wafer-scalable, aligned carbon nanotube transistors operating at frequencies of over 100 GHz. Nature Electron 2:530–539
79. Patil N, Lin A, Zhang J, Wei H, Wong H-SP, Mitra S (2011) Imperfection-immune carbon nanotube VLSI circuits. In: Jha NK, Chen D (eds) nanoelectronic circuit design [Internet]. New York, NY: Springer; 2011 [cited 2019 Nov 25]. p. 277–305. Available from: https://doi.org/10.1007/978-1-4419-7609-3_8
80. Božanić M, Sinha S (2015) Power amplifiers for the S-, C-, X- and Ku-bands: an EDA perspective. Springer
81. Raab FH, Asbeck P, Cripps S, Kenington PB, Popovic ZB, Pothecary N, et al (2003) RF and microwave power amplifier and transmitter technologies—Part 1. High Frequency Electron 2:22–36
82. Preez J du, Sinha S (2017) Millimeter-Wave power amplifiers. Springer
83. Eccleston KW, Smith KJI, Gough PT (2011) A C-band Compact Class-F/class-C Doherty Amplifier. Microwave and Optical Technology Letters. 53:1606–1610
84. Banerjee A, Hezar R, Ding L (2015) Efficiency improvement techniques for RF power amplifiers in deep submicron CMOS. IEEE custom integrated circuits conference (CICC), pp 1–4
85. Božanić M, Sinha S (2017) Millimeter-Wave low noise amplifiers. Springer, Cham
86. Chaturvedi S, Bozanic M, Sinha S (2017) Millimeter wave passive bandpass filters. Microwave J 60
87. Chaturvedi S, Božanic M, Sinha S (2017) Millimeter wave active bandpass filters. Microwave J 60
88. Ismail A, Abidi AA (2004) A 3–10-GHz low-noise amplifier with Wideband LC-ladder matching Network. IEEE J Solid-State Circuits 39:2269–2277
89. Min B, Rebeiz GM (2007) Ka-Band SiGe HBT low noise amplifier design for simultaneous noise and input power matching. IEEE Microwave Wirel Compon Lett 17:891–893
90. Grebennikov A, Kumar N, Yarman BS (2017) Broadband RF and microwave amplifiers. CRC Press, Boca Raton
91. Ludwig R, Bretchko P (2000) RF circuit design: theory and applications. Pearson Education
92. Odyniec M (2002) RF and microwave oscillator design. Artech House
93. Grebennikov A (2007) RF and microwave transistor oscillator Design. Wiley

94. Imani A, Hashemi H (2018) Frequency and power scaling in mm-wave colpitts oscillators. IEEE J Solid-State Circuits 53:1338–1347
95. Momeni O, Afshari E (2011) High Power terahertz and Millimeter-Wave oscillator design: a systematic approach. IEEE J Solid-State Circuits 46:583–597
96. Yang X, Matthaiou M, Yang J, Wen C, Gao F, Jin S (2019) Hardware-constrained Millimeter-Wave systems for 5G: Challenges, Opportunities Solut. IEEE Commun Mag 57:44–50
97. Everard J (2001) Fundamentals of RF circuit design with low noise oscillators. Wiley
98. Yeo KS, Ma K (2018) Low-power wireless communication circuits and systems: 60GHz and Beyond. CRC Press, Boca Raton
99. Chaturvedi S, Bozanic M, Sinha, Saurabh (2017) A 50 GHz SiGe BiCMOS active Band-pass Filter. 2017 IEEE 20th international symposium on design and diagnostics of electronic circuits & systems (DDECS). Dresden
100. Chaturvedi S, Bozanic M, Sinha S (2019) 60 GHz BiCMOS active bandpass filters. Microelectron J 90:315–322
101. Tummala R (2007) System on package: miniaturization of the entire system. McGraw Hill Professional
102. Greig W (2007) Integrated circuit packaging, assembly and interconnections. Springer Science & Business Media
103. Cheema HM, Shamim A (2013) The last barrier: on-chip antennas. IEEE Microwave Mag 14:79–91
104. Wang H, Lin K, Tsai Z, Lu L, Lu H, Wang C et al (2009) MMICs in the Millimeter-Wave regime. IEEE Microwave Mag 10:99–117
105. Goettel B, Winkler W, Bhutani A, Boes F, Pauli M, Zwick T (2018) Packaging solution for a Millimeter-Wave system-on-chip radar. IEEE Trans components, packaging and manufacturing technology. 8:73–81
106. Leong KMKH, Deal WR, Radisic V, Mei XB, Uyeda J, Samoska L et al (2009) A 340–380 GHz integrated CB-CPW-to-waveguide transition for sub Millimeter-Wave MMIC packaging. IEEE Microwave Wirel Compon Lett 19:413–415
107. Jameson S, Socher E (2015) A wide-band CMOS to waveguide transition at mm-Wave frequencies with wire-bonds. IEEE Trans Microw Theory Tech 63:2741–2750
108. Lu D, Wong CP (eds) Materials for advanced packaging [Internet]. Springer US; 2009 [cited 2019 Jun 8]. Available from: https://www.springer.com/gp/book/9780387782188
109. Heinrich W (2005) The flip-chip approach for millimeter wave packaging. IEEE Microwave Mag 6:36–45
110. Wojnowski M, Issakov V, Knoblinger G, Pressel K, Sommer G, Weigel R (2012) High-Q inductors embedded in the Fan-Out area of an eWLB. IEEE Trans components, packaging and manufacturing technology 2:1280–1292
111. Mack W, System in package–how to cope with increasing challenges? Electron Dev Failure Anal 202AD;14:4–11
112. Santagata F, Dong M, Yuan C, Sokolovskij R, Wei J, Zhang G (2015) 3D system-in-package design using stacked silicon submount technology. Microelectron Int 32:63–72
113. Raj PM, Sharma H, Sitaraman S, Mishra D, Tummala R (2017) System scaling with nanostructured power and RF components. Proc IEEE 105:2330–2346
114. Liu Y, Agrawal A, Natarajan A (2016) Millimeter-Wave IC-antenna cointegration for integrated transmitters and receivers. IEEE Antennas Wirel Propag Lett 15:1848–1852
115. Wu P, Liu F, Li J, Chen C, Hou F, Cao L et al (2017) Design and implementation of a rigid-flex RF front-end system-in-package. Microsyst Technol. 23:4579–4589
116. Hagelauer A, Wojnowski M, Pressel K, Weigel R, Kissinger D (2018) Integrated systems-in-package: heterogeneous integration of Millimeter-Wave active circuits and passives in Fan-Out wafer-level packaging technologies. IEEE Microwave Mag 19:48–56
117. Lin Y, Kang C, Chua L, Choi WK, Yoon SW (2016) Advanced 3D eWLB-PoP (Embedded Wafer Level Ball Grid Array—Package on Package) Technology. 2016 IEEE 66th electronic components and technology conference (ECTC), pp 1772–7

118. Dai WW (2016) Historical perspective of system in package (SiP). IEEE Circuits Syst Mag 16:50–61
119. Decrossas E, Glover MD, Porter K, Cannon T, Stegeman T, Allen-McCormack N et al (2015) High-performance and high-data-rate quasi-coaxial LTCC vertical interconnect transitions for multichip modules and system-on-package applications. IEEE transactions on components, packaging and manufacturing technology 5:307–313
120. Lee YC, Kong M, Zhang Y. Microelectromechanical systems and packaging. Materials for Advanced Packaging. Springer; 2017. p. 697–731
121. Lau JH (2010) Design and process of 3D MEMS system-in-package (SiP). J Microelectron Electron Packaging 7:10–15
122. Kim MS, Pulugurtha MR, Sundaram V, Tummala RR, Yun H (2018) Ultrathin High-Q2-D and 3-D RF inductors in glass packages. IEEE Trans components, packaging and manufacturing technology 8:643–652
123. Samanta KK, Robertson ID (2011) Advanced multilayer thick-film system-on-package technology for Miniaturized and high performance CPW microwave passive components. IEEE transactions on components, packaging and manufacturing technology 1, pp 1695–1705
124. Lim K, Pinel S, Davis M, Sutono A, Lee C-H, Heo D et al (2002) RF-system-on-package (SOP) for wireless communications. IEEE Microwave Mag 3:88–99
125. Robertson I, Somjit N, Chongcheawchamnan M (2016) Microwave and Millimetre-Wave design for wireless communications. Wiley

Chapter 5
Visible Light Communications for 6G

Abstract This chapter strives to investigate, in detail, the theory of VLC. Whereas VLC technology is nowhere near as mature as radio, it is potentially much better in various scenarios, which include V2X communications, underwater communication, or communications where low interference or high security is needed. After an introductory section describing the most important aspects of VLC, VLC system architecture and the physics of optical signal transmission and reception will be described in more detail. Then, considering that the VLC channel is different from the electromagnetic channel, VLC channel properties will be discussed. Physics of light generation and reception is also looked at in this chapter. Lastly, this chapter looks into modulation techniques for VLC, hybrid RF/VLC network implementations and some of the applications of VLC and other OWC types.

In the previous two chapters, it was discussed how increasing the frequency of operation can open the electromagnetic spectrum required to reach data rates that will be required by the wireless networks of the future. As one moves even higher in the electromagnetic spectrum, all the way into the optical frequency domain, another opportunity comes alive: optical communication through unguided media, also called free space optics or OWC. Optical frequencies range from 3 THz to 300 PHz (petahertz, 10^{15} Hz) and include infrared, visible, and ultraviolet bands. Between the three bands, communication in the visible band (430–790 THz) is the most appealing, because of the availability of inexpensive LEDs, which can be used as the signal source, and photo diodes that can be used as signal detectors. As an alternative to LEDs, more expensive laser diodes can be used, especially when it comes to outdoor settings where focused light beams are essential. A most important property of LEDs and laser diodes is that they can switch to different light intensity levels very quickly, which allows data to be encoded in emitted light in a variety of ways [1–3]. When comparing VLC with RF communications, many advantages can be identified, including ultrawide bandwidths that support data rates in the order of Gb/s, lower latency, removal of the need for licensing (the spectrum is free), improved security (light does not propagate through opaque objects so it is contained in a small area), lack of signal interference (only other light sources can interfere), low transmitter

M. Božanić and S. Sinha, *Mobile Communication Networks: 5G and a Vision of 6G*, Lecture Notes in Electrical Engineering 751,
https://doi.org/10.1007/978-3-030-69273-5_5

power and high energy efficiency, low cost of transmitter and receiver deployment, as well as health safety [4–8]. At the same time, LEDs enjoy similar benefits to those transistors enjoy from the technology scaling described by Moore's law: every several years, the price of LED technology halves, acting as a stimulus for further research into LEDs, photodiodes and related technologies (such as light modulation techniques). This, in turn, leads to ever-better transmission rate achievements, with Tb/s rates expected to become feasible in the near future.

There are many aspects of 5G and 6G that can benefit from VLC and OWC in general. First, OWC, typically in the infrared frequency range, could be established for communication links over long distances [9]. This could be, for example, for ultra-broadband backhaul (communication between base stations). Secondly, communications from and to satellites, which have been identified as an enabler of ubiquitous coverage, could be established with optical links. Third, light is the most promising option for underwater communications, massively outperforming acoustic and RF communications in this medium. Fourth, V2V and V2X communications can be established by using head and tail lights of various autonomous and semi-autonomous vehicles in combination with traffic lights [10, 11]. In microelectronics, light can be used for communication between chips, replacing RF transmission lines with miniature optical links [12]. However, the immediate application of OWC and specifically VLC would be in in indoor settings, where, according to various published statistics, data users spend about 70–80% of their time [5], as well as in crowded city streets, where light signals could be emitted from lamp posts [13]. VLC can be deployed for establishing communications in places where the electromagnetic spectrum is crowded and no additional data capacity can be added without a massive increase in the system cost, or in environments where the presence of electromagnetic radiation may be dangerous, such as in hospitals or on airplanes. In this context, links can be regarded as short-range (typically limited to several meters only), but the capacity benefits these links would introduce can easily overshadow any other means of wireless communication. In addition, in short-range links (including V2V links), the same LEDs can be used both for illumination and communication. On top of this, harvesting of the light energy emanating from the transmitter can also be considered.

This chapter strives to investigate, in detail, the theory of VLC, with some occasional generalization attempts towards OWC. After an introductory section describing the most important aspects of VLC, VLC system architecture and the physics of optical signal transmission and reception will be described in more detail. Then, considering that the VLC channel is different from the electromagnetic channel, VLC channel properties have to be discussed. Physics of light generation and reception is also looked at in this chapter. Lastly, this chapter also looks into modulation techniques for VLC, hybrid RF/VLC network implementations and some of the applications of VLC and other OWC types.

5.1 Introduction to Visible Light Communications

VLC can be established over the visible light part of the spectrum, spanning from 400 THz ($\lambda \approx 750$ nm) to 790 THz ($\lambda \approx 380$ nm), as illustrated in Fig. 5.1 [11, 13]. As is evident from this figure, the visible light forms only a small portion of the total optical spectrum (which includes infrared and ultraviolet). In OWC systems, modulation and demodulation are direct, as there are no radios, which makes the optical transmission system much simpler than competing RF systems [14]. However, visible light remains the most accessible part of the optical spectrum, owing to the invention of the inexpensive white LED that has the characteristic of a wide field of view.

LEDs were invented in 1960s. Over just a few decades, LEDs went from single-color availability to a multitude of colors, from low luminance to high luminance, and from short-lifespan to long-lifespan devices [15]. The luminance alone has been increasing 20 times per decade, while the price has decreased to 1% of the original price. However, VLC as a concept was only proposed some two decades ago by Japanese scientists, who proposed that the information can be encoded into the light source by fast flickering the light based on the encoded information. Fortunately, this flickering is not detectable by human eyes, as long as the flickering rate is at least 200 Hz [5]. This allows for seamless integration of VLC LEDs with lighting

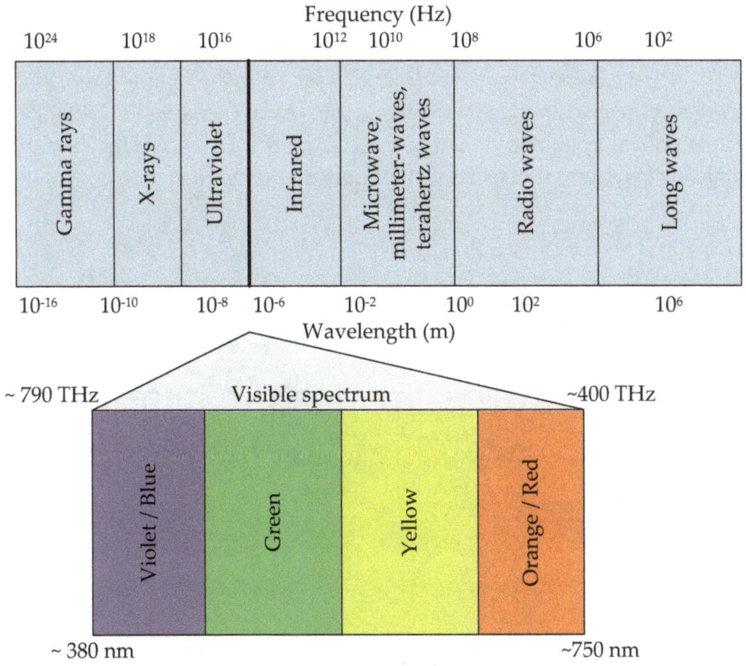

Fig. 5.1 Electromagnetic part of the spectrum used in VLC

LEDs. The first VLC standard, called light fidelity (LiFi), was only revealed in 2011. Since its invention, the VLC concept has come a long way, with data rates in excess of 15 Gb/s reported by 2019. For example, Bian et al. [16] reported a data rate of 15.73 Gb/s over a 1.6 m long link using OFDMA as an access scheme, using just off-the-shelf LEDs. In another study, Xie et al. [17] reported that links with speeds of over 6.5 Gb/s and 1.5 Gb/s can be maintained at distances of 10 m and 20 m respectively.

While there are numerous advantages to VLC, as discussed in the introduction to this chapter, VLC also faces many challenges. It should be noted that the faster the LED flickers, the brighter it gets. This can start becoming a problem for two reasons: first, power consumption increases, and second, if the same LED is used for illumination, the emitted light may be too bright for the required space [11]. In this case, various LED-dimming techniques need to be deployed. Another issue is the interference from other light sources (both artificial, i.e. incandescent lights, and natural, i.e. sunshine). The presence of other light sources reflects as noise in the system, thus degrading the data rate and increasing the BER. Fortunately, bandpass blue filters have been proven to limit the degradation caused by solar irradiance [18]. Lastly, the light intensity drops abruptly with distance, which can be limiting in many deployment scenarios [5].

5.2 VLC System Architecture

The VLC system can most easily be described by using the indoor setup, but various system aspects are also applicable to other deployment scenarios. The block diagram of a simple optical system is shown in Fig. 5.2 [5, 11, 15]. As in the case of the RF communication system, it consists of the transmitter and the receiver. It will shortly

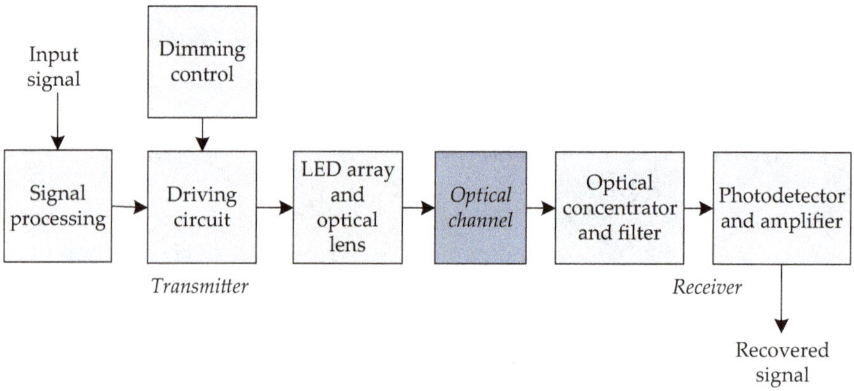

Fig. 5.2 Block diagram of the VLC system. From [11]

be seen that there are many similarities between the VLC and the RF systems, but there are also some major differences, which will be pointed out.

5.2.1 Transmitter

The transmitter, shown on the left of Fig. 5.2, is tasked with mapping the input electrical data to the optical source, which can be an LED (or an LED array) or a laser diode. The modulation is essentially mapping the data signal $m(t)$ to the drive current of the LED (or the laser diode), producing a signal of time-varying light intensity $x(t)$. This process is also known as intensity modulation. This signal is carried over the optical channel to be received by the photodetector, which will be described in the next section. The drive current has to be within the dynamic range of the LED in order not to end up with distortion of the light signal. It is also necessary that the driving signal is real and non-negative. Figure 5.2 shows a dimming control block as well, which is used to regulate the brightness of an LED precisely to the appropriate level of dimming. Dimming is used for at least two reasons. First, it is used to decrease the brightness of the LED for high data rates and place the resulting brightness within the illumination requirements of the intended setting. Secondly, it can help to increase the LED response time.

LEDs are preferred to laser diodes in indoor settings, partially because of their lower costs and large field of view and partially because they can also be used for illumination. LEDs can either be monochromatic (emitting only one color, typically blue, green or red) or white (theoretically containing all the colors simultaneously). VLC is normally established with the of aid white light, which can be generated in three ways:

- In the first method, red, green and blue light (RGB) can all be emitted with precise intensity. The resultant spectrum is shown in Fig. 5.3.
- The second method is slightly more complex and involves using a blue LED with a phosphor coating that discharges yellow light. In such a way, short wavelengths are absorbed, and a light of yellow color is emitted. In the next step of light processing, the wavelength is shifted to red, which is combined with a portion

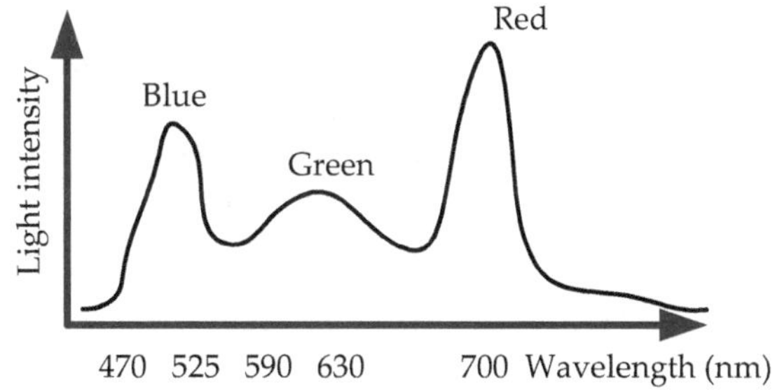

Fig. 5.3 The spectrum of the RGB LED. From [15]

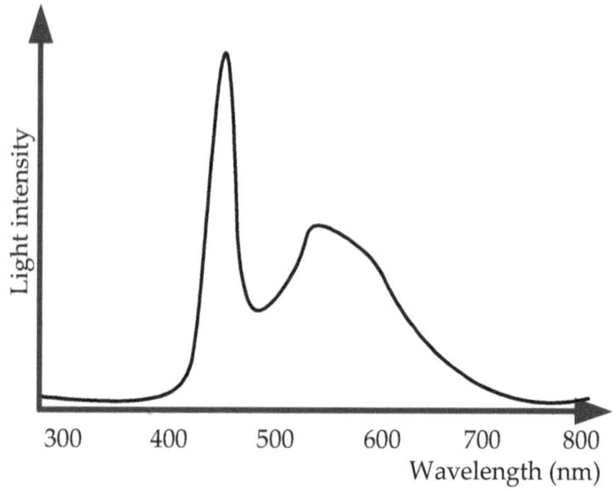

Fig. 5.4 Spectrum of the blue LED with phosphor coating. From [15]

of blue light that is not absorbed. This finally results in white light, with the spectrum shown in Fig. 5.4. This method is more popular than the RGB method, but it results in light with limited bandwidth, owing to the slow absorption and emission of the phosphor layer.

- Lastly, broadband, or full-wavelength radiation can be generated, closely resembling the solar spectrum.

The optical form of the LED is also important. Several forms are typically considered [15]:

- Direct approach, where the LED light is emitted without going through an optical device;
- Diffusion approach, where the light is diffused to reduce surface brightness and expand the luminous range;
- Reflection approach, where the light from the LED is directed via external reflectors;
- Transmission-type approach, where a lens is added in front of an LED so the that the contour map of the light can be changed; and
- A combination of reflection and transmission approaches.

It should be noted that there are several issues associated with LED-based transmitters. These include overcoming nonlineraities in LEDs, the need for precise LED chromaticity control, as well as previously mentioned requirements for dimming control and flicker-free communication in order not to harm human eyes [6]. Power electronics associated with driving of LEDs thus plays a critical role in error-free and controlled VLC transmission [19]. A typical LED driving circuit is shown in Fig. 5.5 [20].

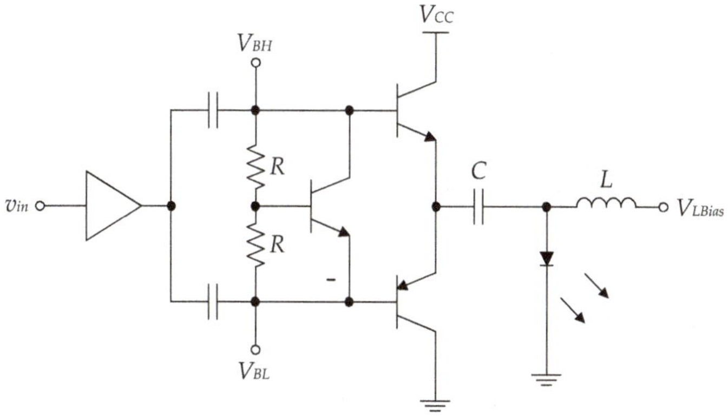

Fig. 5.5 A typical push-pull circuit used for LED driving [20]

5.2.2 Receiver

The optical receiver, shown on the right of Fig. 5.2, is based on the photodetector. The photodetector accepts the transmitted signal $x(t)$ that has passed through the channel with the response $h(t)$, creating the photocurrent $y(t)$. Other important blocks in the receiver are the optical concentrator, which ensures that the received light is reflected and focused on the photodetector, and the optical filter, which filters out unwanted wavelengths. The working principle of the optical concentrator is shown Fig. 5.6, where it is also shown that the optical filter can be included as part of the optical concentrator. An electronic amplifier is also typically needed in the receiver to convert the detected current into electrical quantities suitable for further processing in subsequent receiver blocks.

There are at least three types of photodetectors: photodiodes, imaging and camera sensors, as well as solar panels. Among these, photodiodes are most commonly used. Two types of photodiodes are typically deployed in commercial VLC systems: positive-intrinsic-negative (PIN) diodes and avalanche photodiodes

Fig. 5.6 An optical concentrator with a filter [3]

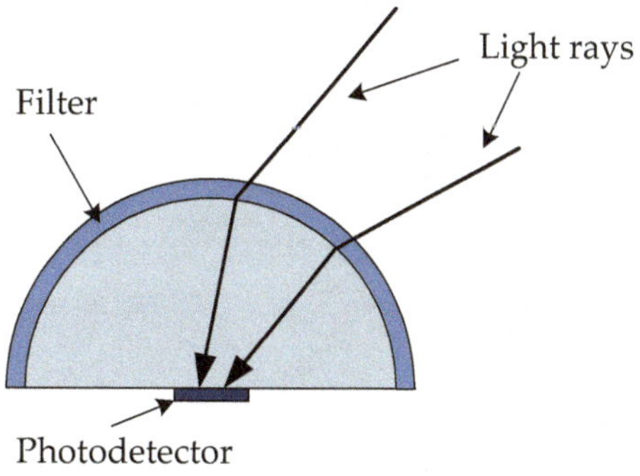

(APDs). Although APDs have higher gains than PIN diodes, they suffer from higher shot noise. On the other hand, PIN diodes are inexpensive and have higher temperature resistance. However, two other photodetector types also have advantages. For instance, solar panels can be simultaneously used for energy harvesting, while camera sensors are usually already available in many devices, such as cellphones, and as a result, their use in VLC would incur only minimal costs.

5.3 Mechanism of Light Generation and Detection

In this section, the mechanism of light generation in LEDs, and light detection in photodiodes, is discussed. Since both devices are diodes, it will be useful in the discussion that follows in this section to note that LED works under a forward bias whereas the photodiode is driven by a reverse voltage.

5.3.1 Light Generation in LEDs

Light generation in LEDs is based on the principle of electroluminescence, which can be defined as the emission of photons in the presence of an electric current or electric field. In order to describe the principle of electroluminescence, recall that the LED is a diode, which is a device consisting of a single *pn*-junction [6, 15]. The operation of the *pn*-junction, a boundary between *p*-type and *n*-type materials, was not discussed in detail in Chap. 4 when other *pn*-junction-based devices and transistors were discussed, with the aim not to digress. Now, this discussion cannot be deferred any longer.

A *pn*-junction in its equilibrium condition is illustrated in Fig. 5.7a. In the *p*-type region, the holes are major carriers and the electrons are minor carriers; the opposite is true for the *n*-type region. At the boundary of the two regions, during the *pn*-junction

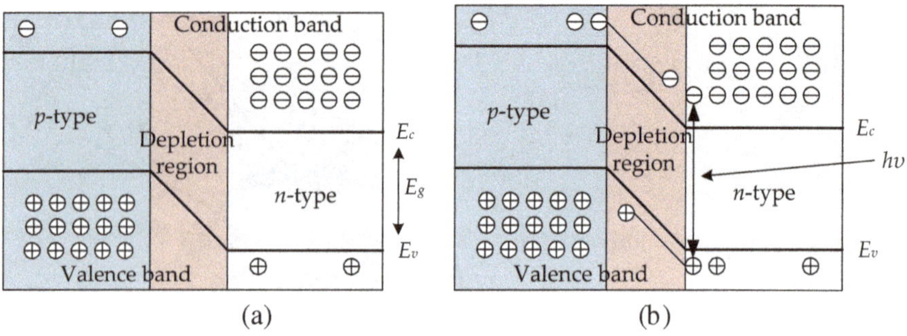

Fig. 5.7 *pn*-junction in equilibrium condition (**a**) and under forward bias (**b**)

fabrication, some electrons from the n-type region diffuse into the p-type region, thus recombining with holes and forming a depletion zone (region). This results in the formation of a built-in potential of the pn-junction, the magnitude of which depends on acceptor and donor carrier concentrations (N_A and N_D respectively), as well as on the intrinsic carrier concertation n_i:

$$V_D = \frac{kT}{q} \ln \frac{N_A N_D}{n_i^2}, \tag{5.1}$$

where other symbols have already been defined in Chap. 4. Without an applied external potential, the equilibrium state is not disrupted, and the electrons in the n-type region are concentrated in the conduction band, whereas holes in the p-type material are concentrated in the valence band. However, if a forward bias voltage V is applied, the holes in the p-type region and the electrons in the n-type region start returning to their original positions, which results in a width decrease of the depletion zone. If V gets sufficiently close to V_D or surpasses it, the depletion region becomes so thin that neutral electrons and holes in the conduction and valence bands can recombine swiftly, as shown in Fig. 5.7b. Because there is an energy band gap between the conduction band and the valence band, the recombination of the electron-hole pair can also cause photon emission. The bandgap energy can be expressed as the difference in conduction band energy (E_c) and valence energy (E_v):

$$E_g = E_c - E_v. \tag{5.2}$$

Furthermore, it can be shown that the emitted photon energy is

$$E_g = h\upsilon, \tag{5.3}$$

where $\upsilon = c/\lambda$ is the frequency of the light emitted and h is Planck's constant, equal to 6.626×10^{-34} J·s. This means that the frequency of light is proportional, and the wavelength is inversely proportional to the bandgap energy of the semiconductor material in which the LED has been fabricated. Therefore, the color of the LED can be changed by modifying the LED semiconductor material composition. For example, InGaN/GaN produces blue light, whereas InGaP produces red light. Bandgap energies of some commonly used semiconductor materials have already been tabulated in Chap. 4.

5.3.2 Detection of Light in Photodiodes

Photodetectors are devices that transform optical radiation signals into electrical signals [15]. A good photodetector should have the following characteristics:

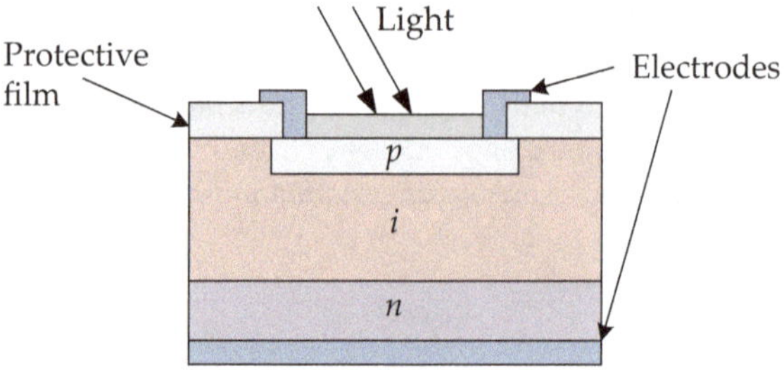

Fig. 5.8 Cross-section of the PIN photodiode

- *Sufficiently high responsivity in the working wavelength.* In other words, the incident power of the appropriate wavelength should be able to generate sufficiently high current for meaningful signal processing in the subsequent receiver blocks.
- *Fast response speed.* This means that the photodetector must be able to support the data rates required by its intended application.
- *Low noise.* Similar to LNAs in RF systems, the photodetector is the first component in the VLC system, and as such, it should add as little noise as possible in order to keep the SNR high and to preserve the integrity of the incoming signal.
- *High linearity.* High linearity ensures there is no distortion during the signal conversion process.
- *Other characteristics.* These include small size, long service life, low cost, etc.

A PIN photodiode is a photodetector that satisfies many of the listed requirements and is thus the most popular detector used in VLC systems. Like the LED, a PIN photodiode is also a diode, but the main difference between a regular diode and a PIN photodiode is that p-type and n-type material are separated by a layer of lightly doped intrinsic material. In this way the width of the depletion layer is increased. A cross-section of a PIN photodiode is shown in Fig. 5.8. To understand the reason for the intrinsic semiconductor layer insertion, one has to look at the principle of photodetection. Photodetection happens in reverse order from photoemission: when a photon with sufficiently high energy hits a depletion layer of the photodiode, an electron moves from the valence band to the conduction band, which results in the generation of an electron-hole pair [6]. This, in turn, causes the current to start flowing. The conversion efficiency of the photodiode depends on the width of the depletion region. Formally, the conversion efficiency is called the quantum efficiency, and is expressed as

$$\eta = 1 - e^{\alpha W}, \tag{5.4}$$

where α ($\alpha(\lambda)$) is the junction absorption coefficient at a specific wavelength and W is the width of the photodiode depletion region. Applying the reverse bias voltage to the photodiode can increase the depletion region width, but a much better effect in depletion region widening can be achieved with an insertion of the intrinsic material, as is the case with the PIN photodiode. In PIN photodiodes, most of the light, in fact, is absorbed by the intrinsic layer. In the alternative device, the APD, the diode avalanche effect is used as a way to improve the diode conversion efficiency. In short, in these devices, the avalanche effect practically acts as a multiplication buffer for the generated current.

The energy of the photon hitting the depletion region of the photodiode should be larger or equal to the bandgap energy of the semiconductor used in the diode fabrication, i.e.

$$h\upsilon \geq E_g. \tag{5.5}$$

This sets the limit on the upper cutoff wavelength of the photodiode:

$$\lambda_c = \frac{hc}{E_g}. \tag{5.6}$$

As a result, the semiconductor material used in the diode composition plays an equally important role to the one it plays in LEDs. Both direct-bandgap semiconductors, such as InGaAs and GaAs, and indirect-bandgap ones, such as Si and Ge, can be used in photodiodes. Note that direct-bandgap materials often have higher absorption coefficients. Practical photodetectors typically use Ge and InGaAs for receiving long-wavelength light in the infrared spectral range, and Si can be used for short-wavelength light in the range between 200 and 1600 nm, which also encompasses visible light. As a result, Si photodiodes are quite common in VLC systems, despite the fact that they might have lower absorption coefficients.

The current generated in the photodiode can be expressed as [20]

$$I_{ph} = \eta P_o \frac{q\lambda}{hc}(1-r), \tag{5.7}$$

where P_o is the incident optical power and r is the reflection coefficient at the air-diode interface.

The bandwidth of the photodiode is one of its most important properties. Two main factors influence the bandwidth: photodiode capacitance and the drift velocity of the carriers drifting through the depletion region. The method of time constants can be deployed to estimate the bandwidth. If the photodiode equivalent circuit can be represented by the circuit diagram in Fig. 5.9, it is easy to conclude that the first time constant influencing the bandwidth is $\tau_{RC} = R_s C_p$. In this case, the influence of the series inductance L_s is ignored, as it becomes prominent only at extremely high

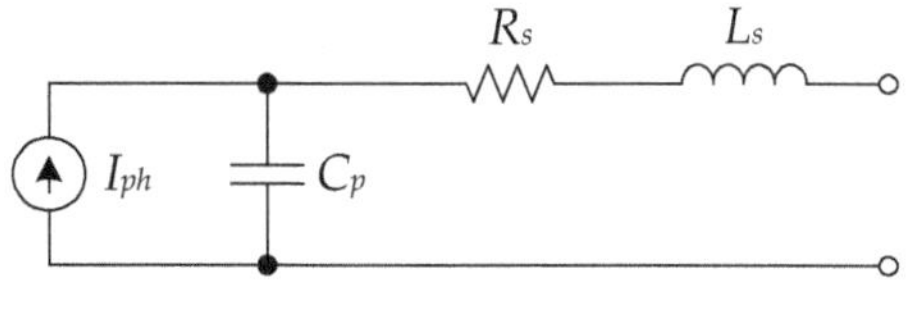

Fig. 5.9 The equivalent circuit model of a photodiode

frequencies. Then, if v_{sat} is the (saturated) carrier velocity, the second time constant, $\tau_{tr} = W/v_{sat}$, viz. the carrier transfer time, can be identified. Then the photodiode bandwidth is simply

$$BW = \frac{1}{2\pi(\tau_{tr} + \tau_{RC})} \tag{5.8}$$

The bandwidth of the photodiode should not be confused with its pulse response time and resulting frequency characteristics, which show its response speed to the high-speed modulated optical signal that is being received. The pulse response and the signal maximum frequency are related to the diode switching speed (its turn-on and turn-off times). For fast switching diodes, however, the maximum signal frequency becomes limited by the carrier transfer time.

Noise in the photodiode is another photodiode characteristic that needs to be considered. Diodes suffer from shot noise, thermal noise and flicker noise (which can be neglected at higher frequencies). However, any currents created from the light emanating from unwanted wavelengths are unwanted signals, and therefore, can also be regarded as noise. Proper optical systems use techniques (filters) for blocking the light with unwanted wavelengths, as already mentioned. Photodiodes also suffer from dark current noise. A dark current is the concept that is used to describe the situation where a small amount of current is flowing out of the diode when no light is present.

5.4 VLC Channel and Propagation

The third component of the optical communication system, in addition to the transmitter and the receiver, is the VLC channel.

5.4.1 Propagation of Optical Waves and the VLC Channel Model

Optical propagation links can be classified into two categories: LOS links and nLOS links [6]. Because of the low reflection rate of light, LOS links have a much greater power efficiency and the amount of power reflected more than once is negligible [5]. However, in indoor settings, usually at least one reflected path can be found, meaning that the signals can be detected if in the presence of blockage. Propagation links can also be classified in terms of the directionality of the transmitter and receiver. Both the transmitter and the receiver can be directional or non-directional. Scenarios where one of the two is directional and the other one is not lead to so-called hybrid links. If the presence of the LOS is simultaneously considered, eight types of links can be identified as a result:

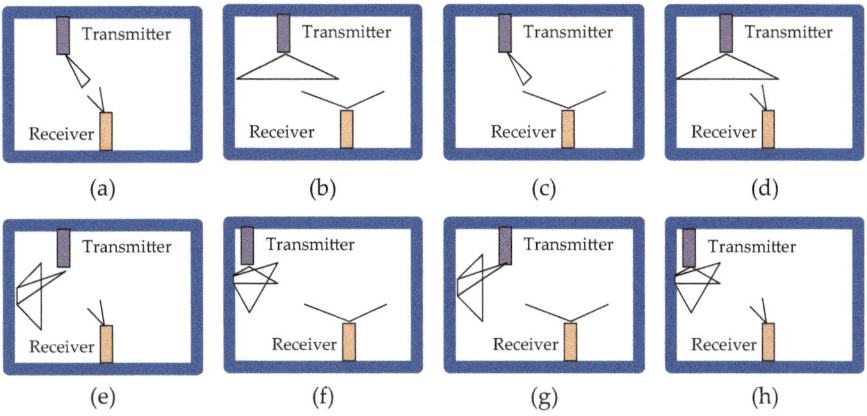

Fig. 5.10 Various types of VLC links: **a** LOS/directed link, **b** LOS/non-directed link, **c** and **d** LOS/hybrid links, **e** NLOS/directed link, **f** NLOS/non-directed link **g** NLOS/hybrid links. From [6]

- LOS/directed link;
- LOS/non-directed link;
- LOS/hybrid link 1, where the transmitter is directional, and the receiver is not;
- LOS/hybrid link 2, where the receiver is directional and the transmitter is not;
- nLOS/directed link;
- nLOS/non-directed link;
- nLOS/hybrid link 1, where the transmitter is directional, and the receiver is not; and
- nLOS/hybrid link 2, where the receiver is directional and the transmitter is not;

These links are shown in Fig. 5.10.

One of the major differences between RF and optical systems is that the RF systems suffer from multipath fading, while the optical systems do not. This is because the physical detection area of the photodiode is much larger than the square wavelength of the incoming light. Optical systems, especially directional outdoor systems, on the other hand, are much more prone to pointing errors and pointing loss RF links, even those operating at millimeter-wave or terahertz frequencies.

The VLC can be modeled by the amount of attenuation the light signal suffers as it propagates from transmitter i to receiver j. As already mentioned, only two paths are feasible: a direct link and a link with a single reflection. Both possible paths are shown in Fig. 5.11. The attenuation in the VLC channel can be modeled using the Lambertian model. The generalized Lambert law can be applied on the side of the transmitter, which indicates that the radiant light intensity is dependent on the cosine of the angle ϕ between the emitted light and the line normal to the LED surface. Then, to work out the link attenuation, first, one needs to define the Lambertian index as [5, 6]

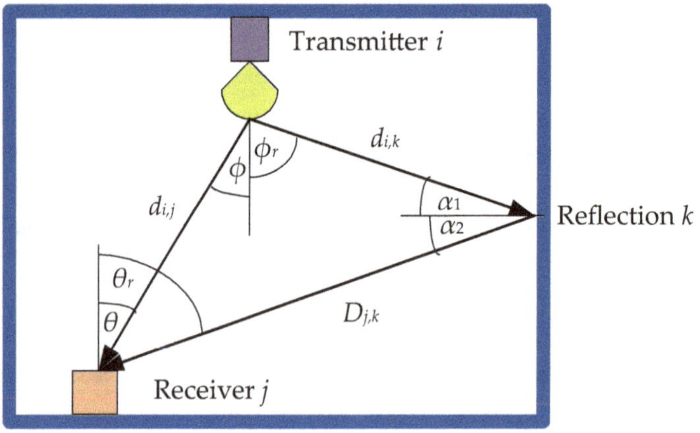

Fig. 5.11 The principle of optical wave propagation, taking into consideration the direct wave and the first reflected link [5]

$$m = \frac{-1}{\log_2(\cos\theta_{1/2})}, \tag{5.9}$$

where $\theta_{1/2}$ is the half-intensity radiation angle of the LED. Second, on the receiver side, the optical concentrator gain is

$$f(\theta) = \begin{cases} \frac{n^2}{\sin^2\Theta}, 0 \leq \theta \leq \Theta, \\ 0, \theta > \Theta, \end{cases} \tag{5.10}$$

where n is the refractive index and Θ is the semi-angle of the field of view of the receiver photodiode. Then, for the direct link, the DC gain (which is smaller than 1 and effectively the attenuation) can be expressed as

$$h_{i,j} = \frac{A_p(m+1)}{2\pi d_{i,j}^2} \cos^m \phi g_{of}(\theta) f(\theta), \tag{5.11}$$

where A_p is the physical area of the photodiode, $d_{i,j}$ is the distance between the transmitter and the receiver and g_{of} is the gain of the optical filter. For $\theta > \Theta$, there is no LOS, and the DC gain is zero. For the reflected link, the DC gain can be worked out from

$$dh_1 = \frac{A_p(m+1)}{2\pi d_{i,k}^2 d_{j,k}^2} \rho d A_s \cos^m \phi_r \cos\alpha_1 \cos\alpha_2 g_{of}(\theta_r) f(\theta_r) \cos\theta_r, \tag{5.12}$$

where $d_{i,k}$ is the distance from the transmitter to the reflecting surface, $d_{j,k}$ is the distance from the reflecting surface to the receiver, ϕ_r and θ_r are the angles corresponding to the reflected link, α_1 and α_2 are incidence and irradiance angles at the reflection point and lastly, ρ and dA_s are the reflection factor and reflective area respectively.

5.4.2 VLC Channel Capacity and Throughput

Radical differences between RF and VLC systems mean that the capacity of the VLC channel does not follow the Shannon channel capacity introduced in Chap. 3 [5]. There seem to be two constraints on the channel capacity: the requirement for the non-negative real-valued intensity signals and the average light intensity for the eye safety standard. In addition, the illumination requirements of the environment in which the VLC system is deployed, and the LED dynamic range must also be considered. Regrettably, there is no analytical expression for channel capacity subject to the average-power and peak power constraints, and only capacity bounds (lower and upper) can be found [6]. The bounds differ depending on the chosen constraints and it is beyond the scope of this book to present different capacity bounds alternatives.

An alternative metric that can be looked at in VLC is the channel throughput, which calculates the actual data rate. The throughput requires the modulation scheme and BER to be known beforehand. In Sect. 4.5, it will be identified that OFDM, widely used for RF, is also doable in VLC. The throughput that is seen by the user j if OFDM is used can be worked out from

$$X_j = \frac{W}{\beta L} \sum_{i=1}^{\beta L - 1} \eta_i \tag{5.13}$$

where W is the modulation bandwidth, L is the number of subcarriers, β is the constant that depends on the kind of OFDM used (typically, $\beta = 0.5$) and η is the subcarrier efficiency constant that incorporates the effect of the modulation scheme, coding scheme and the received SNR.

5.4.3 Influence of Weather on VLC

Outdoor VLC and other optical links, such those used in V2V communications, could be influenced by adverse weather. In general, the earth's atmosphere is not an ideal propagation medium for optical waves [21]. Two main factors can result in the link loss. The first is due to the temperature gradients that induce turbulence, which leads to the loss of coherence because of refraction. The second is interaction with the solid and liquid particles present in the air. For relatively short links, turbulence is generally not a problem, thus this section will focus on the influence of particles present in the air and turbulence will be reconsidered for under-water VLC systems.

Some particles (constituents) are always present. These include gases such as oxygen, nitrogen, and water vapor. Some particles (components) are only present in adverse weather. The probability that a particle will influence the optical wave propagation largely depends on the size of the particle relative to the wavelength of light and the concentration of such particles in the link path. With the assumption that the link is designed to operate in the presence of constituents, it is only the adverse weather that should be considered treated as a potential challenge. High attenuation could be caused by heavy fog, rain, or snow, all generally shortening the transmission range [22]. However, a VLC receiver has better sensitivity than human eyes, meaning that what humans could consider as the loss of visibility may not necessarily be applicable to the VLC system. The following is generally true [21]:

- Fog is the worst possible weather condition and it can produce attenuation as large as 170 dB/km in cases where visibility drops below 100 m.
- Attenuation due to rain is not marginal (30 dB/km for heavy high rain), but this figure is still much lower than for fog.
- Hail is generally not a problem.
- The effect of snow depends on the type of the snow particle, but the attenuation is generally comparable to that of rain.

5.4.4 Increasing the VLC Coverage

Because the signal is practically lost if there is more than a single reflection in the system, the coverage area of a VLC system using just one light source is very small. In large indoor spaces, such as in large open-plan offices, this necessitates placing multiple transmitters with the overlapping coverage area, as shown in Fig. 5.12. In outdoor spaces, the same concept can be conceived, with LED streetlamps used for

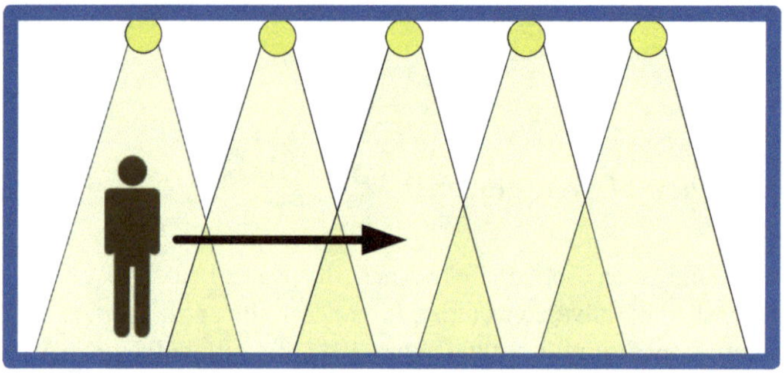

Fig. 5.12 Arrays of VLC transmitters for increasing the VLC coverage

both illumination and data transmission. With the network users constantly moving between different cells, the VLC systems need to be designed for a fast handover.

5.5 Modulation Techniques for VLC

VLC transmitters, like RF transmitters, require the signal to be modulated on order for it to be transmitted over a wireless channel. Modulation and the related concept of channel access were formally defined in Chap. 3. However, there is one fundamental difference between the modulation in the optical domain to that in the RF domain. Whereas in the RF domain modulation implies frequency, phase, and amplitude translation of the signal, in the optical domain it is only possible to vary the light intensity emitted by an LED. The same holds for demodulation. In this case, frequency and phase translation are replaced by direct detection in a photodiode [5, 11].

Several groups of modulation schemes are considered for VLC. General modulation schemes have some similarities with RF modulation schemes and will be discussed first. Power-efficient modulation schemes integrate dimming as a part of the modulation process. Lastly, NOMA schemes are also being considered for VLC.

5.5.1 General Modulation Schemes

The modulation scheme for a particular VLC network is usually selected based on several criteria [11]:

- Power efficiency;
- Bandwidth or spectrum efficiency;
- Transmission reliability;
- Complexity; and
- Robustness in the presence of external ambient light sources.

Additionally, all modulation schemes have a requirement to integrate run length limited (RLL) series to avoid long series of 0 and 1 s, which would cause flickering irrespective of the data rate. Fortunately, RLL also takes care of data recovery and clock synchronization.

Popular modulation schemes for VLC include:

- On-off keying (OOK);
- PAM;
- Pulse position modulation (PPM);
- Pulse-interval modulation (PIM);
- Pulse-width modulation (PWM);
- Pulse-position PWM;

- Pulse dual slope modulation (PDSM);
- Color-shift keying (CSK); and
- Orthogonal schemes, such as discrete multitone (DMT) modulation and OFDM.

Simple Modulation Schemes

OOK is the simplest (and slowest) modulation scheme. In OOK, the LED will be on if the transmitted signal is binary 1, and it will be off if the transmitted signal is binary 0. Sometimes, instead of the complete absence of the signal for state 0, the LED will simply emit a different light intensity. If the data are encoded into several light intensity or light amplitude levels, PAM is achieved. The concept of PAM in the VLC context is very similar to PAM in the RF context described in Chap. 3. PAM has good bandwidth efficiency.

PPM encodes data into pulses of short duration placed at different positions of the pulse train. It is more power-efficient (LED is on for only a small percentage of the total pulse train duration), but it requires larger bandwidth and synchronization on the receiver end. Four symbols of PPM with $M = 4$ are shown in Fig. 5.13. PIM modifies PPM by inserting blank slits between two adjacent symbols.

PWM is similar to PPM, except that the data are encoded into pulses of different widths, as shown in Fig. 5.14. If both PPM and PWM are deployed simultaneously, PPMPPW is achieved. This is an example of a hybrid technique, with the aim to improve data throughput, power efficiency and BER.

In PDSM, 0 bit is denoted by the growing edge and 1 bit is denoted by the dropping edge of the duration of the pulse, with the other edge kept constant. This scheme can

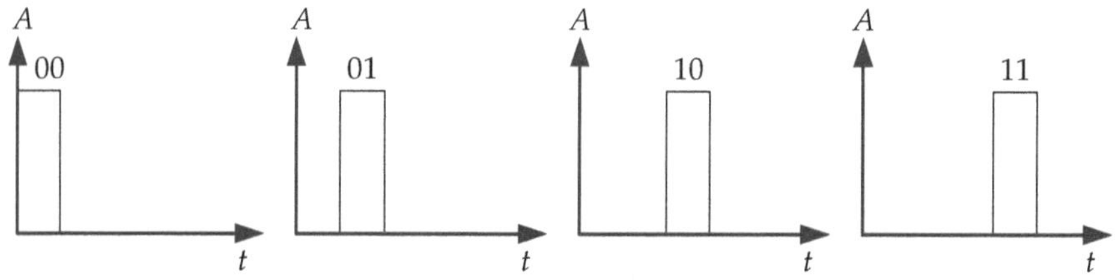

Fig. 5.13 PPM ($M = 4$)

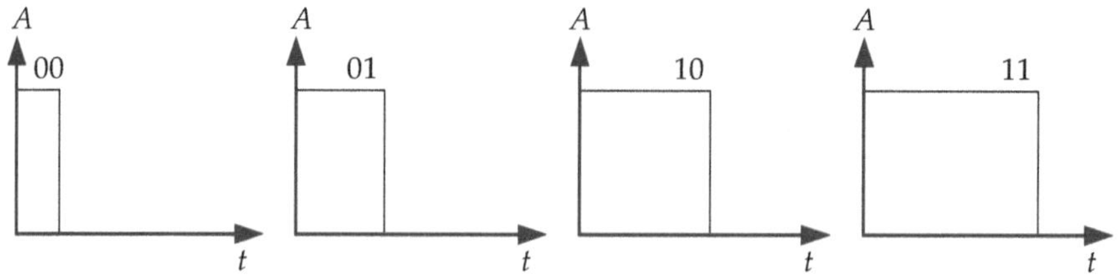

Fig. 5.14 PWM ($M = 4$)

be used to avoid inter-and intra-frame flicker. Symbols for binary 0 and 1 in PDSM are shown in Fig. 5.15.

CSK is a modulation scheme in which a bit or a combination of bits is encoded into a light of a different color [6, 24]. The simplest way to do the encoding is to convert a combination of symbols into one of the seven main wavelength bands, or colors, violet, blue, cyan, green, yellow, orange, and red, as shown in Table 5.1. Each of the seven color bands can be expressed as an (x, y) coordinate on the CIE 1931 color space chromaticity diagram, shown in Fig. 5.16. Higher order CSK modulation is possible, and this involves using typically three or four LEDs of different colors and firing them simultaneously. Not all colors can be combined so that the resultant light can be decoded successfully—legal light combinations for three-color PSK can be seen in Table 5.2. For three different colors, constellations for 4-CSK, 8-CSK and even 16-CSK can be found, each resulting in symbols spaced far apart in the constellation space so that error-free detection is not compromised. Note that in the modulation schemes that use multiple LEDs, such as CSK, light equalization is necessary.

Orthogonal Modulation Schemes

Orthogonal modulation schemes are also considered in VLC systems for at least three reasons [15]:

- High spectral efficiency;
- Ability to resist multipath interference; and
- Dynamic subcarrier allocation to achieve the maximum bit rate.

Fig. 5.15 Symbols in PDSM [23]

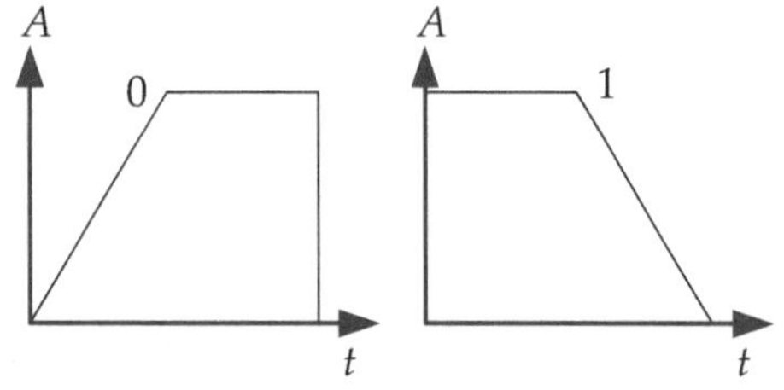

Table 5.1 Encoding of the data into different color bands

Color	Band (nm)	Code	λ (center) (nm)
Violet	380–478	000	429
Blue	478–540	001	509
Cyan	540–588	010	564
Green	588–633	011	611
Yellow	633–679	100	656
Orange	679–726	101	703
Red	726–780	110	753

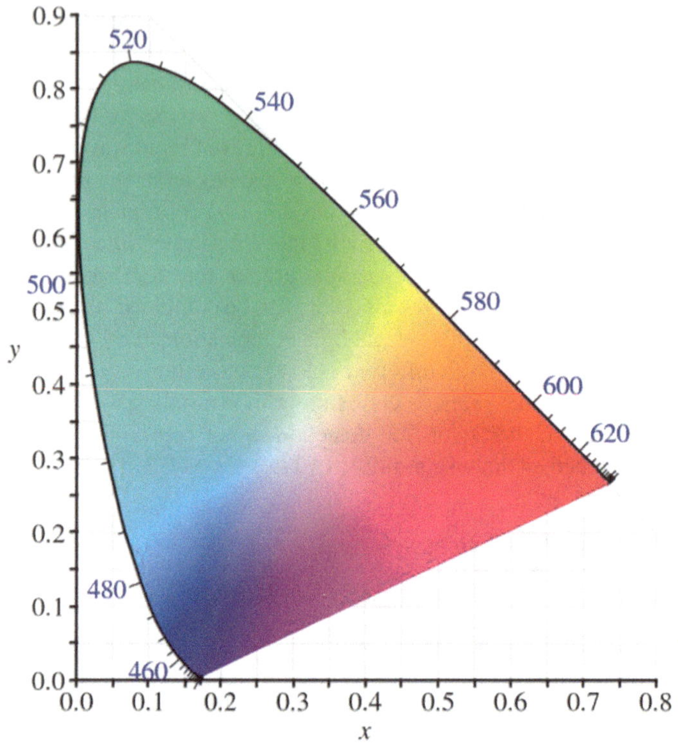

Fig. 5.16 The CIE 1931 color space chromaticity diagram

Table 5.2 Legal three-color CSK combinations [6]

Combination	Light source 1	Light source 2	Light source 3
1	110	010	000
2	110	001	000
3	101	010	000
4	101	001	000
5	100	010	000
6	100	001	000
7	011	010	000
8	011	001	000
9	010	001	000

Many orthogonal modulation schemes have a signal constellation that is the same as that of QAM, shown in Chap. 3 in Fig. 3.16. Examples of orthogonal modulation schemes are OFDM and DMT.

OFDM in the VLC domain is similar to the OFDM in the RF domain, described in Sect. 3.4.1. QAM mapping, S/P conversion, IDFT, the addition of a CP, and a

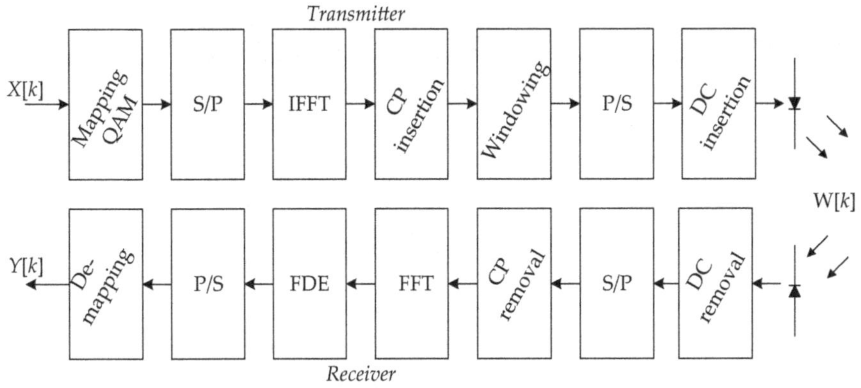

Fig. 5.17 Block diagram of the OFDM transmitter and the receiver modified for VLC (DCO-OFDM). Adapted from [15]

P/S conversion step are all present in the optical OFDM [15]. The main difference between the OFDM in the VLC domain from the OFDM in the RF domain is that instead of broadcasting the signal with the aid of antenna, an additional step of DC insertion is added, which allows for the electrical signal to become positive so that it can drive the LED. On the receiver side, optical signals are detected by the photodetector, then passed through the DC removal step, S/P conversion, CP removal step, FFT converter, FDE, P/S conversion and QAM de-mapping. The modified block diagram of the OFDM then looks as shown in Fig. 5.17.

The OFDM technique described is just one of the many different optical OFDM variants and is known as DC-biased optical OFDM (DCO-OFDM). On the other hand, if only odd subcarriers are used to modulate the data, asymmetrically clipped optical OFDM (ACO-OFDM) is achieved [5]. Other types of optical OFDM are also possible but will not be mentioned here.

DMT can be regarded as the baseband implementation of the OFDM [11]. Instead of two-dimensional mapping (QAM), the subcarriers are usually modulated by a one-dimensional modulation, such as PAM or PWM. PAM-DMT, for example, uses asymmetrical clipping, which is characteristic of ACO-ODFM [24]. Data are modulated only on the imaginary components of the subcarriers, which come in pairs of complex conjugates. Thus, there are 2 N symbols, satisfying this relationship [15]:

$$C_{2N-n} = C_n^*. \tag{5.14}$$

The first three modulation steps (mapping, S/P conversion and IFFT) are illustrated in Fig. 5.18 for a PAM-DMT system. DMT is a good option for a noisy setting, as it does not add additional noise, because clipping of the noise is contained in the real domain, whereas the mapping is in the imaginary domain.

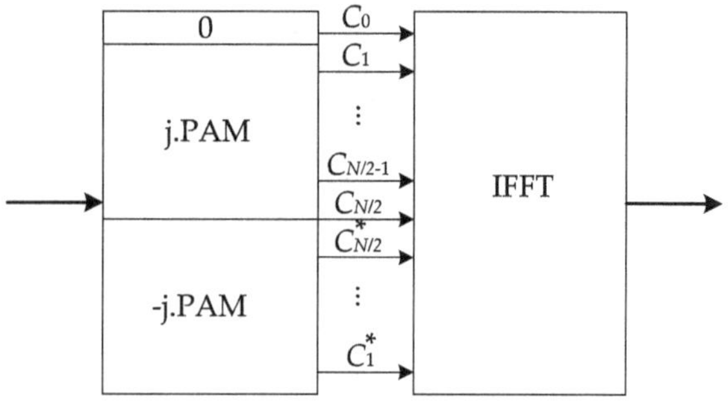

Fig. 5.18 Mapping, S/P conversion and IFFT in a PAM-DMT system [24]

Other orthogonal modulation schemes that have been researched for VLC include wavelength division multiplexing and polarization division multiplexing. The discussion of these modulation schemes will be omitted.

5.5.2 *Power-Efficient Modulation Schemes*

Power-efficient modulation schemes need to be deployed in the systems in which the VLC light sources are also used for illumination [11]. Several modulation techniques already described can be regarded as power-efficient techniques. These include CSK and ACO-OFDM.

However, several modulation schemes have been specifically designed for dimming control. These include variable OOK (VOOK), variable PPM (VPPM), multiple PPM (MPPM), variable pulse amplitude and position modulation (VPAPM) and overlapping PPM (OPPM) [11, 25]. As can be derived from the names of most of these modulation schemes, a degree of digitally controlled variability is added to the non-power-efficient counterparts of these modulation schemes to achieve the effect of dimming and power efficiency.

5.5.3 *Non-orthogonal Channel Access in VLC*

The concept of the NOMA explored in RF communications has also triggered optical communication researchers to consider it as option for VLC. The main goal of NOMA is to boost the achievable data rates. Note that techniques such as optical OFDM and use of MIMO have already been designed with this goal in mind. However, traditional modulation schemes do have their disadvantages, which could potentially be corrected with NOMA. In OFDM, for example, the achievable data rates are inevitably reduced by spectrum partitioning [26].

NOMA, on the other hand, can simultaneously serve multiple users in the same degrees of freedom by splitting them in the power domain (PD-NOMA), as explained in the context of RF systems in Sect. 3.4.2. Recall that in PD-NOMA, each user gets to operate at a different power level [5]. Consequently, indoor VLC systems with a large number of users can benefit from NOMA, because the sum throughput of the multiuser VLC system is increased when compared to other access schemes [27]. What is more, NOMA can bring about significant performance gains in high SNR situations, which are likely in indoor VLC systems that deploy links over short distances. However, implementing the NOMA in VLC networks requires considering the unique properties of VLC networks, including limited bandwidth of LEDs, the maximum allowed transmit power, the massive influence of blockages and much higher deterioration in the VLC channel than in RF systems, particularly for long-distance links [5].

It was also suggested that NOMA can be used with other techniques for data link improvement. For example, in [26], Chen et al. argue that using NOMA in conjunction with MIMO can bring improvements over PD-NOMA, and as such, can be considered promising for future high-speed multi-user VLC systems.

5.6 Green Aspects of VLC

In Chap. 1, it was indicated that one of the key aspects of 6G planning is energy efficiency, with a serious move towards aspects such as energy conservation and energy harvesting, that can potentially make the 6G network a "green" network. A full chapter (Chap. 6) will be dedicated to the discussion of this topic; however, there are several aspects in which VLC can contribute to this regard, and these will be discussed in this section.

First, it should be noted that VLC networks are more power-efficient than RF networks. This is because LEDs, by nature, are energy-efficient devices [5]. Secondly, since the same LED can, in general, be used for data transmission as well as for illumination, as long as the illumination of the transmitting LED is within specified illumination requirements, data transmission adds virtually no energy overhead. However, this saving is not possible during daytime in outdoor environments or indoor environments where natural lighting is possible. Bear in mind that with the VLC capability of high data transfer rates, it is easy to increase power consumption up to a point where energy efficiency is lost. In those cases, the system should rather be optimized for several aspects, in which energy efficiency is only one of several considerations.

Energy efficiency can also be achieved with the addition of energy harvesting into the system. In this case, it is fine if the transmitter does not meet certain energy efficiency requirements, as long as this can be offset by harvesting some of the transmitted energy by one or more receivers in the network. This is possible if the receiver is equipped with a solar panel that can convert the received modulated light signal into an electrical signal. The light, which is modulated, needs to be separated into AC and DC parts: the AC part is demodulated, and the DC part is used for powering up the system.

The energy that can be harvested by a user from a single LED is

$$E = f I_{DC} V_{oc}, \tag{5.15}$$

where f is a fill factor of approximately 0.7, I_{DC} is the received DC part of the photodiode current I_{ph} described in Eq. (5.7) and V_{OC} is the open-circuit voltage,

$$V_{oc} = V_T \ln\left(1 + \frac{I_{DC}}{I_0}\right), \tag{5.16}$$

with I_0 being the dark saturation current of the photodiode. Since the value of I_{ph} depends on the incident optical power at the receiver, the fact that the amount of energy that can be harvested if the system uses multiple LEDs increases, does not warrant any further explanation.

Energy harvesting systems are, naturally, not only limited to harvesting the energy from the transmitter light source, but also from any other available ambient light (e.g. daylight). With that in mind, the topic of energy harvesting becomes a much wider discussion, applicable not only in the VLC context but also for other systems that might be used in future wireless networks (e.g. to RF). As such, it will be discussed in more detail in Chap. 6.

5.7 Hybrid VLC/RF Networks

Hybrid VLC and RF, or in the case of 6G, a millimeter or terahertz wave system, introduces a robust communication solution, which combines the best aspects of both VLC and radio transmission regimes. Hybrid systems are introduced for two reasons: (1) ubiquitous coverage and (2) increasing the SNR of the VLC system in a setting where there are large amounts of light, which leads to inevitable SNR loss, thus allowing such a system to operate in a wide range of channel conditions [28].

In the indoor setting, for example, the hybrid VLC/RF system could be organized as described below. As the primary link, the VLC part of the system would provide for extremely high data rates, while simultaneously providing illumination. However, the VLC may not be able to provide full coverage (despite a large number of cells, as shown earlier in Fig. 5.12), which is why the secondary RF system is required, albeit providing slower data rates (though sufficiently high in the case of millimeter-wave or terahertz links), but introducing ubiquitous coverage [29]. Similarly, for outdoor directional optical links, there could be interruptions in the optical transmission, for example in adverse weather, such as fog. Again, the slower radio link could kick in as a secondary backup link. In both cases, users' experience is significantly improved from the case where either of these technologies is deployed in isolation.

A block diagram of the hybrid VLC/RF system is shown in Fig. 5.19 [30].

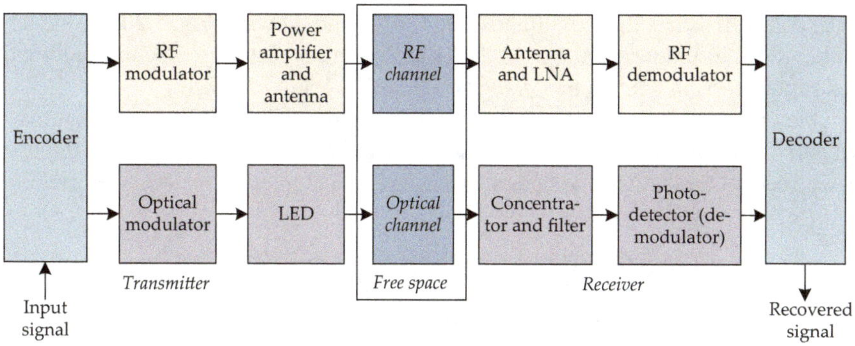

Fig. 5.19 Block diagram of the hybrid VLC/RF system [30]

One of the issues with the hybrid RF/VLC network that needs to be considered is how to optimize the network for the best possible performance. In other words, the question is, how to distribute the users among the access points, both RF or VLC, to improve the overall system performance while retaining acceptable fairness of the system [5]. The idea is the minimize the number of handovers between VLC and RF segments. Research on this, and similar topics, continues.

5.8 Applications of Optical Wireless Communications

Verticals enabled by VLC will introduce unique flavor to the future wireless network generations beyond 5G. In addition to being able to support wireless links of a remarkably high information rate, thereby complementing traditional RF links (inclusive of millimeter-wave and terahertz variants), there are certain aspects that may only be achievable with optical technologies. These include, among others, V2X links, indoor localization, underwater communication, etc. Of course, general communication links, for example in indoor environments or on street level, discussed throughout this chapter, should not be forgotten.

Consequently, the most promising applications of VLC and OWC in general will be considered in this section.

5.8.1 Wireless LAN and Li-Fi

Wireless LAN and LiFi are terms that various sources use to describe the VLC communication networks that could be deployed in indoor and in outdoor environments, where VLC base stations form part of indoor lighting or streetlights, respectively. As already mentioned in a few discussions in this chapter, the idea is that the VLC piggybacks off the existing LED lighting infrastructure. This idea is not far from reality. In 2020, LED lightbulbs, such as the one shown in Fig. 5.20, are already widely used in indoor lighting, while LED lighting is slowly replacing traditional gas street lamps in many cities in the world, as shown in Fig. 5.21.

One thing that can be stressed here is that the optical communication in this scenario can typically only be established for the DL. This is because (1) the user device cannot transmit in the optical domain if placed in one's pocket or a briefcase

Fig. 5.20 An LED lightbulb used for indoor lightning. The glass enclosure has been broken, revealing an array of LEDs. Lightbulb provided by Azoteq (Pty) Ltd

Fig. 5.21 LED streetlights along Atterbury road in Pretoria, South Africa, June 2020

[9], or (2) the light emitted by the user device could potentially be harmful to users' eyes. Thus, the wireless LAN is the scenario in which the hybrid RF/VLC scheme discussed in Sect. 5.7 is in fact more applicable.

5.8.2 Vehicular Communications

In recent years, research efforts to develop autonomous or self-driving vehicles have been increasing rapidly [31]. At present, autonomous vehicles are in the process of field testing and it is expected that it will still take several years (if not tens of years) before self-driving becomes a reality. Today, experimental vehicles use numerous interrelated systems, including radars, lidars and cameras, in combination with GPS, to observe their environment, avoid obstacles and plot a course of their movement. In addition, cars become smarter if they can communicate with each other (V2V), as well with stationary objects, such as the roads on which they are driving, traffic lights, information boards, or other possible providers of information (V2X). VLC has been suggested, in addition to the millimeter-wave and terahertz technologies, as one of the promising candidates to provide high data rates that would support a high volume of information that self-driving vehicles would need. Of course, even human-driven vehicles can benefit from vehicular communication, where the collected information can be used to improve the overall driving experience and safety of the driver and passengers in the vehicle. Benefits that could be brought about by vehicular communications include emergency brake assistance, lane change assistance and blind spot warning, forward collision warning, traffic condition information, upcoming traffic light information and many others.

Various advantages make VLC a more attractive solution for vehicular communication than radio links [22]:

- *Low complexity and cost of VLC systems.* This becomes clear if it is noted that most modern vehicles are already fitted with LED lights.
- *The localization or positioning capability of VLC systems.* VLC is more capable of accurately predicting the position of the vehicle in question as well as other vehicles on the road.
- *Scalability.* VLC can operate safely if many or few vehicles are present.
- *Security.* As already discussed, VLC links are inherently more secure than radio links.
- *Operation in adverse weather.* In most weather conditions, except perhaps in heavy fog, VLC links are more stable than radio links.

5.8.3 Underwater Communications

When it comes to underwater communication, VLC is by far the most promising alternative, given that light propagates much better under water than radio or even

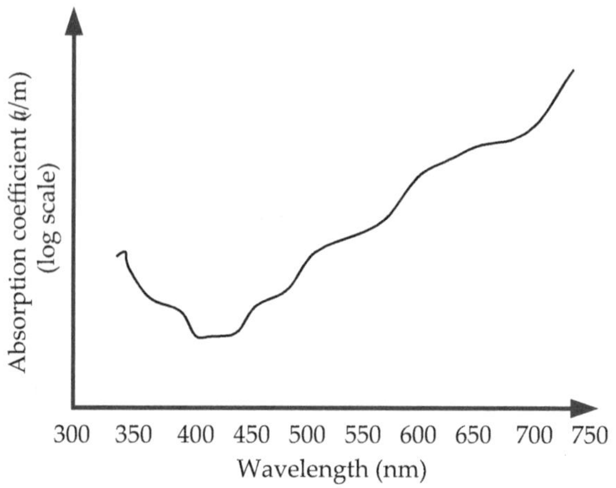

Fig. 5.22 Absorption coefficient of visible light of different wavelengths (adapted from [33])

acoustic signals, allowing for data rates in the order of hundreds of Mb/s or even Gb/s to be achieved underwater [32, 33]. The demand for UOWC is increasing owing to various new applications, such as environmental monitoring and maintenance, underwater exploration, offshore oil field exploration, as well as port security, tactical surveillance, and warfare.

UOWC is influenced by the following factors [33]:

- *Absorption.* Overall absorption combines the absorption of pure water, as well as other particulates in the water, for example colored organic matter and phytoplankton. Absorption of pure water is dependent on the wavelength, as shown in Fig. 5.22. From this figure, it can be deduced that blue/green light has the best propagation properties in pure water.
- *Scattering.* Scattering is essentially the altering of the path of the photon after interaction with an object in the water. Scattering can be either molecular scattering or scattering by large particles. Molecular scattering is scattering by objects that are much smaller than the wavelength of the light. This group also includes microorganisms such as bacteria. Large particle scattering is caused by objects larger than roughly 10 times the wavelength and is responsible for most of the scattering under water.
- *Turbulence.* In the context of underwater communications, turbulence refers to the mechanism in which the refractive index of the wafer is changed as light travels through waters of different temperature, salinity, or pressure. The refractive index is also dependent on the wavelength of light. In general, it has been found that the refractive index decreases with increasing temperature or wavelength and increases with increasing pressure or salinity.

The overall result of all these effects can be modeled by the approximate attenuation of water. The attenuation will depend on the location of the underwater systems. Typically, a distinction can be made between the open ocean, coastal ocean, and harbor waters. Given that harbor waters are the dirtiest and most turbulent, it is no

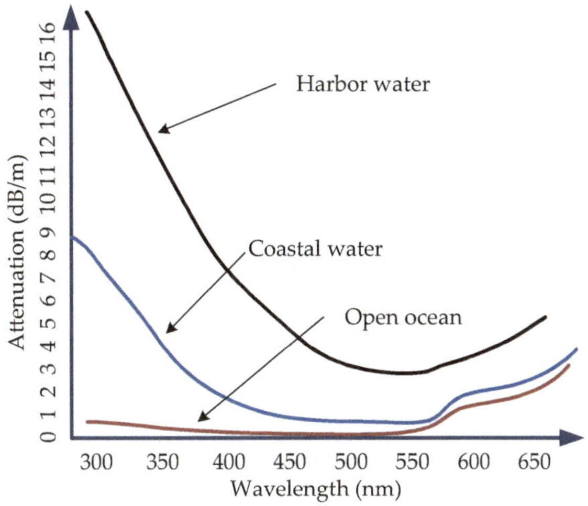

Fig. 5.23 Average attenuation of light in ocean water (adapted from [33])

surprise that the highest amount of attenuation is found in these waters. Conversely, the lowest attenuation is observed in open waters. The attenuation remains dependent on the wavelength, but it should be noted that the window of lowest attenuation shifts from shorter to longer wavelengths, as shown in Fig. 5.23.

Lastly, note that communication underwater is not only limited by spatial factors, but also by temporal factors, that is, that the channel conditions underwater are not constant, but rather change constantly in response to factors such as the wind, currents, the passage of animals or man-made disturbances.

5.8.4 Indoor Localization

VLC is one of the most promising technologies for indoor localization. The principle behind indoor localization is simple and can be related to GPS, which is used for outdoor localization [34].

Basically, a VLC network uses LED lights as wireless access points and the user can only communicate to each access point via either a direct link, or via one reflection. In indoor environments, such as large multi-story office buildings or shopping malls, the location of every LED light would be known precisely, meaning that the location of the user could be pinpointed to within a distance of a few meters, by simply determining to which base station the user is connected and knowing the area the VLC link can cover. Then, a more accurate position would be found using triangulation: if the user can establish a link with at least three base stations simultaneously, the user is found in the area where the coverage of the three base stations overlaps, as shown in Fig. 5.24. Lastly, if the distance from each base station can be determined (e.g. by computing the strength of the signal), a location specific to even a few centimeters can be found.

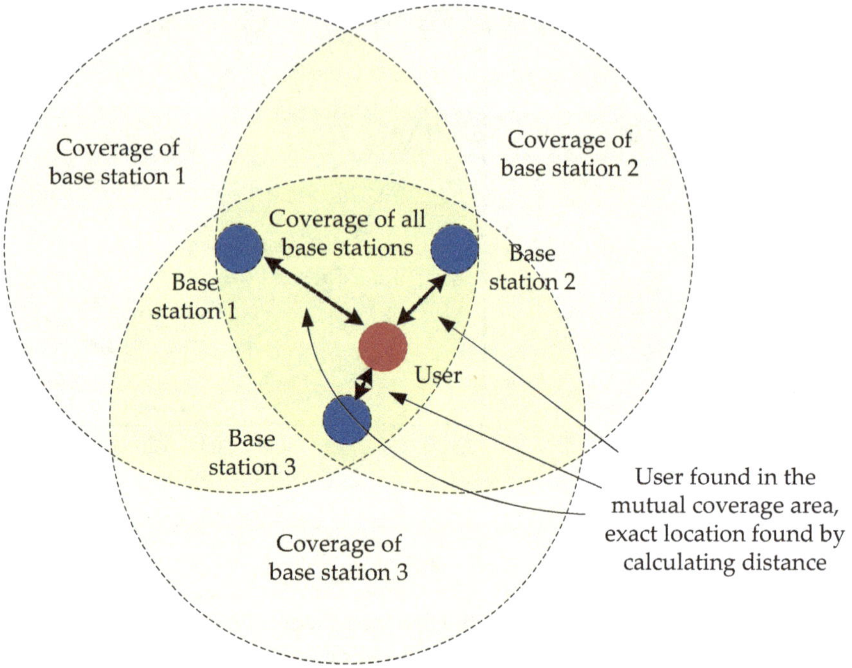

Fig. 5.24 Principle of triangulation

Indoor localization based on VLC works better than radio-based localization, because RF waves can propagate further than light waves, can be diffracted around and reflected from objects, and radio transceivers are usually spaced much further apart than VLC transceivers. Also, in multi-story buildings, radio waves can propagate at least to an adjacent floor, whereas this is not possible with light and as a result, the user's location is always pinpointed to a specific floor when VLC localization is used.

5.8.5 *Ultra-Short-Range and Short-Range Communications*

There are several scenarios where VLC could be useful in ultra-short-range and short-range communications [14]. For example, in various environments where high-speed chip-to-chip or intra-chip communication is required, but electromagnetic interference or power consumption is an issue, links could be established using VLC. In another example, VLC could be used between wearables and implanted sensors in a concept called wireless body area networks.

5.8.6 Light-Based Internet of Things/Internet of Everything

In recent years, the concepts of IoT and IoE have been receiving a lot of research attention. In short, in IoT (becoming IoE in practice), virtually everything, beyond people, vehicles and machines, is connected. One of the limitations of practical IoT networks is the difficulty of providing the energy supply to various "things" that need to receive and transmit information. This is where the concept of light-based IoT (LIoT) fits in [35].

In LIoT, the light is exploited in multiple ways, specifically with light harvesting for energy in mind. The LIoT concept becomes particularly interesting if it is combined with printed electronics, including among others printed solar cells, printed optical components (photodiodes, LEDs, lenses) and printed displays, although the energy efficiency of printed solar cells is considerably less than the efficiency of conventional solar cells.

5.8.7 Long-Range and Ultra Long-Range Communications

On the opposite side of the distance scale, VLC and other types of optical communications in general can be used in various scenarios where communication over several kilometers is required, which can be classified as long-range communications. One example is the use of optical (infrared) connectivity in last-mile (backhaul) links, which implies provision of high-speed internet to areas that cannot be reached via alternative technologies, such as copper or fiber-optic cabling. This specifically includes rural areas, particularly in undeveloped countries [13, 14]. Another scenario includes connecting LANs between buildings in office parks and on university campuses. In yet another scenario, optical and VLC links could be used in the military, for communication between aircraft, ships, submarines and various other military vehicles.

When it comes to ultra-long-range communications (thousands of kilometers), the specific application that is considered is in satellite optical feeder links. The success of such communication depends on the development of incredibly powerful and focused lasers. As a reminder, seamless satellite integration in 6G networks is one of the main conditions for establishing ubiquitous access.

5.9 Other Types of VLC

Lastly, before concluding this chapter, it must be noted that VLC established by utilizing the LED and a photodiode is not the only way of establishing communication using the visible spectrum. Recall that even the optical telegraph mentioned in Chap. 2 can be considered a type of VLC. Reading a bar code is also a type of optical communication, where the information from paper or from an electronic screen is transferred via the barcode reader—or even a camera—onto another device. 3D barcode reading has been becoming increasingly popular around the world for exchanging personal information or even for payments (e.g. WeChat Pay and Alipay, originating in China).

One type of VLC area that should be specifically mentioned is the area of optical camera communications (OCC) [6]. OCC implies using the camera image sensor, either in standalone cameras or in smart devices, such as smartphones and tablets, to receive the transmitted information. The information that is transmitted by arrays of LEDs (such as those within an information board) would be received by the image sensors and captured at a specific frame rate by the image sensor and subsequently processed within the device itself. Some of the advantages of this type of communication include pervasiveness (transmitters and receivers are already available) and multi-color capability. Disadvantages typically include difficulty in synchronization, the presence of shot noise, perspective distortion, misalignment, blur, and susceptibility to ambient light. Another issue that needs to be considered is that the limited frame rate of the receiver (typically 60 fps) would require the transmitter to transmit images that change at lower speeds than those at which an LED can be flickered in traditional VLC, which in turn could re-introduce flickering. Thus, although OCC is an interesting concept, it is clear that much more research is needed to implement it practically.

One should, however, not confuse OCC with optical imaging, where cameras are used to sense the environment and make conclusions based on images. For example, as already mentioned, technologies such as self-driving cars already use cameras to capture images of the road ahead, and the technology is also used in applications such as lane-change assistance or blind-spot assistance in modern non-autonomous vehicles.

5.10 Concluding Remarks

In this chapter, the concept of VLC communication was considered as an alternative to, or even a way to complement traditional electromagnetic (radio) communications. VLC systems can introduce incredible data rates while remaining extremely power-efficient and inexpensive to deploy, which makes them suitable for future wireless networks beyond 5G. Specifically, VLC can add to the ubiquitous coverage that is envisioned by 6G.

VLC technology has not yet made a breakthrough and it is still largely waiting for deployment outside laboratories [35]. This is attributed to the dominance of radio technologies, which have a history that surpasses that of VLC by about 100 years. More research is needed before VLC can be commercialized: VLC channel modeling and characterization of various deployment scenarios are still in their infancy [9]. Moreover, there is space for improvement upon VLC modulation schemes, as well as for improvements in new materials and optoelectronic devices. However, as the world moves to 6G and concepts such as underwater communications, indoor localization and V2X communications and communication in settings where radio communication is not desirable (e.g. in hospitals) become important, VLC technology will undoubtably find its place.

The concept of light ties in closely with energy harvesting, seeing that the energy from light has been harvested for decades. Energy harvesting, and green networking as a whole, is the topic of the next chapter, Chap. 6.

References

1. Huang T, Yang W, Wu J, Ma J, Zhang X, Zhang D (2019) A survey on green 6G network: architecture and technologies. IEEE Access 7:175758–175768
2. Nawaz SJ, Sharma SK, Wyne S, Patwary MN, Asaduzzaman Md (2019) Quantum machine learning for 6G communication networks: state-of-the-art and vision for the future. IEEE Access 7:46317–50
3. Green R, Leeson M (2012) The optical wireless channel. In: Arnon S, Barry J, Karagiannidis G, Schober R, Uysal M (eds) Advanced optical wireless communication systems. Cambridge University Press, pp 240–72
4. Calvanese Strinati E, Barbarossa S, Gonzalez-Jimenez JL, Ktenas D, Cassiau N, Maret L et al (2019) 6G: the next frontier: from holographic messaging to artificial intelligence using subterahertz and visible light communication. IEEE Veh Technol Mag 14:42–50
5. Obeed M, Salhab AM, Alouini M-S, Zummo SA (2019) On optimizing VLC networks for downlink multi-user transmission: a survey. IEEE Commun Surv Tutor 21:2947–2976
6. Wang Z, Wang Q, Huang W, Xu Z (2017) Visible light communications: modulation and signal processing. John Wiley & Sons
7. Soderi S (2020) Enhancing security in 6G visible light communications. 2020 2nd 6G wireless summit (6G SUMMIT), pp. 1–5
8. Yesilkaya A, Cogalan T, Erkucuk S, Sadi Y, Panayirci E, Haas H, et al (2020) Physical-layer security in visible light communications. 2020 2nd 6G wireless summit (6G SUMMIT), pp 1–5
9. Rajatheva N, Atzeni I, Bjornson E, Bourdoux A, Buzzi S, Dore J-B, et al (2020) White paper on broadband connectivity in 6G. arXiv:200414247 [eess] [Internet]. [cited 2020 Sep 16]. http://arXiv.org/abs/2004.14247
10. Tariq F, Khandaker M, Wong K-K, Imran M, Bennis M, Debbah M A speculative study on 6G. IEEE Wirele Commun 27:118–25
11. Memon NK, Umrani FA (2017) Efficient modulation schemes for visible light communication systems. Networks of the future: architectures, technologies, and implementations. CRC Press, pp 355–74
12. Božanić M, Sinha S (2020) Methodologies for design of Millimeter-wave and terahertz integrated circuits: an lna case study. In: 2020 30th international conference radioelektronika (RADIOELEKTRONIKA). Bratislava, pp 1–6
13. Lambrechts W, Sinha S (2019) Last mile internet access for emerging economies. Springer
14. Ghassemlooy Z, Uysal M, Khalighi MA, Ribeiro V, Moll F, Zvanovec S, et al (2016) An overview of optical wireless communications. In: Uysal M, Capsoni C, Ghassemlooy Z, Boucouvalas A, Udvary E (eds). Optical wireless communications: an emerging technology [Internet]. Cham: Springer International Publishing [cited 2020 Jun 18], pp 1–23. https://doi.org/10.1007/978-3-319-30201-0_1
15. Chi N (2018) LED-based visible light communications. Springer, Cham
16. Bian R, Tavakkolnia I, Haas H (2019) 15.73 Gb/s visible light communication with off-the-shelf LEDs. J Lightwave Technol 37:2418–24

17. Xie E, Bian R, He X, Islim MS, Chen C, McKendry JJD et al (2020) Over 10 Gbps VLC for long-distance applications using a GaN-Based series-biased micro-LED array. IEEE Photonics Technol Lett 32:499–502
18. Islim MS, Videv S, Safari M, Xie E, McKendry JJD, Gu E et al (2018) The impact of solar irradiance on visible light communications. J Lightwave Technol 36:2376–2386
19. Sebastián J, Lamar DG, Aller DG, Rodríguez J, Miaja PF (2018) On the role of power electronics in visible light communication. IEEE J Emerg Sel Topics Power Electron 6:1210–1223
20. Alves LN, Rodrigues L, Cura JL (2017) Lighting and communications: devices and systems. In: Ghassemlooy Z, Alves LN, Zvánovec S, Khalighi MA, (eds) Visible light communications: theory and applications. CRC Press, Boca Raton, pp 9–70
21. Nebuloni R, Capsoni C (2016) Effects of adverse weather on free space optics. In: Uysal M, Capsoni C, Ghassemlooy Z, Boucouvalas A, Udvary E (eds) Optical wireless communications: an emerging technology [Internet]. Springer International Publishing, Cham [cited 2020 Jun 23], pp 47–68. https://doi.org/10.1007/978-3-319-30201-0_3
22. Luo P, Tsai H-M, Ghassemlooy Z, Viriyasitavat W, Minh HL, Tang X (2017) Car-to-car visible light communications. In: Ghassemlooy Z, Alves LN, Zvanovec S, Khalighi M-A (eds) Visible light communications: theory and applications. CRC Press, pp 253–82
23. Oh M (2014) Pulse dual slope modulation for VLC. KSII Trans Internet Inf Syst 8
24. Cseh T, Rajbhandari S, Fekete G, Udvary E, Ghassemlooy Z, Alves LN, et al (2017) Modulation schemes. visible light communications: theory and applications. CRC Press, pp 97–143
25. Lee K, Park H (2011) Modulations for visible light communications with dimming control. IEEE Photonics Technol Lett 23:1136–8. IEEE
26. Chen C, Zhong W-D, Yang H, Du P (2018) On the performance of MIMO-NOMA-based visible light communication systems. IEEE Photonics Technol Lett 30:307–310
27. Yang Z, Xu W, Li Y (2017) Fair non-orthogonal multiple access for visible light communication downlinks. IEEE Wirel Commun Lett 6:66–69
28. Dang S, Amin O, Shihada B, Alouini M-S (2020) From a human-centric perspective: What might 6G be? Nat Electron 3:20–29
29. Basnayaka DA, Haas H (2017) Design and analysis of a hybrid radio frequency and visible light communication system. IEEE Trans Commun 65:4334–4347
30. Letzepis N, Fàbregas AG (2012) Hybrid RF/FSO communications. In: Arnon S, Barry J, Karagiannidis G, Schober R, Uysal M (eds) Advanced optical wireless communication systems. Cambridge university press
31. Božanić M, Sinha S (2019) Emerging transistor technologies capable of terahertz amplification: a way to re-engineer terahertz radar sensors. Sensors 19:2454
32. Khalighi M-A, Gabriel CJ, Pessoa LM, Silva B (2017) Underwater visible light communications, channel modeling and system design. Visible Light Communications. CRC Press, pp 337–72
33. Cochenour B, Mullen L (2012) Free-space optical communications underwater. In: Arnon S, Barry J, Karagiannidis G, Schober R, Uysal M (eds) Advanced optical wireless communication systems. Cambridge Univ. Press, Cambridge, pp 201–239
34. Fehér G, Udvary E (2016) VLC-based indoor localization. In: Uysal M, Capsoni C, Ghassemlooy Z, Boucouvalas A, Udvary E (eds) Optical wireless communications: an emerging technology [Internet]. Springer International Publishing, Cham [cited 2020 Jun 27], pp 609–22. https://doi.org/10.1007/978-3-319-30201-0_28
35. Katz M, Ahmed I (2020) Opportunities and challenges for visible light communications in 6G. 2020 2nd 6G Wireless Summit (6G SUMMIT), pp 1–5

Chapter 6
6G: The Green Network

Abstract One of the primary visions associated with 6G is the vision of the green 6G network. In this chapter, the ongoing efforts to achieve energy efficiency in future network generations will be discussed. At present, energy efficiency is already starting to take shape in 5G, but it is expected that it will only be with 6G that a truly green network will come into existence. The discussions in this chapter start with an expanded overview of the green communications concept. This is followed by an investigation of the historical wireless network consumption figures, research into present and envisioned techniques for energy efficiency in 5G and 6G networks respectively, as well as into the energy-saving contribution of the VLC that is proposed as an alternative to electromagnetic communication for 6G. The remainder of this chapter then looks at the most promising energy-harvesting, energy transfer and energy storage techniques for future networks.

One of the primary visions associated with 6G is the vision of the green 6G network.

In today's world, green is associated with sustainability, pollution reduction, carbon footprint reduction, deterring of climate change and related aspects. Simply put, being "green" means that one is using available energy efficiently. However, recall that starting with 5G, the amount of information that is transmitted and the data rate at which transmission occurs are enormous. Data rates of 5G are at least 100 times higher than those achievable in 4G and the data rates of 6G will be a few orders of magnitude higher than even those of 5G. At the same time, the number of users increases with every wireless generation, with many of the users nowadays being machines rather than humans, as seen in Chaps. 1 and 2. Hence, the energy consumption required to generate and transmit all the expected information is increasing dramatically with each wireless generation. As a result, future wireless networks are on track to become anything but green. In fact, future wireless networks could become the highest energy consumers (and carbon generators) of the future.

It is obvious that this scenario is not practical or sustainable. The solution is, however, simple: the energy required to transmit a piece (bit) of information must

M. Božanić and S. Sinha, *Mobile Communication Networks: 5G and a Vision of 6G*, Lecture Notes in Electrical Engineering 751,
https://doi.org/10.1007/978-3-030-69273-5_6

be decreased so that the overall power[1] consumption of future networks, despite the increased traffic flow, mostly stays on par with that of previous generations. Mathematically looking, energy per bit, which can be described with a metric that has units of joules per bit, needs to decrease at least at the rate at which the data throughput increases. In general, this has not been the case to date and the rate of efficiency improvement has been slower than the traffic growth rate [1]. However, it is widely accepted that if future wireless networks maintain the status quo of the said energy consumption, they would be considered "green". Naturally, the concept of "energy-per-bit" is fairly abstract. Instead, the idea is that the energy required to transmit and receive chunks of information is decreased, and, if possible, effectively eliminated, leading to an average energy consumption decrease in the whole network. In practice, this is achieved typically in two ways: (1) reducing the conventional energy consumption by energy saving within the architecture and the hardware of the network, both on the side of the base station and on the side of the remote user, and (2) harvesting the freely available (renewable) energy (e.g. from the sun) to reach a net effect of low or even zero energy consumption [2]. In a hybrid approach, the overall net effect of energy reduction can also be achieved if some of the energy that would normally be wasted during information generation or transmission is harvested elsewhere, typically on the side of the receiver.

Whereas energy efficiency has been the focus of wireless technology researchers from the inception of wireless communications, the energy-harvesting concept is somewhat newer and is becoming ever more important with every wireless network generation. This is because the number of devices that are connected to the network is constantly increasing. Many of these devices are battery-powered, for example cellular telephones or remote sensors for IoT, and the capacity of batteries is only increasing at a rate of 1.5 times per decade [3]. As the number of remote nodes increases, a question that comes up is: when does it become impractical to recharge batteries from the mains in the case of cellphones, or to replace batteries in the case of wireless sensors? The belief of 6G researchers is that in many instances, batteries could be recharged by harvesting the available ambient energy, or even completely eliminated by purposefully transferring the wireless energy directly from the base station to the node. Thus, wireless energy transfer is steadily becoming a particularly important subcategory of energy harvesting.

Energy saving and energy harvesting are the primary steps towards becoming energy-efficient. However, if the scope of energy consumption analysis is expanded on the verticals that have been envisioned for 6G, a bigger energy picture can be built and secondary energy-saving techniques could be identified. For example, 6G will become the backbone of the smart city concept, which itself is imagined as a green or energy-saving concept, where various networked sensors detect the changes in the environment and adjust heating, cooling, lighting and other energy-consuming aspects optimally, leading to lower energy consumption in homes, offices and whole cities. On that note, even the heat energy generated as a side product of signal transmission could be reused elsewhere in a smart city, which would result in less

[1]Recall that $P = E/t$, where P is power, E is energy and t is time.

strain on the base-station air-conditioning. In another example, in V2X networks, energy-consuming vehicles receive traffic information and communicate with other vehicles and as a result, they can move at optimal speeds and via optimal routes, with their acceleration, deceleration and idling time minimized, which has the net effect of optimal fuel usage and lower energy consumption [4].

Moreover, in emerging economies, many of which can be found in Africa, the energy to run future wireless networks may not even be readily available in the first place. Even South Africa, which is typically considered one of the more developed countries in Sub-Saharan Africa, experiences rolling blackouts regularly [5]. In these scenarios, green, sustainable energy solutions with energy-harvesting initiatives could be the only be scenario that would work in practice for deployment of networks beyond 5G.

To put everything discussed so far into perspective, in this chapter, the ongoing efforts to achieve energy efficiency in future network generations will be discussed. At present, energy efficiency is already starting to take shape in 5G, but it is expected that it will only be with 6G that a truly green network would come into existence. The discussions in this chapter start with an expanded overview of the green communications concept. This is followed by an investigation of the historical wireless network consumption figures, research into present and envisioned techniques for energy efficiency in 5G and 6G networks respectively, as well as into the energy-saving contribution of the VLC that is proposed as an alternative to electromagnetic communication for 6G. The remainder of this chapter then looks at the most promising energy-harvesting, energy transfer and energy storage techniques for future networks.

6.1 Overview of the Green Communications Concept

Consider a network that is designed for the best utilization of available spectrum resources, but with green communications in mind. According to the discussions in Chap. 3, the maximum achievable data rate, or the capacity of the channel, is governed by the Shannon capacity limit equation that relates the channel capacity, available channel bandwidth, transmit power and the channel noise. The Shannon equation is repeated here for convenience:

$$C = W \log_2\left(1 + \frac{P}{N_0 W}\right). \tag{6.1}$$

Recall that the units of capacity are bits/Hz. For a network that effectively uses the available spectrum resources, the capacity equation can be expanded to include three factors, namely the spectrum, spectrum efficiency and spectrum reuse [6]:

$$C = \text{Spectrum} \times \text{spectrum efficiency} \times \text{spectrum reuse}. \tag{6.2}$$

In Eq. (6.2), spectrum is a quantity in Hz, spectrum efficiency is in bits/s/Hz and spectrum reuse is a factor that has no units. According to [7], the spectrum efficiency above is quantified as

$$\eta_{SE} = \log_2\left(1 + \frac{P}{N_0 W}\right). \tag{6.3}$$

However, in green communications, researchers are rather interested in energy efficiency, which is expressed in bits/s/W (bits/J):

$$\eta_{EE} = W\frac{\log_2\left(1 + \frac{P}{N_0 W}\right)}{P}. \tag{6.4}$$

With some manipulation, a relationship between energy efficiency and spectral efficiency can be found. Their relationship is independent of the transmit power but shows dependence on AWGN power:

$$\eta_{EE} = \frac{\eta_{SE}}{(2^{\eta_{SE}} - 1)N_0}. \tag{6.5}$$

The result of Eq. (6.5) is important for green communications. This equation states that for a given level of noise power, there is a fundamental trade-off between spectral efficiency and the energy efficiency that can be achieved in a channel, as illustrated in Fig. 6.1. The energy efficiency trade-off applies both on the side of the base station and that of the user equipment, because the user equipment relies on the network configuration [8]. It should be noted here that Fig. 6.1 is applicable only to an ideal case when various other factors such as the transmitter circuit power consumption are disregarded [7]. However, in order to achieve better energy efficiency (move left in the graph), the requirement for the more complex transceiver architecture, among others, necessitates more complex circuitry. If the influences of circuit complexity are included, the energy efficiency versus spectral efficiency curve loses monotonicity and looks something like that shown in Fig. 6.2. In this case, there is only one point or one region where optimum energy usage is possible.

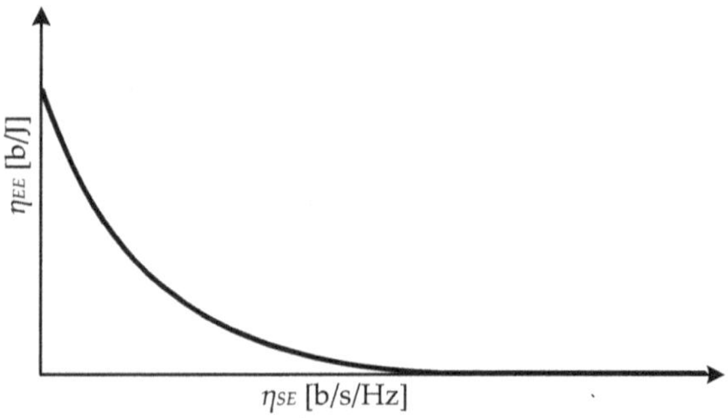

Fig. 6.1 Fundamental trade-off between spectral efficiency and energy efficiency in wireless networks

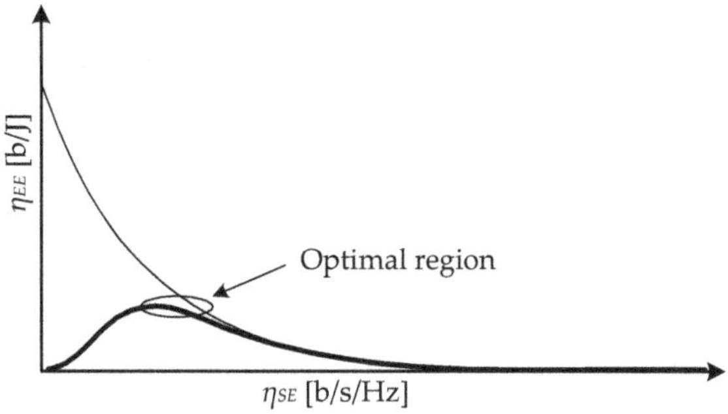

Fig. 6.2 Trade-off between spectral efficiency and energy efficiency in wireless networks when non-ideal effects are taken into consideration

In summary, the Shannon capacity limit states that energy efficiency cannot be improved indefinitely by only applying spectrally efficient communication technologies [9]. Note that this is an additional finding to that in Chap. 3, which was that the Shannon capacity limit stated that the only feasible way to increase the data throughput was to increase the transmit power and bandwidth simultaneously. Consequently, the demand for an increased data rate can easily diminish any gains that are achievable by spectral and energy efficiency optimization. As a result, improving the energy efficiency of a network is only one of several steps towards solving the overall power consumption problem to maintain sustainable wireless technologies. Simultaneously, a shift towards more efficient transmitter and receiver topologies is also necessary, given that a lot of energy is wasted in signal processing and amplification, particularly power amplification, as discussed in Chap. 4. Another possible step may be a move towards VLC, which is not hindered by the Shannon capacity limit, as a way to complement RF communications, wherever such a solution is possible (for example in indoor environments, as discussed in Chap. 5). Lastly, it is not too early to start exploring the availability of green energy and energy conservation.

The relationship between these different green network enablers is shown in Fig. 6.3, which assists in building an overview picture of all the aspects discussed in this chapter. Starting from Sect. 6.3, various green network enablers shown in this figure will be discussed in more detail. Before this is done, however, a brief review of typical power consumption figures of various network components, as well as those of different transceiver system blocks, in past and present network generations, will be completed.

6.2 Power Consumption Analysis of Past and Present Network Generations

The power consumption of wireless networks has been increasing steadily with every generation. This trend has been summarized quantitatively by Abrol and Jha [3] and is partially reproduced here in Table 6.1, where the average power density of the network, as well as the power usage of the base station and a typical mobile user

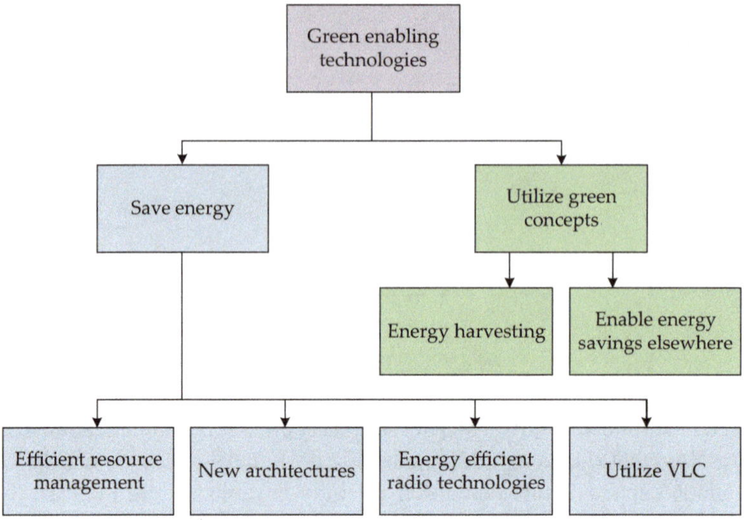

Fig. 6.3 Main avenues leading to the green 6G network concept

Table 6.1 Power trends observed with the wireless network evolution (adapted from [3]

Generation	Frequency band (MHz)	Power density (W/m^2)	Base station power level	User device power level
1G	800	4	Low	Low
2G	900/1800	4.5–9	14–46 dBm	24–39 dBm
3G	900/1800/2100	4.5–10	24–38 dBm	21–33 dBm
4G	1800/2600	10	43–48 dBm	23 dBm

device, is shown. In principle, as expected, the analog 1G network had the lowest power density and the lowest power consumption on the side of the base station as well as on the side of the user device. The exact power transmission levels associated with 1G are not shown in Table 6.1, since they are irrelevant to this discussion owing to the fully analog nature of this network. Moving away from 1G, in 2G, 3G, and 4G networks the largest increase occurred in the power density, which jumped from 4 to 10 W/m^2. The power at the base station also increased: earlier 2G setups consumed as little as 14 dBm, whereas new 4G deployments can utilize as much as 48 dBm. What is interesting to see is that the user devices have become more power-conservative over time, with 4G devices consuming 23 dBm, while older 2G devices consumed 39 dBm. This was necessitated by the fact that power had to be conserved so that it could be used to power up other components that were included in the mobile device with the network evolution, such as large screens, cameras and other networking capability such as Bluetooth and WiFi, while still using batteries with limited power capacity. With all the additional features typically associated with user devices, it

Table 6.2 Comparison of power consumption of 4G and 5G base stations (approximate figures, from [4])

Network	Bandwidth (MHz)	Transmission power (W)	Baseband power (W)	Radio unit power (W)	Total power (W)	Max data rate (Mb/s)
4G LTE	20	40	150	950	1100	120
5G NR	100	240	220	4080	4300	2000

makes sense to focus the power consumption analysis on the side of the base station initially.

In the power analysis presented above, the power consumption figures for 5G have not been included. A comparison of the power consumption between 4G and 5G can be found in the paper by Chih-Lin et al. [4] and is shown in Table 6.2. This analysis shows a massive jump in power consumption that is observed with the move from 4G LTE to 5G NR. With the increase in bandwidth from 20 to 100 MHz, the transmission power increases from 40 to 240 W. The total amount of power required to reach 240 W of transmission power is some 18 times higher than that and is almost 4.3 kW. Thus, it is not surprising that the radio unit is the largest power consumer of the whole base station. However, the baseband processing power consumption contribution was still significant in 4G, with about 14% contribution, whereas the power consumption of the radio unit in 5G has increased so much that the baseband processing power contribution in 5G is now almost negligible at 5%. In the radio itself, the largest energy consumer is the power amplifier. The second largest energy consumer is the air-conditioning required to keep the base station equipment cool enough to operate properly. Specifically, whereas about four-fifths of the total radio power is attributed to power amplification, at least one-fifth of the total power is spent for air-conditioning [10, 11], which by far surpasses the power required in other parts of the base station, for example in the circuitry needed for baseband signal processing.

Note that one important factor was omitted in the power analysis in Table 6.2. From this table, it appears that the total power consumption of the 5G base station is about four times that of the 4G. However, 5G base stations need to be packed much closer together. Roughly, three 5G base stations are needed to cover the area of one 4G station, meaning that the overall power consumption of the 5G network is at least 12 times that of the 4G network [4].

What may not be very obvious is that the power consumption in the receiver is also increasing with the network evolution. Although these power figures may seem absurd when comparing them to the numbers presented above for the signal transmission path, they could become important for future battery-operated user devices (such as remote sensors). In [12], Skriponis et al. experimentally compared the receiver power usage for operation at two different frequencies: 28 GHz, a frequency feasible for 5G, and 140 GHz that may become feasible in 6G. This comparison is shown in Table 6.3. From this analysis, it is evident that if a fully digital architecture is used, a

Table 6.3 Power consumption estimates for digital and analog receiver architectures at two different operating frequencies (adapted from [12])

Component power consumption (mW)	Fully digital		Analog	
	28 GHz	140 GHz	28 GHz	140 GHz
LNA	9.05	325	180	6515
Mixer and LO	80	6272	20	196
ADC	8.18	261	16.4	65.4
Filtering	57.52	1022.8	22.72	45.44
Total	155	7881	239	6822

50-fold increase in power consumption is observed if the operation is shifted from 28 to 140 GHz for a fully digital system, which can be decreased to a 30-fold increase if an analog architecture is deployed. In the fully digital implementation, at 140 GHz, the largest amount of power is lost in the mixer, whereas in the fully analog implementation, it is the LNA that is the highest power consumer. In either case, authors argue that the power consumption at these frequencies is still prohibitive for small devices and that further research is needed if operation at such high frequencies is to become feasible in future.

The overall picture then is that the energy consumption path at present is following the black curve (line a) in Fig. 6.4 despite the energy-saving efforts that have been deployed in previous wireless network generations. This is determined by the requirement for a high data rate and unless there is a paradigm shift towards the green approach discussed in the opening of this chapter, it will become impossible to keep

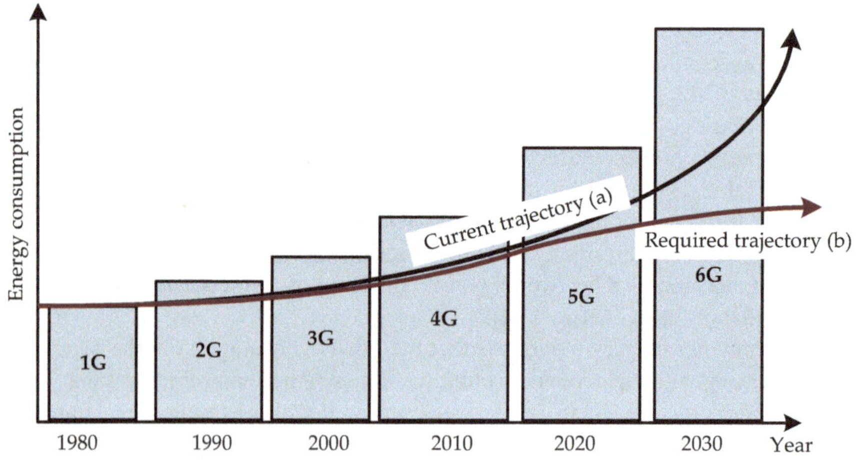

Fig. 6.4 Current trajectory of power consumption of wireless networks as predicted by network traffic and other factors (line marked a) and the required trajectory (line marked b) (adapted from [3])

up with this energy increase trend. Thus, there is no other direction than to make every effort rather to stay on the light curve (line b) [3], as detailed in the remainder of this chapter.

6.3 Energy-Saving Methods in 5G and 6G Networks

Starting with 5G, research into the energy efficiency of wireless networks has received a massive boost. This is evident from a large number of research articles on this topic, published over the last several years [1–4, 6, 8, 9, 11, 13–19]. Before expanding on seminal findings, it should be noted that this section considers only the energy-saving methods applicable to electromagnetic (radio) communications, and that the reader will be remined of the VLC as an effective method for energy saving only in the following section. Furthermore, any discussions on wireless energy transfer or energy harvesting will be omitted at present and will be handled late in Sect. 6.5.

In general, most researchers agree that the techniques for energy saving can be classified into several categories. Abrol and Jha [3] argue that three such categories are:

- deploying energy-efficient architecture;
- utilizing energy-efficient resource management; and
- introducing energy-efficient radio technologies.

Energy-efficient architectures could include strategies such as optimization of the cell size, relaying or cooperative communications. Energy-efficient resources may include techniques such as joint power and resource allocation, utilization of mMIMO or packet scheduling. Energy-efficient radio technologies could include heterogeneous network deployment or the utilization of higher frequency (millimeter-wave and terahertz) bands. However, the distinction between different types of energy-saving techniques is not always completely obvious and there are techniques that can easily belong in multiple categories or stand outside of the three mentioned categories. Hence, in this section, various techniques will be discussed in the order of their importance for the evolution of 5G and for 6G and irrespective of their classification.

6.3.1 New Device and Circuit Research

The first place to begin looking for power saving is inside the transmitter and receiver circuits themselves.

As seen in Sect. 6.2, the largest amount of power is lost in the transmitter, specifically for power amplification and the removal of the generated heat. This is due to, typically, the low efficiency of power amplifiers, as discussed in Chap. 4. As new, faster devices, improved device technologies or even innovative circuit topologies become available, it may be possible to decrease some of the unnecessary losses that occur in the power amplifier itself, as well as in any other circuitry used by any transmitting and receiving device in these networks. As faster devices become available as a result of technology scaling, the complexity of high-frequency (millimeter-wave and terahertz) circuits also has a tendency to decrease (e.g. harmonic generation is not required anymore), resulting in more power in optimal circuits. Furthermore, research into more optimal packaging that minimizes losses and into better, more efficient and more focused antennas is another avenue of exploration to achieve energy efficiency.

An approach that has been proven to work well for energy saving in wireless networks, at least when it comes to power-hungry amplification in the transmitter, is the utilization of Doherty and envelope tracking approaches in building power amplifiers [1]. These advanced power amplifier topologies give the amplifier the capability to transmit different powers at maximum efficiency most of the time during which the amplifier is used. Some of these techniques have been discussed in [20] and [21]. To extend the level of energy saving further, it is also a good idea to keep the entire system in the sleep (low-power) mode for a portion of the time when the traffic demand is low, as described in Sect. 6.3.6. Note that the cooling of the power amplifier is also an energy-demanding affair, meaning that new and innovative cooling and air-conditioning techniques may contribute to energy efficiency as well.

It should not be forgotten either that the receiver circuitry also plays its part in energy efficiency. Having more sensitive receivers or receivers that are less prone to the influence of noise (e.g. owing to advanced LNA topologies) means that the signal may be transmitted at a lower power level than if less sensitive and less noise-sensitive receivers are used. This consequence of the link-budget equation (Eq. (3.10)) alone may introduce significant power savings in any wireless link.

6.3.2 Utilization of Higher Frequency Bands

The utilization of higher frequency bands—the millimeter-wave band for 5G and terahertz band as envisioned for 6G—has proven to introduce an abundance of bandwidth into the picture. What may not be that clear is that the transmitter operation at these high frequencies can also contribute to the power-saving efforts of future green networks.

The explanation for the above statement must be sought in the propagation characteristics of millimeter and terahertz waves. Recall the two related characteristics of these waves: narrow beam widths and directivity (Fig. 3.10). The narrow beam width that is brought about by the frequency increase means that beamforming is necessary to establish a link between the receiver and the transmitter. Such narrow beams tend to consume less power than when the power is radiated in all directions, which is a characteristic of wide beams operating at lower frequencies [3]. Furthermore, directional beams tend to decrease signal interference (in ideal conditions), which also ensures that energy is not lost elsewhere in the signal path. According to [2, 9], hybrid beamforming (a combination of the analog and digital beamforming approach) is the most efficient type of beamforming because such beamforming can reduce the number of RF chains in the system.

Utilization of higher frequencies calls for deployment of large antenna arrays, as well as utilization of the mMIMO concept. Fortunately, these architectures can also contribute to power saving, as discussed next.

6.3.3 *Massive MIMO and Large Antenna Arrays*

MIMO is the key technology used to increase network throughput, spectrum efficiency and reliability and has been used since 4G [3, 6]. In MIMO, the same signals are sent through different paths between transmitter and receiver antennas. MIMO also involves multiplexing of the gain by transmitting independent signals in parallel through spatial channels. The use of MIMO necessitates the use of large antenna arrays. The complexity of the array increases in 5G and 6G as more and more multiple beams are added, leading to the more advanced mMIMO concept. The concept of MIMO is illustrated in Fig. 6.5 and will be described in more detail in Chap. 7.

The bulk of energy saving brought about by mMIMO is on the side of the user device, rather than on the side of the base station, although both the base station and the user device have to be equipped with a large number of antennas. By densifying the number of antennas on both sides, a large number of users can be served with the same time-frequency resource [2]. In addition, better angular resolution is achieved and interference is reduced [18]. The overall result of this is a significant reduction in the transmission power of the user device. Programable metasurfaces, also discussed in Chap. 7, can contribute to power saving efforts in a similar manner.

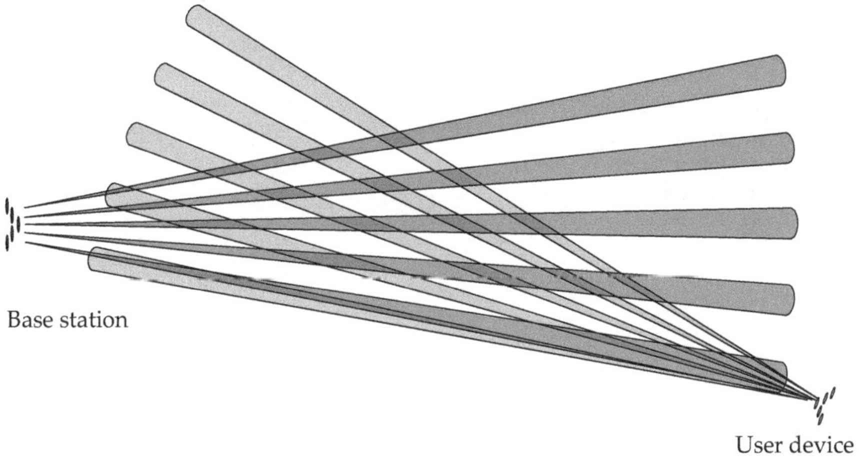

Fig. 6.5 Base station and user device communicating via MIMO

Note, however, that there is a very important trade-off that needs to be considered in this approach: the increased circuit power and processing power needed to feed different antennas could increase to the point where instead of improving energy efficiency, energy efficiency is deteriorated instead [4]. Thus, for MIMO to be effective in energy saving, cheap but power-efficient amplifiers are often needed to ensure that this trade-off remains feasible [6]. However, this is generally not an issue, since the parallelization of power amplifiers is known to result in more effective power amplification than when just a single high-power power amplifier is utilized [1].

6.3.4 High Base Station Density, Relaying and D2D

Increasing the base station density (and thus reducing the cell size), introducing relaying into the network and exploiting the D2D concept (also relaying, but using user devices) are all means of shortening the link between the transmitter and the receiver [9]. If the transmitter needs to overcome a shorter distance, whether going from the user device to a base station that is nearby, from the device to a relay, from relay to relay or from D2D, it can transmit the signal at the lower power level, which results in overall energy saving. The concept of relaying is shown in Fig. 6.6. Here, once again, a trade-off needs to be found between energy saving that can be accomplished by decreasing the link distances against the power cost that is introduced by operating more cells or relays within the same area. 6G, which will be designed predominantly as an ultra-dense network, is expected to benefit greatly from the energy efficiency that can be achieved by using shorter links [16].

An additional potentially useful benefit from the D2D concept is active user cooperation. In addition to acting as relay points, D2D-networked devices can serve as local clusters for multicast transmission or as local caching devices for content exchange [9].

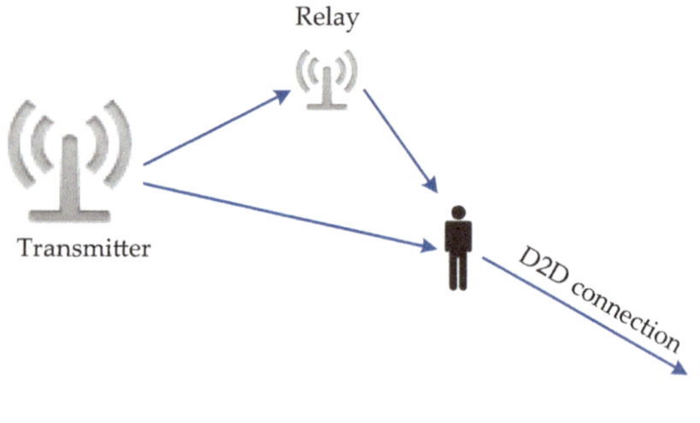

Fig. 6.6 Relaying concept in wireless networks, including signal retransmission via a dedicated relay node as well as via another user device

6.3.5 Non-orthogonal Multiple Access and Other Resource-Sharing Types

As seen in Chap. 3, NOMA is a promising candidate for improving the overall throughput in the network under the condition of a large number of users sharing the same resources. NOMA is seen as a technique that improves spectrum efficiency by allowing a certain degree of multiple-access interference at the receiver [6]. It has been speculated that energy efficiency can also be dynamically improved in NOMA. This would be achieved by dynamically regulating the bandwidth to comply with both constraints of spectral efficiency and energy efficiency simultaneously [13], which is possible because the extra user interference, while degrading the performance, decreases the energy usage [22].

NOMA is not the only resource-sharing approach that has been suggested in recent years. Energy-efficiency improvement can also be observed if the network resources and infrastructure are shared by different network operators [13]. The degree of sharing may vary from partial to full sharing, but the effectiveness of the sharing may of course depend on security and dependence on the proprietary equipment. However, by collaboration, operators are empowered in planning and executing future network deployments based on the network demands in a specific area. Lastly, the problems of either over-provisioning or under-utilization of the deployed network infrastructure that would arise if all operators deployed their equipment separately are avoided.

6.3.6 Advanced Network Sleep Modes

The introduction of so-called sleep modes into the network architecture is another way of saving energy [19]. It has been determined that the base station is lightly loaded about 80% of the time, but despite this, it almost always consumes peak power—this is true at least for elements such as air-conditioning and power-amplifying circuits [9]. This calls for various strategies to be implemented to switch off the base station for periods of time when traffic is light. For this to happen, the dynamics of the traffic load across time and geographical locations need to be exploited, whether by machine learning or otherwise. For example, in dense base station deployment, some base stations can be sent to sleep during off-peak times but could be switched on during peak traffic demand times.

A more efficient approach is keeping the nodes on only when needed and keeping them asleep otherwise [13]. As much as 60% of energy saving can be accomplished in this manner. To achieve this, however, interactive monitoring of power consumption, traffic load and other parameters is needed, which then allows for activation and deactivation of nodes based on a number of pre-set rules [1]. With flexible reference signal design, it is even possible to introduce different levels (depths) of sleep into the base station design [8], each with different duration and different activation and deactivation profiles, collectively lasting from 100 μs to several seconds [1]. This

Fig. 6.7 Different energy-saving levels for different sleep modes and different traffic demands

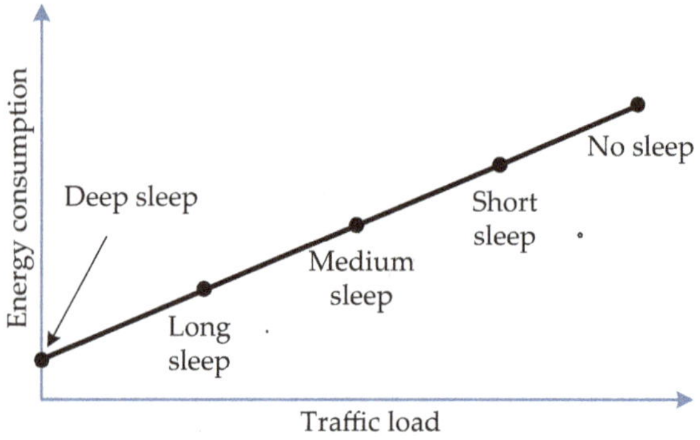

leads to different energy-saving levels for different data traffic demands, which can be pictured as illustrated in Fig. 6.7.

6.3.7 *Energy-Efficient Architecture*

Many aspects of energy-efficient architecture for future wireless networks have been discussed already. For example, the energy-efficient architecture may be the one that utilizes relaying to shorten the distance over which a user device needs to communicate. However, there are other examples of architectural approaches for energy saving. One that has been specifically mentioned in the literature is to separate indoor and outdoor network cells [3]. This separation is accomplished by introducing one or more base stations inside a building where there are potentially many network users, as illustrated in Fig. 6.8. This is a scenario that can work for a large office building, a

Fig. 6.8 Outdoor base station feeding indoor base stations as a way to optimize network architecture for power efficiency

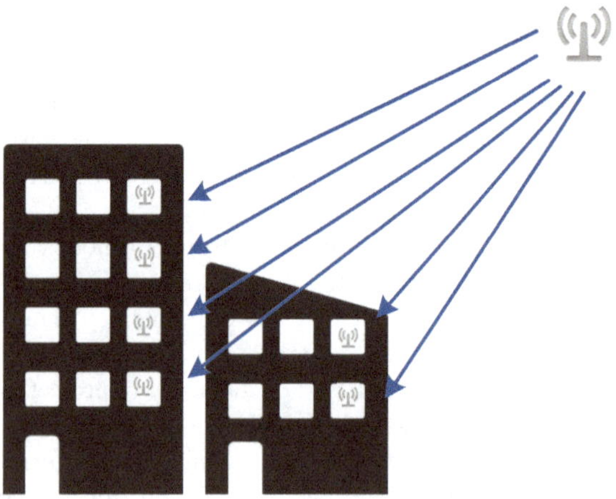

hotel or a sports hall. The energy saving is brought about by noting that building walls tend increasingly to block or heavily attenuate radio signals, especially as frequency increases (i.e. when millimeter-wave or terahertz frequencies are utilized). With an indoor base station, mobile network signals are exposed to less blockage, and neither the base station nor the user device has to transmit at high power levels to overcome obstacles.

6.3.8 Smart Energy Resource Management

Smart energy resource management is a mechanism that could be deployed by future networks to optimize the balance between energy demand and energy availability dynamically [16]. It makes sense to include this technique when the network relies heavily on energy harvesting of resources that provide energy only when a particular type of renewable energy is available. Considering that the energy comes from green resources and the network communicates with what is known as the smart grid, a network that relies on energy resource management can be called a smart grid-enabled network [2]. A smart grid-enabled network simultaneously has to manage factors such as random traffic arrival, dynamic link quality, on-site power storage, wireless energy transfer, on-grid energy price and grid load, among others.

6.3.9 Traffic Offloading

Traffic offloading has already been used to alleviate bottlenecks and balance the load across different wireless access technologies such as 3G, 4G and Wi-Fi [2]. This concept will be inherited by 5G and 6G to achieve a similar set of effects. In the context of green communications, however, it may be more important to note that it can also be used to improve energy efficiency. Traffic offloading rules would be used to reduce the traffic by certain network functions or verticals in instances when a fair amount of energy saving can be accomplished without compromising the QoS. Alternatively, the traffic could be redirected via less energy-consuming network nodes, or more advanced packet-scheduling approaches could be considered [1, 2].

6.3.10 Edge Computing

The last approach to achieving energy efficiency of future wireless networks that will be discussed in this section is edge computing.

A typical approach to computing in the network is to utilize the higher capabilities of cloud/centralized data centers, which allows for sophisticated and detailed data analysis at the expense of transport and processing delays [17]. On the other hand, in

Chap. 1 it was deduced that a data latency reduction can be achieved if computing in 6G is moved from a central location to the edge of the network. One of the inherent advantages of such an approach is that data do not need to be transferred to and back from the central location; instead, pushing of the data happens only locally. As a result, the overall energy consumption of the network reduces, making edge computing an important promoter of the green network concept [2, 13]. Furthermore, data caching can happen at the edge of the network, thus achieving a similar effect.

However, computing at the edge of the network may not be fast or accurate enough for certain applications. Rather, the network will need to be optimized in such a way as that, at the edge, fast and localized data processing, analytics, and content caching serve critical applications, thus saving energy, while more in-depth data analysis is still carried out at the central location at the expense of increased energy consumption [17].

6.4 Reflection on Visible Light Communications

Section 6.3 looked at the energy-saving techniques in 5G and 6G networks, but only considered the energy saving that could be accomplished if all network traffic is exchanged via radio links. However, a whole chapter in this book (Chap. 5) was dedicated to VLC as an alternative to radio communication. In this chapter, the hardware that makes VLC possible was analyzed and it was seen that transmitters in optical networks are in general much more energy-efficient than radio transmitters. They do not transmit high power levels; consequently, much less power is lost via heat or in the process of heat removal. Moreover, the capacity or the maximum data rate that is achievable via any form of optical communication is not bound by the traditional Shannon capacity limits, which is the primary reason for promoting VLC.

On the other hand, the propagation of visible light waves is blocked by any object, and therefore deployment of VLC is limited to only certain scenarios, such as inside offices or in well-lit streets. However, if radio communication in 6G is replaced with VLC wherever possible, massive energy savings are possible. The result is a hybrid network, where both types of communication co-exist. While the primary goal of such a network is to unlock the ultimate data rates, and as such, efficient network resource management is one of the critical requirements, the network should be designed such that energy efficiency is considered in resource management as well. As a result, many of the energy-efficiency strategies discussed in Sect. 6.3 remain applicable, with some exceptions. The following is possible if VLC is included:

- A new device and circuit research paradigm can be extended towards device and circuit research in the context of VLC, e.g. towards more sensitive photodiodes and photodetectors, or more efficient LEDs.
- High base station density continues to promote energy savings even when light as a communication medium is utilized.
- Resource sharing remains applicable to both light and radio infrastructure.

- Energy-efficient architecture can take into account deployment scenarios that include both VLC and radio communications, not only radio communication as originally discussed in Sect. 6.3.
- Smart energy resource management is easily extended to VLC.
- Traffic offloading may now include paths that utilize both radio and VLC links.
- Lastly, the edge computing paradigm is as applicable to VLC as it is to radio communications.

Thus, the ultimate green future wireless network will be the network that efficiently utilizes both types of communication—electromagnetic and VLC. However, additional room for improvement, on both accounts, lies in the addition of energy harvesting to offset the excess energy usage that seems unavoidable as traffic increases. Energy harvesting may also include the wireless energy transfer that can be achieved in both RF communications and VLC. It is discussed next.

6.5 Energy Harvesting

Energy harvesting is expected to play a major role in future wireless generations as an aid to overall energy saving, not only in wireless sensor networks, but in society as a whole. Energy is harvested either from the ambient environment (from renewable sources), or from a dedicated power transmitter in a cableless way [22]. There are several benefits of energy harvesting, many of them already discussed earlier in this book, but it does not hurt to mention them again. First, with energy harvesting, the extent of reliance on batteries or mains power, especially in user devices such as smartphones or sensors in IoT, is decreased. This adds to sustainability of such device deployment and to convenience (e.g. no replacing of batteries at remote sensor nodes is needed). Complementing traditional power supplies with energy that was harvested is not a new concept—recall that, for example, calculators that can be powered by both batteries and solar power have been on the market for decades. Secondly and more importantly for this chapter, energy harvesting has the best possibility of becoming the most important energy-saving contributor in 6G.

The major hinderance to the wide adoption of energy harvesting is the reliability of each of the green energy sources. For example, if solar energy is harvested, harvesting can only happen during daytime, and the amount of energy that can be harvested in every 24-h period depends on the latitude and the season (length of the day) and present weather conditions. The same can be said for wind, except that its appearance is even more unpredictable. Harvesting of kinetic energy, on the other hand, depends on the movement of an object, which is also usually random in nature. A notable exception can be made for RF energy harvesting: RF energy in the 21st century is omnipresent and could be harvested universally, by day or at night. However, the problem with RF energy harvesting is the amount of energy that can be harvested with this approach, unless the RF energy harvester is part of a system deliberately designed for wireless energy transfer. A plausible scenario for this is a base station that transmits both RF energy and RF information to the user. In 5G, the RF energy could be in the form of millimeter-waves and in 6G, even terahertz wave harvesting could be considered. Alternatively, energy could also be transferred in VLC systems.

In general, renewable energy sources can be classified into:

- Uncontrollable and unpredictable;
- Uncontrollable but predictable; as well as
- Controllable and partially predictable.

Mechanical vibration or wind are examples of uncontrollable and unpredictable energy sources. Solar energy can be regarded as uncontrollable but predictable for the most part. Wireless energy transfer either by electromagnetic radiation or light could be classified as controllable and partially predictable. A question that may be asked here is why there is no controllable and fully predictable category in energy source classification. This is because there is always at least partial uncertainty about the availability of the energy source—the energy source can be lost intermittently, for example because of random blockages or by channel fading.

According to the type of energy that is harvested, the energy sources are classified into

- Mechanical energy sources;
- Solar or light energy sources; and
- Electromagnetic energy sources.

Mechanical or kinetic sources include sources such as wind, piezoelectric and electrostatic sources. Light energy sources can come from the sun or from artificial light energy sources. Lastly, the most viable electromagnetic energy source is the RF source. Table 6.4 shows the typical amount of power than can be harvested from each energy source. It is clear from this table that the highest amount of green energy comes from the sun.

In many scenarios, it is possible to deploy multiple energy-harvesting techniques in parallel. There are even materials that are capable of simultaneous energy harvesting from multiple sources [23]. Wireless networks in general should never rely on only a single source of energy (renewable or otherwise) in the first place;

Table 6.4 Various energy sources and typical amount of power harvested from each source [22]

Source	Energy type	Typical amount of power harvested
Solar	Light	100 mW/cm^2 (sunny day)
Indoor light	Light	0.01–0.1 mW/cm^2
Wind (5 m/s)	Kinetic	0.38 mW/cm^3
Piezoelectric	Kinetic	0.2–0.4 mW/cm^3
Electrostatic	Kinetic	0.05–0.1 mW/cm^3
Ambient RF	Electromagnetic	0.2 nW–1 $\mu W/cm^2$

powering should rather always be possible from multiple sources, e.g. from solar energy, wind, batteries or even from the grid [9].

This section will focus on a discussion of the principles behind light and RF energy harvesting, given that these are the two most feasible energy-harvesting techniques for 5G and 6G. (The latter is owing to the fact that deliberate wireless energy transfer is possible.) The remainder of the section will then look at other promising energy-harvesting techniques that may be considered in the near future.

6.5.1 Harvesting Energy from Light

Energy from light is the most commonly harvested type of energy [22] and as such remains appealing for future wireless networks. As described in Chap. 5, electricity is generated from photons from the light source by using photovoltaic cells, which use the principle of photoconductivity. Some of the advantages of photovoltaic energy harvesting include large achievable power densities (especially if the light comes directly from the sun with no atmospheric obstacles such as clouds), easy integration with circuits that require power to be supplied to them, and a steady working status without noise or emissions [23]. When it comes to solar energy harvesting alone, the main disadvantage is the fact that this energy is available only during daytime, and that the amount of energy that can be harvested is heavily dependent on weather conditions. This puts solar energy into the category of uncontrollable but partially predictable renewable energy. Moreover, this energy is only available outdoors, whereas indoors it is typically only possible to harvest light energy from artificially made light sources. The amount of energy, or energy density, that can be harvested indoors is in the region of microwatts per m^2, whereas the energy density of energy harvested outdoors is in the milliwatts per m^2 range. However, light energy harvested indoors has the advantage of being relatively controllable, but it still depends on obstacles and the distance from the light source. A special case of indoor energy harvesting is wireless energy transfer in VLC, which was already described earlier in Sect. 5.6 in Chap. 5 as a way to minimize energy wastage when communicating via light waves.

The principle of photoconductivity was, similarly, described in Chap. 5 and will not be repeated here in detail. This principle of electricity generation is still applicable here because most mainstream solar cells are still of the *pn*-junction type [24]. An important relationship that can be taken over from the discussion in Chap. 5 is that a photocurrent

$$I_{ph} = \eta P_o \frac{q\lambda}{hc}(1 - r), \tag{6.1}$$

is generated if a light appears at the *pn*-junction of the photodiode (Eq. (5.7) repeated here for convenience). If a load is connected to the photodiode, the net current will cause a usable voltage to be generated [22]. The relationship between the voltage and the current flowing out of the photodiode as the current source can be expressed by an I-V relationship represented by the relationship.

$$I = I_{ph} - I_0\left(e^{\frac{V}{V_T}} - 1\right), \tag{6.7}$$

where I is the output current of the solar cell, I_0 is the reverse saturated current of the diode and V_T is the thermal voltage. This means that for a loaded photodiode, the current that is generated is always smaller than the unloaded current I_{ph}. Equation (6.7) is known as the ideal photovoltaic cell equation, but a more realistic current value can be found if factors such as the leakage current and the internal solar panel resistances are also taken into consideration. There are several different approaches to include the non-idealities into Eq. (6.7), but perhaps the simplest is to include the diode "ideality factor" A:

$$I = I_{ph} - I_0\left(e^{\frac{V}{AV_T}} - 1\right). \tag{6.8}$$

In energy research, the important metrics for energy harvesting are the amount of power that can be harvested and the efficiency of the solar cell. First, the power can either be expressed as a P-V curve or as a P-I curve. Power as a function of diode voltage is simply $P = VI$, which, if $A = 1$ results in:

$$P = \left[I_{ph} - I_0\left(e^{\frac{V}{V_T}} - 1\right)\right]V. \tag{6.9}$$

To get to the P-I relationship, some manipulation is needed, resulting in

$$P = V_T I \ln\left(1 + \frac{I_{ph} - I}{I_0}\right). \tag{6.10}$$

Inspection of Eq. (6.10) shows that an optimal point where the power is at maximum (P_{max}) exists. The current I that sets P_{max} can be found by solving the following equation:

$$\ln\left(1 + \frac{I_{ph} - I}{I_0}\right) = \frac{I}{I_0 + I_{ph} - I}. \tag{6.11}$$

In solar installations, tracking techniques can be deployed to tune the parameters of the photovoltaic cell for maximum power delivery at any given time. Secondly, the practical value of solar cell efficiency can be found by dividing the received power by the incident optical power P_o:

$$\eta = \frac{V_T I \ln\left(1 + \frac{I_{ph} - I}{I_0}\right)}{P_o}. \tag{6.12}$$

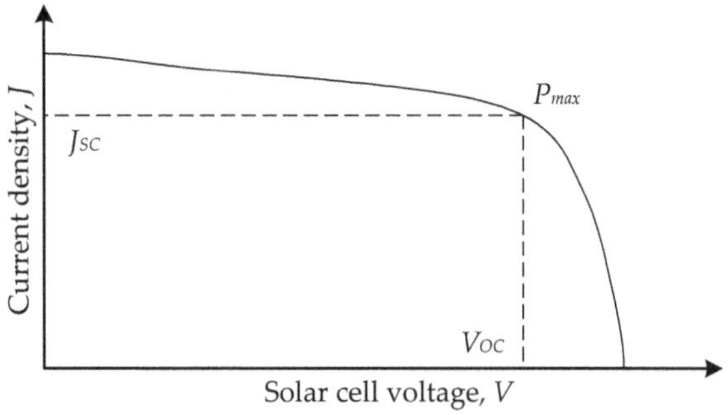

Fig. 6.9 Relationship between current density and solar cell voltage

Some of the other characteristics of the solar cell include [24]:

- Short-circuit current density, J_{SC}, which is current that flows through the external circuit when the electrodes of the solar cell are short-circuited;
- Open-circuit voltage, V_{OC}, which is the voltage when no current flows through the external circuit, given earlier in Eq. (5.16); and
- Fill factor, which is the ratio between P_{max} and the product of J_{SC} and V_{OC}:

$$FF = \frac{P_{max}}{J_{SC}V_{OC}}. \tag{6.13}$$

The relationship between the current density and the solar cell voltage is shown in Fig. 6.9, with points where peak power, J_{SC} and V_{OC}, can be found also indicated on the graph.

The equations in this section are applicable to both *pn*-junction and crystalline Si solar cells [24]. These devices have been used commercially for decades. Over the years, however, research has diverted to other types of solar cells, typically with the aim of better integration into small or monolithic devices. Other types of solar cells that have been considered include amorphous silicon solar cells, $Cu(In_{1-x}Ga_x)Se_2$ semiconductors (CIGS), cadmium telluride (CdTe) thin film solar cells, polymer solar cells and dye-synthesized solar cells (DSSC). The practicality of these devices is investigated in Table 6.5, showing that crystalline silicon remains the first contender in the race for the best material for light harvesting, but that other types of materials are trailing closely behind [24].

A photovoltaic cell cannot operate on its own. At minimum, the photovoltaic system needs a power-management circuit, voltage-regulating unit and an energy-storage unit, such as a battery or a supercapacitor (Sect. 6.6). The energy-management circuit regulates the storing of the energy and the utilization of the stored energy. Voltage regulation is required to deliver a constant voltage from the photodiode voltage that varies over time. Other circuitry, for example that for tracking of the maximum power point, may also exist in the system. A microvoltaic system that deploys all of the mentioned blocks with the aim to supply a device in a wireless

Table 6.5 Performance parameters of different solar cell types [24]

Solar cell type	Power density in the laboratory (mW/cm^2)	Practical power efficiency (mW/cm^2)	Possibility of monolithic integration
Crystalline Si	24	18	Limited
Amorphic Si	14	10	Yes
CIGS	20	12	Yes
CdTe	18	10	Yes
Polymer solar cell	9	5	Yes
DSSC	12	9	Yes

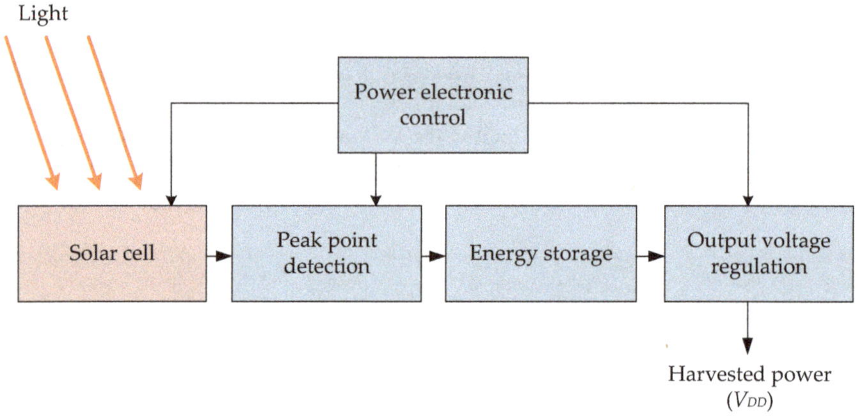

Fig. 6.10 A microvoltaic system for a device in a wireless sensor network

sensor network may be described with the aid of a block diagram like the one shown in Fig. 6.10.

6.5.2 Harvesting RF Energy and Wireless Power Transfer

An alternative to solar power in 5G and 6G networks is the harvesting of RF power. Although the amount of power that can be harvested is extremely small (typically 1 μW/cm^2 [23]) and the RF-to-DC conversion efficiency is very limited [6], this is one of the most promising concepts for future wireless sensor and IoT networks where a reliable batteryless power supply is required, but the amount of energy required may not be too great (i.e. the device is already designed for the optimal usage of energy). This may be true in various instances; for example, sensors that record and transmit their information only occasionally are already common. Since the ambient electromagnetic (RF) power is man-made, even energy harvesting of ambient RF energy can be considered to form part of wireless power transfer (WPT) [9]. If the

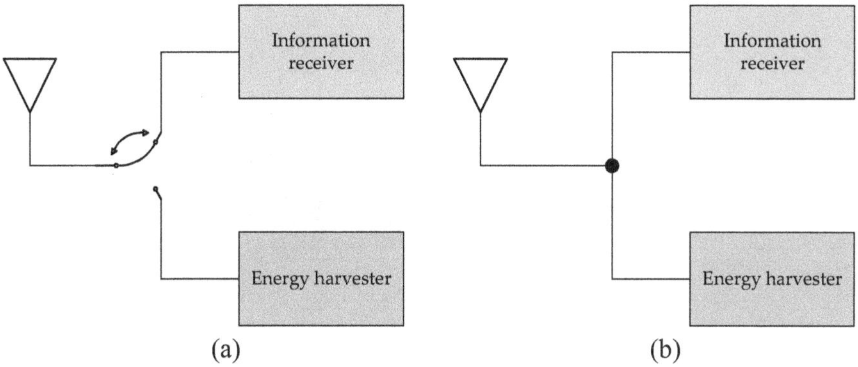

Fig. 6.11 WIT and WPT as separate switched processes (**a**) SWIPT, where the information and power are received simultaneously [15]

excess power is purposefully sent alongside information, with the power beacon and the information transmitter on one side, and the energy harvester and the information receiver, respectively, on the other side, and all form part of a common system, then simultaneous wireless and information power transfer (SWIPT) is achieved [3, 9].

Traditionally, WPT and wireless information transfer (WIT) are two different processes and one device cannot receive both the information and power simultaneously. At the very least, some switching between the regimes is needed, as shown in Fig. 6.11a. In multiband-based wireless communications (5G and 6G), one sub-band can be allocated for power transfer with another sub-band dedicated for information transfer, resulting in true SWIPT, as shown in Fig. 6.11b [15]. In the latter system, WPT design cannot be independent from WIT, requiring a joint optimization of both processes, while considering their particular characteristics [17]. SWIPT is most feasible for far-field low-power applications, similar to wireless sensor networks mentioned earlier. Two more scenarios may be possible: If highly directional transmitting antennas are used for beamforming in the far-field scenario, with highly efficient horn antennas at the receiver, it may be possible to transfer milliwatts of power across distances of several meters [3]. If the distances are limited to about 1 m, the least practical scenario is achieved, where even watts of power may be transferred. The high loss of power with the distance increase is a consequence of the Friis formula (presented in Chap. 3 and repeated here for convenience),

$$P_R = \frac{P_T G_T G_R \lambda^2}{(4\pi r)^2}, \tag{6.14}$$

that indicates that the power decreases at any distance from the transmitter by a square of that distance.

Figure 6.12 shows a simplified block diagram of an RF energy harvester, indicating the three main components: antenna, matching network and a rectifier [22]. In many cases, an energy storage unit is included, also indicated in the figure. Whereas in

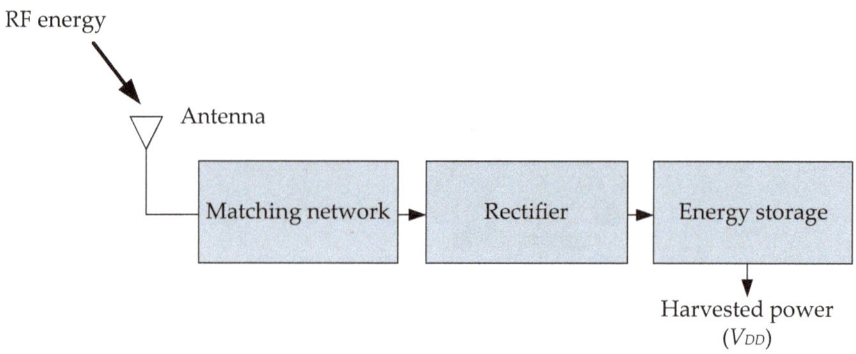

Fig. 6.12 Block diagram of an RF energy harvester

many instances it is possible to remove the energy storage completely, it should be included wherever possible, even if it is a low-capacity storage device with the sole purpose to cater for scenarios where an unexpected break in power transmission occurs (e.g due to blockage, bad weather, etc.).

The antenna is the first component of the RF harvester. Its role is to capture electromagnetic radiation from the air, as in the case of receivers used in information transfer, as described in Chap. 3. Since the signal power reaching the antenna is typically extremely low, the antenna must be designed with efficiency in mind, so that additional power is not wasted in the antenna. As with regular (transceiver) antennas, multiple antennas can be combined to boost antenna efficiency. Usually, antennas are tuned to a single frequency, but it is also possible to harvest energy from multiple frequencies simultaneously to collect more energy.

The second part of the energy harvester is a matching network. The matching network is tuned for the maximum power transfer between the antenna (usually 50 Ω) and the rectifier. Various matching networks are possible, including *R-C*, *L-C* and transformer-matching networks.

The rectifier is the last component in the RF harvester chain, with the role to convert the captured AC signal into a DC voltage. It consists mainly of diodes and capacitors. A very simple rectifier circuit would consist of a single series diode (typically a Schottky diode) that results in half-wave rectification [25], but a better circuit is the full-wave rectifier consisting of four diodes and a voltage regulation capacitor, as shown in Fig. 6.13 [26].

Fig. 6.13 Commonly used rectifier circuit in energy harvesting

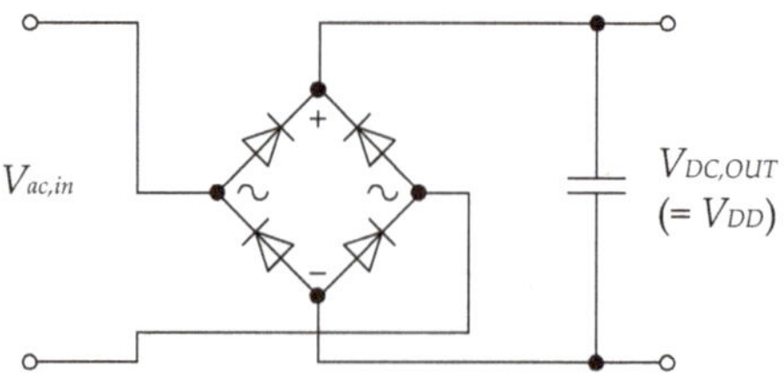

Another block that is not uncommon in the RF energy harvester is the RF voltage-boosting circuit. The RF voltage-boosting circuit can, alternatively, be replaced with a DC-to-DC converter. Either of the two is used to boost or upconvert the typically low harvested voltage to a level that is sufficiently high to power up the required circuitry. The RF boosting circuit is inserted between the antenna and the rectifier, whereas the DC-to-DC converter operates on the other side of the rectifier, on the rectified (DC) voltage. These circuits are usually complex, requiring the use of several components, including large and lossy inductors (in case of an RF boost converter) or multiple capacitors and switches (in the case of a DC-to-DC converter). Switching in the latter case is done with the aid of MOSFETs. Examples of both circuits are shown in Fig. 6.14.

Lastly, unless the device can afford interruptions in the energy supply, the RF energy harvester is equipped with an energy-storage block. Energy storage also requires an energy-management circuit to be included, with the role of controlling the charging or discharging of the battery or the supercapacitor.

The most important metric of the RF harvester is its conversion efficiency [22]. It is expressed as the inverse of the ratio between the incident RF power P_i and the output DC power P_o that exits the energy harvester:

$$\eta = \frac{P_o}{P_i}. \tag{6.15}$$

The achievable conversion efficiency usually differs depending on whether the energy installation is in the near field on in the far field. For near-field energy transfer, efficiency in the region of 80% is achievable, but when it comes to far-field energy harvesting, anything between only 5% and 60% is possible.

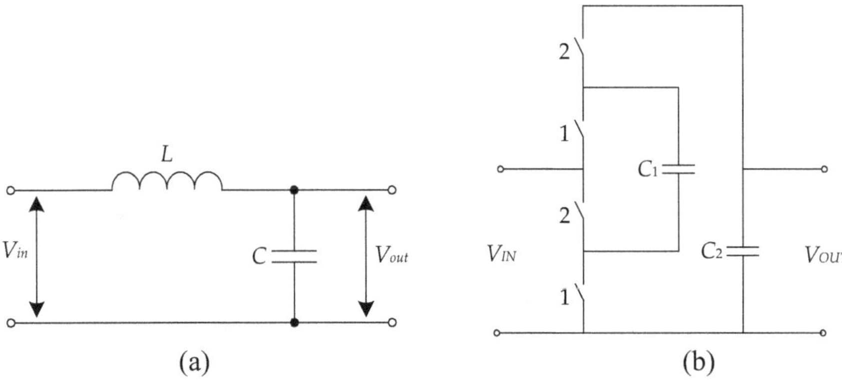

Fig. 6.14 RF voltage boost circuit (**a**) and capacitive voltage converter (**b**). From [25]

6.5.3 Other Harvesting Possibilities

Photovoltaic and RF energy harvesting are ideal energy-harvesting techniques for future wireless networks; however, they are neither the only harvesting possibilities nor universally reliable energy sources. In fact, as the demand for energy increases from network generation to network generation, it makes sense to explore a multitude of alternative energy sources. This includes exploiting principles such as electromagnetic, piezoelectric, triboelectric, electrostatic, electrostrictive, thermoelectric, pyroelectric and magnetostrictive energy harvesting. The explanation of each of these terms, some with a few examples, is given in Table 6.6. The energy build-up allowing for these harvesting principles to be utilized may come from the movement of humans or other objects (e.g during exercising or when vehicles are in motion), fluid flow, atmospheric condition changes (wind or temperature changes) and many other types of influences.

An excellent recent review of different energy-harvesting principles has been completed by Bai et al. [23], indicating that not only is it possible to harvest energy from all of these sources, it is also often possible to combine different techniques to

Table 6.6 Description of various energy-harvesting principles

Energy conversion principle	Type of energy	Description
Triboelectric	Kinetic	Harvesting the energy from the effect where a material becomes electrically charged owing to friction
Electrostatic	Kinetic	Similar to triboelectric harvesting, but the term is used when no friction is required (contactless charge build-up)
Piezoelectric	Kinetic	Harvesting of the electrical potential that is generated across a material when it is strained owing to an applied force
Electromagnetic	Kinetic	Principle of electromagnetic induction used in dynamos, such as those used in wind and water turbines
Electrostrictive	Kinetic	Harvesting energy from electrostrictive polymers (e.g., elastomers) to which applied DC bias electric fields induce statics or polarizations within materials
Magnetostrictive	Kinetic	Harvesting of the energy that is generated by effect where ferromagnetic materials change their shape during magnetization
Thermoelectric	Thermal	Harvesting energy with the aid of materials that can utilize the Seebeck effect (phenomenon in which a temperature difference between two materials produces a voltage difference)
Pyroelectric	Thermal	Harvesting of the energy generated by temperature gradient and fluctuation

harvest energy simultaneously from multiple sources, owing to the fact that in reality a variety of energy sources coexist.

While many of the energy-harvesting techniques have been around for a very long time (e.g dynamos for electromagnetic harvesting), energy-harvesting research remains active and some state-of-the-art results can be found in Table 1 in Ref. [23]. Some of the findings in that table have been summarized here Table 6.7; however, the reader is welcome to consult Ref. [23] for the exact details of the papers where the respective findings have been presented. Note that even some of the state-of-the art energy harvesters presented here are still far from being able to provide energy levels that would be able to find practical use without further research.

Table 1 in Ref. [23] also summarizes multisource or hybrid energy harvesters. Table 6.8 can serve as an indication of what types of hybrid energy harvesters are deemed feasible at present, but possibilities of hybrid energy harvesting are endless, limited only by the system and circuit complexity (see, for example, [27] for some more recently reported hybrid energy-harvesting ideas). Furthermore, this research is still in its very infancy and regrettably, single-source harvesters (for example wind turbines or solar cells) still perform far better than any multi-source harvesters.

Table 6.7 Some of the single-source state-of-the art energy harvesters as summarized in [23]

Energy source	Energy conversion principle	Material/ configuration/ other description	Output power density	Original reference in [23]
Solar	Photovoltaic	GaAs	38.8 mW/cm^2	Ref. 612
Kinetic	Triboelectic	6 kgf pushing force at 5 Hz	0.71 mW/cm^2	Ref. 140
Kinetic	Electrostatic	750 rpm rotary	0.42 mW/cm^2	Ref. 155
Kinetic (vibration)	Electromagnetic	Magnet suspended by other magnets on both sides in a vertical tube	0.61 mW/cm^3	Ref. 614
Kinetic (impact or movement)	Piezoelectric	Semi-transparent flexible PZT ribbon-based nanogenerator, PZT thin film, PET substrate and graphene electrode	88 mW/cm^3	Ref. 616
Thermal	Thermoelectric	Temperature difference of 525 °C	1.02 W/cm^2	Ref. 617
Thermal	Pyroelectric	Temperature fluctuation from −173 to 27 °C	47.4 J/cm^3 per cycle	Ref. 411

Table 6.8 Some of the hybrid energy harvesters summarized in [22]

Energy source	Energy conversion principle	Output power density	Original reference in [23]
Kinetic	Piezoelectric + electromagnetic	2.7 μW/cm^3	Ref. 492
Kinetic	Piezoelectric + triboelectric	13 mW/cm^3	Ref. 619
Kinetic	Electrostrictive + electrostatic	1.8 μW/cm^3	Ref. 496
Kinetic	Electromagnetic + triboelectric	13.8 μW/cm^3	Ref. 509
Kinetic	Triboelectric + electrostatic	0.15 mW/cm^2	Ref. 515
Kinetic	Piezoelectric + electromagnetic + triboelectric	0.75 μW/cm^3	Ref. 307
Thermal	Thermoelectric + pyroelectric	1.5 μW/m^2	Ref. 520
Solar + thermal	Photovoltaic + thermoelectric	11.29 mW/cm^2	Ref. 523
Solar + kinetic	Photovoltaic + triboelectric	–	Ref. 528
Solar + kinetic	Photovoltaic + electromagnetic + triboelectric	–	Ref. 510
Solar + kinetic	Photovoltaic + piezoelectric	34.5 μW/cm^2	Ref. 530
Thermal + kinetic	Pyroelectric + piezoelectric + triboelectric	–	Ref. 538
Solar + thermal + kinetic	Photovoltaic + pyroelectric + piezoelectric	–	Ref. 539

Lastly, the hybrid energy harvesters presented in this table have to be distinguished from multisource energy harvesters that are made from multifunctional materials, that is, materials that are able to harvest energy from multiple energy sources on their own. Only piezoelectric, pyroelectric, and photovoltaic effects can be integrated into the same multifunctional material with the present knowledge.

6.6 Energy Storage

Some form of energy storage—either batteries or supercapacitors—are sometimes added to systems that rely on energy harvesting, which allows those systems—typically wireless sensor devices—to maintain a certain level of QoS [22]. Alternatively, energy storage can be used in other scenarios: as a form of backup power supply, or as an enabler of mobility. Some important differences between batteries and supercapacitors should be highlighted.

Energy is stored in batteries in electrochemical form. One of the characteristics of batteries is that their efficiency for energy storage is not 100% [22]. In other words, some energy is lost during the process of charging the battery, as well as when that energy is used to power up the device. Secondly, batteries suffer from leakage, meaning that the amount of stored charge on the battery, even if the battery is not used, always decreases. Batteries have a certain capacity that is expressed by the number of energy units they can store. This capacity, however, decreases each time the battery is charged and discharged. Furthermore, different types of batteries have different charging profiles, requiring the proper modelling of batteries and the inclusion of intricate power electronics for the control of the charging and discharging process. Most common batteries are lead-acid, nickel-cadmium (NiCd) nickel-metal-hydride (NiMH) and lithium-ion (Li-ion) batteries. Their differences are shown in Table 6.9 [28].

On the other hand, supercapacitors are basically capacitors that are small in physical size yet are able to hold a much larger charge than regular capacitors [22]. Their storage capacity is calculated in the same way for a regular capacitor, using the parallel-place capacitance equation, which states that if the surface area A of the plates is large compared to the separation distance d, the capacitance can be calculated by

$$C = \varepsilon_0 \varepsilon_r \frac{A}{d}, \tag{6.16}$$

where ε_0 is the absolute permittivity and ε_r is the relative permittivity of the dielectric material used. Usually, supercapacitors are based on double-layer capacitor technology, which increases the capacitance to the order of farads [29]. Their resemblance to regular capacitors means that they can be charged and discharged for almost an unlimited number of times, they last longer and require only simple charging circuits. They can also deliver the charge much faster than a battery can. In other words, supercapacitors can store a lot of instantaneous power, whereas batteries store a lot of energy. The main disadvantage of using supercapacitors over batteries is exactly the above: the small energy storage capacity (the best supercapacitors can store about one hundredth of the storage of comparable Li-ion batteries). Also, like batteries, they suffer from leakage, the effect of which can be regarded as more serious, because their storage capacity is already much lower. Nevertheless, supercapacitors are a fairly good choice for wireless sensors, where the amount of energy that needs to

Table 6.9 Differences between commonly used batteries [28]

Battery type	Lead-acid	NiCd	NiMH	Li-ion
Anode	Sponge metallic lead	Cadmium hydroxide	Rare-earth or nickel alloys	Carbon compound, graphite
Cathode	PbO_2	Nickel oxyhydroxide	Nickel oxyhydroxide	Cobalt-oxide/lithium-iron-phosphate/manganese-oxide
Electrolyte	H_2SO_4	KOH	KOH	$LiPF_6$, liquid lithium salts in an organic solvent (inflammable)
Historical applications	Cars, trucks, standby/back-up systems	Calculators, digital cameras, laptops, flashlights, medical defibrillators, electric vehicles, space applications	Today satellite applications, formerly cellular phones, emergency backup lighting, power tools, laptops, portable electronics, electric vehicles	Laptops, cellular phones, electric vehicles
Specific energy density	35–50Wh/kg	45–80 Wh/kg	60–120 Wh/kg	150–200 Wh/kg
Volumetric density	60–70 Wh/l	50–150 Wh/l	150–300 Wh/l	250–550 Wh/l

be supplied at any given time is small to begin with, but a fast and reliable supply is required. In addition, it is possible to place several supercapacitors in parallel to increase the storage capacity.

6.7 Concluding Remarks

As wireless networks evolve, energy consumption becomes a pressing issue. The principle is simple: the more information that needs to be pushed back and forth, the larger the effort to generate and transmit this information, which is reflected in increased energy consumption. Whereas it seems that there is still no indication of when the trend in the information increase is going to slow down, it is clear that it will soon become difficult to provide all the required power to keep these intricate systems running. Thus, now is the time to make every effort to save energy on all fronts, whether by network design, information flow optimization, energy conservation or even complementing the "lost" energy by harvesting the energy from renewable sources. In this chapter an attempt was made to put all the recent efforts to achieve this cause together, as well as to provide some ideas that may be useful in future.

Some of the concepts that aid in energy saving, such as umMIMO and programmable metasurfaces, have been proposed for 6G networks for the purpose of achieving faster data rates. The detailed operation of these technologies, as well that of some other futuristic 6G ideas, is discussed in the next chapter, Chap. 7.

References

1. Gati A, Salem FE, Serrano AMG, Marquet D, Masson SL, Rivera T, et al (2019) Key technologies to accelerate the ICT Green evolution–An operator's point of view. arXiv:190309627 [cs] [Internet]. [cited 2020 Jul 10]. http://arxiv.org/abs/1903.09627
2. Zhang S, Cai X, Zhou W, Wang Y (2018) Green 5G enabling technologies: an overview. IET Commun IET Digital Libr 13:135–143
3. Abrol A, Jha RK (2016) Power optimization in 5G networks: a step towards GrEEn communication. IEEE Access 4:1355–1374
4. I C-L, Han S, Bian S (2020) Energy-efficient 5G for a greener future. Nat Electron 3:182–184. Nature Publishing Group
5. Sustainable energy critical to SA's new dawn [Internet]. [cited 2020 Oct 28]. https://www.bizcommunity.com/Article/196/704/209781.html
6. Bohli A, Bouallegue R (2019) How to meet increased capacities by future green 5G Networks: a survey. IEEE Access 7:42220–42237
7. Chen Y, Zhang S, Xu S, Li GY (2011) Fundamental trade-offs on green wireless networks. IEEE Commun Mag 49:30–37
8. Li Y-NR, Chen M, Xu J, Tian L, Huang K (2020) Power saving techniques for 5G and beyond. IEEE Access 8:108675–108690
9. Wu Q, Li GY, Chen W, Ng DWK, Schober R (2017) An overview of sustainable green 5G networks. IEEE Wirel Commun 24:72–80

10. Lorincz J, Garma T, Petrovic G (2012) Measurements and modelling of base station power consumption under real traffic loads. Sensors 12:4281–310. Molecular Diversity Preservation International
11. Yao M, Sohul MM, Ma X, Marojevic V, Reed JH (2019) Sustainable green networking: exploiting degrees of freedom towards energy-efficient 5G systems. Wirel Netw 25:951–960
12. Skrimponis P, Dutta S, Mezzavilla M, Rangan S, Mirfarshbafan SH, Studer C, et al (2020) Power consumption analysis for mobile MmWave and Sub-THz receivers. 2020 2nd 6G Wireless Summit (6G SUMMIT), pp 1–5
13. Usama M, Erol-Kantarci M (2019) A survey on recent trends and open issues in energy efficiency of 5G. Sensors 19:3126. Multidisciplinary Digital Publishing Institute
14. Chughtai NA, Ali M, Qaisar S, Imran M, Naeem M (2020) Energy efficiency maximization in green energy aided heterogeneous cloud radio access networks. 2020 IEEE 91st Vehicular Technology Conference (VTC2020-Spring), pp 1–6
15. Liu X, Zhang X, Jia M, Fan L, Lu W, Zhai X (2018) 5G-based green broadband communication system design with simultaneous wireless information and power transfer. Phys Commun 28:130–137
16. Huang T, Yang W, Wu J, Ma J, Zhang X, Zhang D (2019) A survey on green 6G network: architecture and technologies. IEEE Access 7:175758–175768
17. Mahmood NH, Alves H, López OA, Shehab M, Osorio DPM, Latva-Aho M (2020) Six key features of machine type communication in 6G. 2020 2nd 6G Wireless Summit (6G SUMMIT), pp 1–5
18. Davaslioglu K, Gitlin RD (2016) 5G green networking: enabling technologies, potentials, and challenges. 2016 IEEE 17th Annual Wireless and Microwave Technology Conference (WAMICON), pp 1–6
19. Mowla MM, Ahmad I, Habibi D, Phung QV (2017) A green communication model for 5G systems. IEEE Trans Green Commun Netw 1:264–280
20. Božanić M, Sinha S (2015) Power amplifiers for the S-, C-, X- and Ku-Bands: an EDA perspective. Springer
21. Preez J du, Sinha S (2017) Millimeter-wave power amplifiers. Springer
22. Chen Y (2019) Energy harvesting communications: principles and theories. John Wiley & Sons
23. Bai Y, Jantunen H, Juuti J (2018) Energy harvesting research: the road from single source to multisource. Adv Mater 30:1707271
24. Li S, Wang W, Wang N, O'Mathuna C, Roy S (2015) Micro photovoltaic module energy harvesting. In: Briand D, Yeatman E, Roundy S (eds) Micro energy harvesting. Wiley
25. Visser HJ, Vullers R (2015) Far-field RF energy transfer and harvesting. In: Briand D, Yeatman E, Roundy S (eds) Micro energy harvesting. Wiley
26. Szarka GD, Stark BH, Burrow SG (2012) Review of power conditioning for kinetic energy harvesting systems. IEEE Trans Power Electron 27:803–815
27. Codd DS, Escarra MD, Riggs B, Islam K, Ji YV, Robertson J, et al (2020) Solar cogeneration of electricity with high-temperature process heat. Cell Rep Phys Sci 100135
28. Williams BW (2003) Principles and elements of power electronics. Barry W Williams
29. Chau KT (2014) Pure electric vehicles. In: Folkson R (ed) Alternative fuels and advanced vehicle technologies for improved environmental performance. Woodhead Publishing, pp 655–84

Chapter 7
Futuristic Technological Aspects of 6G Networks

Abstract There is a multitude of emerging or proposed technologies that may for the first time be used in upcoming network generations. This chapter will cover some of the advanced, possibly even futuristic, technological aspects that are expected to be unique to 6G. Advanced aspects include ultra-massive umMIMO, beamforming and channel control with programmable (reflective) metasurfaces. Futuristic concepts include holographic radio, quantum and molecular communication, aerial networks established with the help of drones and satellites, as well as the use of blockchain technologies in network security.

In four previous chapters, this book looked at some of the defining technologies of 6G. Chapter 3 looked at the use of the additional (millimeter-wave and terahertz) radio spectrum and the fundamentals of the channel coding theory as a means of increasing the bandwidth and data rate of transmission. In Chap. 4, transistor device technologies and device packaging were examined a means of miniaturization of entire electronic systems, which are some of the main aspects in efforts to enable future mobility required in 6G. Chapter 5 considered the VLC as an alternative technology to radio transmission, discussed in Chap. 2, with the same if not better potential to achieve the throughput goals of 6G. Finally, Chap. 6 investigated the technologies behind energy saving and energy harvesting for 6G networks. What is common to all those technologies is that they have been known or available before 6G, or even 5G networks, were conceived. This means that these technologies are not really unique to 5G and 6G concepts. On the other hand, there is a multitude of emerging or proposed technologies that may for the first time be used in upcoming network generations. Some of these technologies have only been proposed, while proofs of concept of others have already been demonstrated in the laboratory. Whereas some of the ideas are just advanced implementations of existing concepts (e.g. umMIMO), others may still feel futuristic at the moment (e.g. quantum communications). Thus, it will not come as a surprise that some ideas that are to be discussed in this chapter may not be fully ready before initial 6G deployment, and could only appear in later 6G releases, if at all.

M. Božanić and S. Sinha, *Mobile Communication Networks: 5G and a Vision of 6G*, Lecture Notes in Electrical Engineering 751,
https://doi.org/10.1007/978-3-030-69273-5_7

The chapter will start with the discussion of various approaches to beamforming. Sect. 7.1 considered different approaches to beamforming. umMIMO beamforming is the technique that is already used to improve the throughput in 5G, and its continued role in 6G networks is discussed first. Two closely related concepts, viz. holographic communications and channel control using programmable metasurfaces, are discussed thereafter, followed by a discussion on reconfigurable antennas. Sect. 7.2 discusses aerial networks enabled by autonomous drones, as well as satellite networks, both envisioned as technologies that will allow for 3D coverage in 6G, by forming part of 6G as an integrated satellite and terrestrial network. This space-air-ground network could be expanded to include underwater communication networks as well, discussed in Sect. 7.3. Quantum communications, discussed in Sect. 7.4, and molecular communications, discussed in Sect. 7.5, are still quite some time from becoming a reality, but are interesting concepts that may come to life during the lifetime of 6G and deserve to be mentioned. Sect. 7.6 discusses nanopthotonics, another emerging field of research. Lastly, blockchain technologies that may be required to increase the security of 6G networks, among other uses, are discussed in Sect. 7.7, before chapter conclusions are drawn.

7.1 6G Beamforming Techniques

In this section, the concepts of umMIMO, holographic radio, IRSs and configurable antennas are discussed. While holographic radio is not strictly a beamforming technique, it is a concept that is related to the other three concepts and deserves to be discussed together with these.

7.1.1 Ultra-Massive MIMO

MIMO is a technology equipping access points with a large number of antennas, which are utilized for spatial multiplexing of many data streams per cell to one or multiple users [1]. The concept of MIMO is not new. In fact, MIMO was introduced in the 1990s already, under the name spatial division multiple access [2]. Introducing spatial division multiple access to wireless networks at that point, however, did not make much financial sense, particularly because it used a small number of antennas and it did not actually make for major data throughput gains. This situation has changed over time, with MIMO first being introduced in 4G, and mMIMO being introduced in 5G. Currently, 5G base stations use 64 fully digital transceiver chains (each with multiple MIMO antennas) in several countries. It is envisioned that 6G will use antenna structures with counts beyond those of mMIMO, usually defined as umMIMO. MIMO in the VLC domain, involving illuminating LED arrays and multiple photodetectors, is also possible [3, 4].

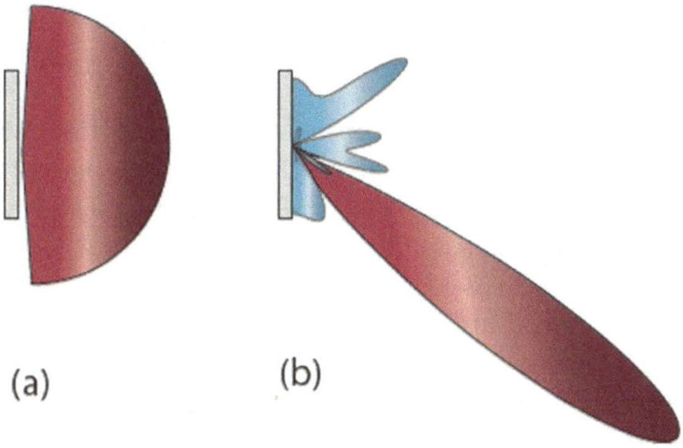

Fig. 7.1 MIMO and beamforming: **a** Radiation pattern of the classical base station, **b** MIMO assisting in formation of multiple signals focused at specific receivers. Reproduced with permission from [2]

The MIMO transmitter assists with beamforming. MIMO can most easily be understood with the help of Fig. 7.1 [2], which compares the signals radiated by the classical base station (Fig. 7.1a) with the beamformed signal by a MIMO antenna array Fig. 7.1b. The antenna array consists of M antennas. Basically, instead of radiating a single uniform signal with wide coverage (which is the case with a classical base station, e.g. as used in 2G and 3G), the antenna array can focus the signal in a single direction, towards a user. Beamforming and beam steering are typically done by controlling the amplitude and phase shift of the signal at each antenna element in analog, digital or hybrid fashion [5]. If M antennas are used, users observe the strength of the signal that is M times the signal that they would observe with just one antenna. The drawback is, naturally, that the signal strength in any direction other than the direction of the beam is virtually zero. On the other hand, each antenna is controlled separately, meaning that K users, where $K \leq M$, can be serviced at the same time, with each user seeing the signal strength of M/K generated by what is called an antenna subarray. However, the gain and beam width also depend on the size of the subarray, hence the scenario where $M >> K$ is the preferred operating regime in MIMO for keeping the received signal strength high. To retain this condition as the number of users (K) increases, M should be increased accordingly. With the increase in M, MIMO becomes mMIMO and eventually umMIMO. For even higher user count, multiple transceivers (each with multiple antennas) are added forming transceiver chains—for example, umMIMO scenarios with 1 024 transceivers with 1 024 antennas each are deemed possible for 1 THz transmissions [6].

mMIMO and umMIMO are appealing for 5G and 6G, respectively, for several reasons [2]:

- IC and packaging technologies have reached the stage where the cost of building antenna arrays is not prohibitive anymore.
- Information theory has evolved to the point where selection of the right operating regime is no longer an issue.
- The signal-processing complexity required for building and controlling MIMO systems has become manageable in a small form factor.

- A large number of antennas allow for unprecedented spatial resolution, robustness towards small-scale fading and the ability to suppress interference.

Specifically for 6G, it is envisioned that umMIMO will be implemented using fully digital arrays with hundreds or thousands of phase-synchronized antennas [1]. The implementation is equally feasible in sub-6 GHz bands and in the millimeter-wave and terahertz bands. In addition, for the millimeter-wave/terahertz case, umMIMO will leverage new materials and advanced chip-processing technologies, making on-chip compact ultra-massive arrays possible, for use not only in base stations, but also in user devices. This will, in essence, mean that a large number of antennas will be present on both the transmitting and receiving sides of the communication link, allowing double-umMIMO to become a reality as well. Recall, from Chap. 6, that this double-MIMO link also allows for MIMO to be considered as an energy-efficient concept. With the increase in frequency, however, beam tracking and acquisition progressively become the key challenges of this concept [7].

Note that the higher the antenna count in the array, the higher the beam accuracy will become. This leads to consideration of the use of a continuous aperture antenna, which is the ultimate configuration for beamforming. One way to build such an antenna is to integrate tiny antennas into a small space, forming a metasurface. Regrettably, in this scenario, tightly coupled, ultra-dense antenna deployment prohibits these antennas from being actively driven. On the other hand, passive structures, capable of relaying transmitted signals, can easily be realized in practice, and are some of the most promising technologies for 6G. Another solution would be holographic radio technology, which expands the concepts of optical holography into the RF domain and eliminates the need for beamforming completely. Each of these concepts will be discussed in the two sections that follow, starting with holographic radio.

7.1.2 Holographic Radio

Holographic radio is one of the methods proposed to create a spatially continuous electromagnetic aperture to achieve ultra-high resolution spatial multiplexing [1]. Most readers will be familiar with optical holography; it is however interesting to note that the principle of holography can be extended to a wave of any frequency, including those in the electromagnetic (radio) domain [2, 8, 9]. As already mentioned, holographic radio is the next step in the evolution of the mMIMO concept; thus, various sources also refer to it as the holographic RF system, holographic beamforming and holographic MIMO/mMIMO.

In holography, the electromagnetic field in space is recorded based on the interference principle of electromagnetic waves [1]. Thus, in addition to the signal wave, the reference wave, assisting in capturing the wave on the transmitter side and reconstructing the wave on the receiver side, must be generated. With this, two core differences between regular radio transmission and holographic transmission define the holographic principle: (1) in addition to the amplitude of the electromagnetic wave, the phase is captured, and this must be done in continuous fashion, and (2) the reference wave must be strictly coherent.

To describe the principle of RF holographic transmission and reception, one can draw a parallel to optical holography. Optical holography differs from regular (2D) photography in the sense that photography only responds to the intensity of the electromagnetic field, in this case, light. The intensity of the illuminating wave $a(x, y)$ in the $x-y$ coordinate system is $|a(x, y)|^2$ [2]. With the addition of the reference wave $e^{j(\alpha x+\beta y)}$ to wave $a(x, y)$, the combined wave becomes

$$b(x, y) = a(x, y) + e^{j(\alpha x+\beta y)}. \tag{7.1}$$

The photographic medium will still record the intensity of the incident wave,

$$|b(x, y)|^2 = \left|a(x, y) + e^{j(\alpha x+\beta y)}\right|^2. \tag{7.2}$$

However, after some manipulation of complex numbers, it can be seen that the phase term of the original wave $a(x, y)$ is actually retained in the second term in the following equation:

$$|b(x, y)|^2 = |a(x, y)|^2 + 2Re\left\{a(x, y)e^{-j(\alpha x+\beta y)}\right\}. \tag{7.3}$$

The phase in the wave is thus inherently recorded.

The captured holographic image is reconstructed by illuminating a transparent display with the replica of the reference wave. This results in

$$c(x, y) = e^{j(\alpha x+\beta y)}|b(x, y)|^2 = a(x, y) + a^*(x, y)e^{j2(\alpha x+\beta y)} + \left[|a(x, y)|^2 + 1\right]e^{j(\alpha x+\beta y)}. \tag{7.4}$$

Careful choice of the reference wave allows for the first component of wave $c(x, y)$ to be separated from its other two components, thus allowing for $a(x, y)$ to be reconstructed in full. The same effect can be achieved by using a mirror instead of the transparent display in conjunction with the complex conjugate of the reference wave ($e^{-j(\alpha x+\beta y)}$) instead of its exact replica. Now, returning from the optical domain to the RF domain, the analysis may remain the same, but with these replacements:

- The illuminating wave becomes the waveform emitted by transmitting user devices.
- The reflecting object (mirror) becomes the propagation environment.
- A reference wave is generated inside a surface containing an array of tightly coupled antennas. These surfaces are sometimes called holographic MIMO surfaces, or HMIMOS [8].

Consequently, in holographic communication, there is no concept of beams, just the concept of the holographic radio space. The holographic radio space, as envisioned in [2], would be created by integrating HMIMOS that can receive and emit electromagnetic waves into walls, windows or even fabrics.

Holographic radio is still in conceptual phases of development, and the exact implementation remains a wide-open research area [1]. However, it is clear that its implementation will be a multidisciplinary problem involving optical processing, spectrum computing, large-scale photon integration, electro-optical mixing, and analog-digital photon hybrid integration technologies. This will require a new physical layer to be designed for 6G to make use of these new methods. It will also require researchers in this field to have a thorough understanding of both communication and electromagnetic theories. Currently, such researchers are a scarce resource.

Much more progress has been made in IRS research, which is another alternative when it comes to utilization of a spatially continuous electromagnetic aperture. This is discussed next.

7.1.3 Intelligent Reflective Surfaces

IRS or programmable metasurfaces[1] are innovative technology allowing for significant improvement in the performance of wireless communication networks, by smartly reconfiguring or manipulating the wireless propagation environment with the aid of massive low-cost passive sub-wavelength reflecting elements integrated within a planar surface [10–12]. The principle of the operation of IRS is shown in Fig. 7.2,

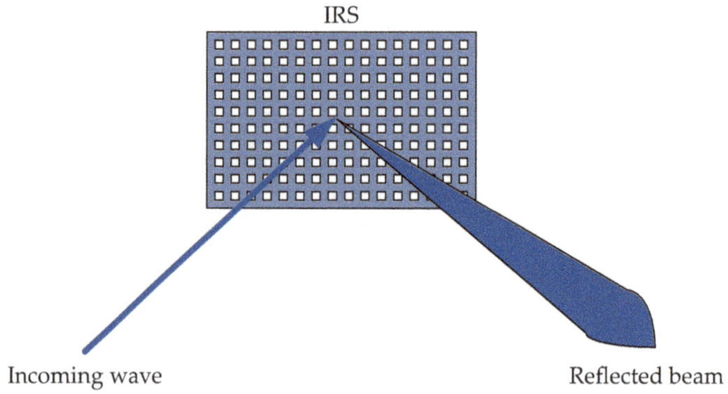

Fig. 7.2 Principle of operation of the IRS

[1]Different researchers use different naming for this technology. The terms intelligent reflective surfaces or programmable metasurfaces are used most widely; however, other sources give these structures other names, among others reconfigurable intelligent surfaces, reconfigurable reflectarrays, smart reflecting arrays, software-controlled metasurfaces, passive intelligent mirrors, artificial radio space, etc.

illustrating that the IRS is basically a reflective surface (similar to a mirror), but with the difference that it can manipulate the incoming (incident) radio wave as it reflects from it. The building block of the IRS is a meta-atom [13], which is a controllable conductive structure that can alter EM properties. Meta-atoms or surface elements are repeated periodically across a rectangular tile, illustrated with white squares in Fig. 7.2. Each element can be configured independently to exhibit unique electromagnetic properties, such as customized permittivity and permeability levels, and negative refraction. As a result, the surface can perform wave focusing, absorption, polarization, scattering, beam steering, beam splitting and phase control (time-delay) of the reflected wave, among other manipulations. Sometimes, the amplitude of the signal can also be altered. On the system level, use of IRS may lead to the joint optimization of certain system performance metrics, such as transmit power, achievable rate, energy efficiency, and received SNR, thus maximizing the efficiency of the end-to-end links. Some of the negative, destructive and uncontrollable effects of the wireless channel are thus overcome [14].

One of the most important properties of programable metasurfaces is the fact that they do not require active amplification elements, such as power amplifiers, which are characteristic of the amplify-and-relay schemes used in telecommunications traditionally [15]. Active devices may still be needed to activate individual surface elements, but their energy footprint is much lower than if full transceiver blocks were used. As such, IRSs are passive, low-cost devices, and although they can grow very large as the number of antenna elements increases, their low complexity could potentially make them invisible to the human eye. As such, they can be integrated in almost any object that has a large area—such as walls, ceilings and furniture (indoors) and building facades (outdoors) [1]. Note that the fact that IRS is a passive structure means that the signals are effectively received and transmitted simultaneously, making these structures full-duplex devices [16], whereas MIMO systems, described earlier in this chapter, are only half-duplex systems.

The most obvious use of IRS is in situations where the LOS link is blocked or not strong enough for a reliable communication path to be established. In these cases, IRS can assist in establishing additional transmission paths by reflecting the incoming waves towards the user in the low-coverage area [10]. This is shown as the top-left scenario in Fig. 7.3. There are, however, many other possible use scenarios. For example, metasurfaces can also assist in secrecy and security. If an eavesdropper is detected, IRS can be tuned to cancel out the non-IRS-reflected signal from the base station in the direction of the eavesdropper (top center scenario in Fig. 7.3). In the third scenario, IRS can boost the signal for the user at the cell edge, as seen in the top right of Fig. 7.3. Similarly, it can assist with D2D communication (bottom left). The last use that will be mentioned here is the use of IRS in SWIPT for devices that require external power, such as IoT devices. IRS can focus the power beam on the device and thereby assist in energy harvesting. This is shown in the bottom right part of Fig. 7.3.

The employment of an IRS can also be a dominant factor in the received signal power even in the presence of a direct (LOS) communication link. To illustrate this, one must expand on the signal propagation theory described in Chap. 3. Recall that

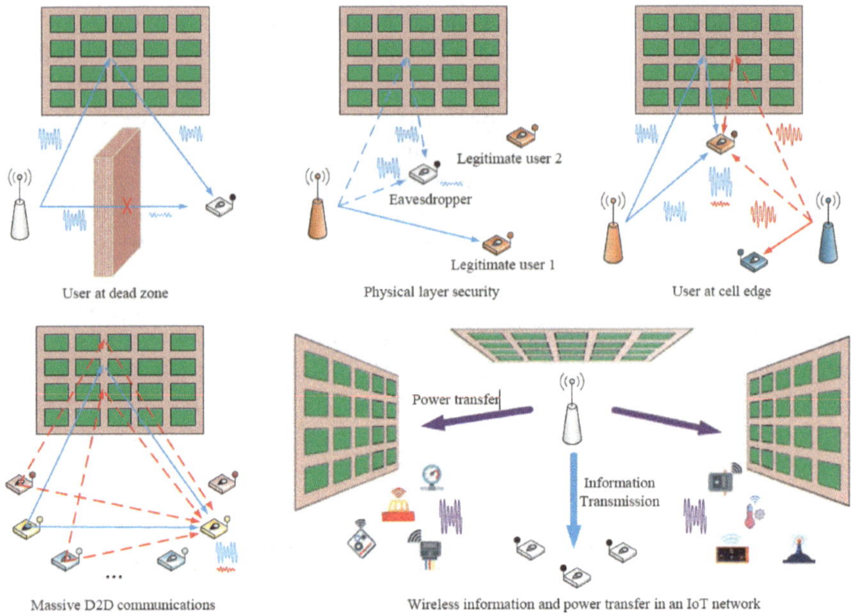

Fig. 7.3 IRS use cases in 6G networks. Reproduced with permission from [10]

the received power, with the LOS present, at the distance r between the transmitter and receiver is

$$P_R = P_T\left(\frac{\lambda}{4\pi r}\right)^2, \tag{7.5}$$

with unity antenna gains. Another way the wave travels is possibly by reflecting off the ground at point G, as shown in Fig. 7.4. The total received power in that case is the power received via both the LOS and nLOS paths [15]:

$$P_R = P_T\left(\frac{\lambda}{4\pi}\right)^2\left|\frac{1}{r} + \frac{Re^{-j\Delta\phi}}{r_1 + r_2}\right|^2, \tag{7.6}$$

where $\Delta\phi = \frac{2\pi(r_1+r_2-r)}{\lambda}$ is the phase difference between the two paths. In a special case when $R \approx -1$, Eq. (7.6) becomes

$$P_R = P_T\left(\frac{\lambda}{4\pi}\right)^2\left|\frac{1}{r} - \frac{e^{-j\Delta\phi}}{r_1 + r_2}\right|^2. \tag{7.7}$$

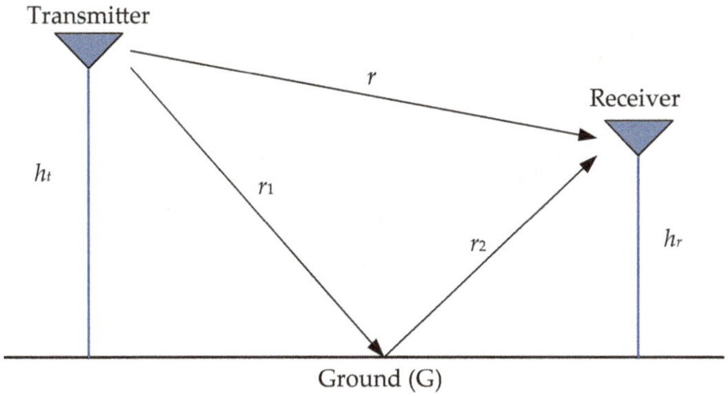

Fig. 7.4 Two different signal-transmission scenarios: LOS (path r) and nLOS (path r_1–r_2)

Complex manipulation will show that

$$P_R \propto P_T \left(\frac{1}{r^2}\right)^2, \tag{7.8}$$

if in the last step of manipulation, r is large enough so that the assumption $r \approx r_1 + r_2$ can be used. The result of Eq. (7.8) shows that there is a scenario in which a single ground reflection has an extremely destructive effect on the signal propagation, causing it to be perceived as if it was decaying with the fourth power of distance, and not with the square of the distance, which is the case with no ground reflection Eq. (7.5).

If the ground is coated with the metasurface, the metasurface can adjust the phase of the reflected wave so that it sums up coherently at the receiver. This requires R to be set to $e^{j\Delta\phi}$, resulting in

$$P_R = P_T \left(\frac{\lambda}{4\pi}\right)^2 \left|\frac{1}{r} + \frac{1}{r_1 + r_2}\right|^2. \tag{7.9}$$

If assumption $r \approx r_1 + r_2$ still holds,

$$P_R \approx 4 P_T \left(\frac{\lambda}{4\pi r}\right)^2. \tag{7.10}$$

The result of Eq. (7.10) is that not only does the surface aid in allowing the wave always to propagate by attenuating with the square of the distance; it also boosts the received gain by a factor of 4, despite being a passive element. This finding, based on a two-ray system, is one of the main findings of the metasurface theory that triggered research activity on this topic. Another favorable mathematical expression must be noted here: as the number of independently configurable metasurface (IRS) elements

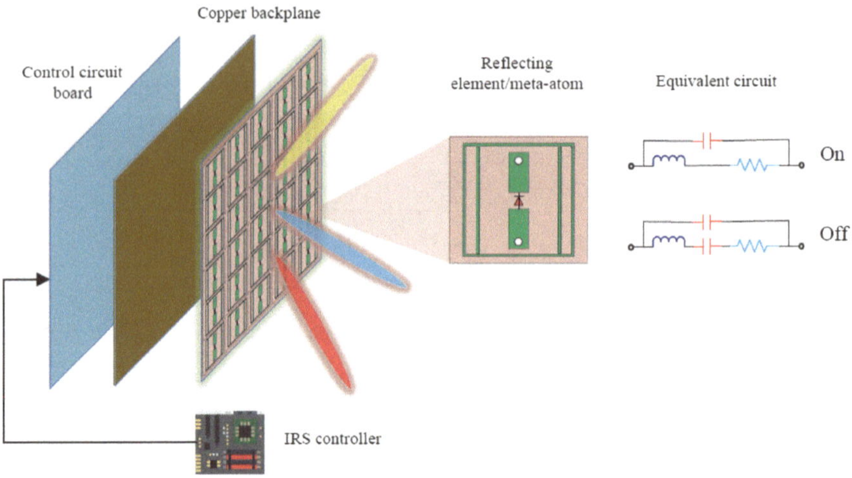

Fig. 7.5 Architecture of IRS. Reproduced with permission from [10]

increases to some arbitrary number N, the received power will continue increasing even further, scaling with the second power of $N + 1$:

$$P_R \approx (N+1)^2 P_T \left(\frac{\lambda}{4\pi r}\right)^2. \tag{7.11}$$

The main question that remains to be answered is how to construct the IRS in order to have many elements, or meta-atoms, independently configured [10, 13]. Independent control of meta-atoms calls for the use of devices such as PIN diodes, FETs or MEMS switches. PIN diodes are typically used to modify the phase shift, whereas resistors, for example, can be added to modify the amplitude of reflected signals. Multiple PIN diodes would be needed to be able to achieve a finer resolution of phase shifts. A typical IRS structure is shown in Fig. 7.5 [10]. It requires three layers and a smart controller. The outer layer consists of a large number of metallic patches (elements) that are printed on a dielectric substrate to interact directly with incident signals. The middle layer is the copper plate, which is used to avoid signal energy leakage. Lastly, the inner layer is the control printed circuit board responsible for adjusting the reflection of each element. Practical control in the past has been achieved with the help of field-programmable gate arrays, but future networks may also use machine learning to assist in metasurface control.

The number of elements in the IRS depends on the size of the surface and the availability of advanced technologies for meta-element construction, which dictate their form factor. However, with the indefinite increase in the number of elements, the received power probably does not continue to scale with the second power of N + 1 indefinitely. In other words, Eq. (7.11) does not hold for large values of N. In some transmitter-IRS-receiver configurations, Eq. (7.11) may hold for large values of N

and the scaling of the surface may indeed continue indefinitely, but there are often other factors that must be taken into consideration when predicting system behavior (e.g. whether the operation is in the near-field or far-field). It is for this reason that a calculation of the feasible element count limit should rather be performed [16, 17].

7.1.4 Fluid Antennas

Whereas IRSs are an excellent solution for the manipulation of wave propagation, they are in essence passive relay systems. However, they cannot manipulate the wave propagation on the transmitter or receiver end of the link. Fortunately, this can be achieved with reconfigurable antennas, which can dynamically adapt their beam patterns upon receiving commands from the network software [18–20]. Traditional reconfigurable antenna systems use multiple antennas for beam control and beam steering, or, in the case of single antennas, mechanical mechanisms. Whereas multiple antennas are an integral part of the mMIMO concept, mechanical steering may not be the optimal solution for 6G owing to the size and complexity of mechanical steering mechanisms. Thus, research into advanced reconfigurable antenna solutions continues. Some of the recently proposed reconfigurable antenna alternatives include antennas built with flexible membrane materials, as well as fluid antennas.

Fluid antennas may be the low-cost solution for 6G mobile devices [21]. Fluid antennas are made of conductive fluid, metal fluid or ionized liquid that can be shaped into a multitude of different forms to suit the propagation environment [19, 21]. The fluidic structure sits on the boundary between the antenna hardware and the output stage of the transmitter (or the input stage of the receiver), allowing for optimization of antenna position and shape, thereby changing antenna gain and beam direction. It has been suggested that a single software-defined fluid antenna can provide performance comparable only to mMIMO antenna arrays. However, research into these structures is still in its infancy.

7.2 Aerial and Satellite Components of 6G Networks

Everything that was discussed in this book and this chapter up to now could apply to communication between various points on the ground or close to the ground. In other words, all the systems discussed considered mounting transmitters, receivers and repeaters on masts, street poles or inside and outside of man-made structures (buildings). However, one of the mission-critical requirements of 6G is ubiquitous coverage in all three dimensions, meaning that the coverage should be able to leave the confines of the earth's surface. This is where the non-terrestrial (aerial and satellite) components of 6G networks come into play. The overreaching result is an integrated space and terrestrial network, or an air-ground-space network, to be more precise, which consists of a smart combination of both terrestrial and non-terrestrial base

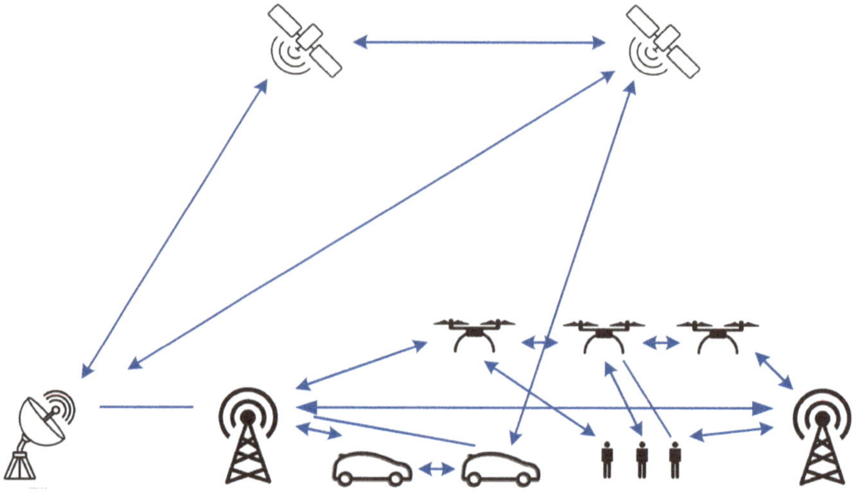

Fig. 7.6 6G pictured as an integrated space and terrestrial network (space-air-ground network)

stations, something that can be illustrated as in Fig. 7.6. Of course, aerial and satellite networks would not only enable 3D coverage (e.g. for a user on board a plane), but also provide coverage to remote areas or areas that are not easily accessible. This includes, for example, providing coverage in remote rural areas, as well as providing coverage in open sea/ocean waters.

Integrated terrestrial and satellite networks are organized in three network layers. Ground base stations are still the main access points for most users in these integrated networks, and they will form the conventional ground-based network layer [1]. Ground base stations are complemented by airborne base stations, which are made of UAVs or drones, forming the most flexible, airborne network layer. Drones will allow for network coverage in areas that are difficult to supply with the signal otherwise. Finally, the spaceborne layer consists of various satellites, deployed at different earth orbits, as described in Sect. 1.4.5 in Chap. 1. The LEO satellites may prove to be the most useful in future, because the SNR and the latency that can be achieved when sending the information to and from this type of satellite could still be in the acceptable range for most use cases of 6G, because of their relative proximity to the surface of the earth.

With the rest of the book covering the traditional ground-based network layer in detail, the remainder of this section will focus on the new airborne and spaceborne network layers.

7.2.1 Airborne Network Layer

The airborne network layer is the intermediate network layer between the ground-based and spaceborne network, and is formed by a network of UAVs, also known as drones [1, 18]. Other airborne vehicles, including balloons and airships, may be considered alongside drones [6]. High-altitude platform stations, which would be stationed in the stratosphere, at an altitude of about 20 km, would be able to provide better coverage than drones, and could also be considered to form part of the airborne network layer. Although, like that of drones, their deployment is not costly, they might suffer from difficulties such as refueling, as well as stabilization [18], and will not be discussed further here.

Drones introduce the advantage of great mobility and flexibility into the space-air-ground network [11]. In the past, drones have already proven to have their place in uses such as search and rescue missions and package delivery, yet providing network coverage by drones is a novel concept. Other advantages of drones may include low cost of deployment (with proper design), high scalability, improved network capacity and the characteristic that they are less affected by nature and terrain.

Cellular-enabled drones would require high-speed connectivity to ground feeder links and would be deployed to areas that ground base stations cannot cover. Integration of cellular-enabled drones with the network of ground stations is one of the new key challenges and design considerations that must be addressed in the 6G network design. Typically, millimeter-wave links, which use adaptive antenna arrays and high power, possibly used in conjunction with umMIMO and full-duplex operation, are mentioned as the most promising answers to this challenge. Usually, advanced antenna solutions are mentioned in addition to these technologies as enablers of drone-based communications.

Another primary concern associated with UAVs is the cost-effectiveness of such systems. While their deployment is not costly, their operation is, meaning that their use may be limited to scenarios where the quality of the network coverage needs to be improved for a relatively short period of time (on demand), e.g. during disasters or for the duration of special events with a large number of attendees (such as concerts and sports events). Alternative uses of cellular drones are found in relaying of signals between base stations (airborne or otherwise), and in aerial backhaul.

Certain challenges associated with UAVs, which will require more research, should be pointed out. These include the limited operation range of UAVs, channel mobility issues and related channel-modeling difficulties, energy constraints imposed by battery operation, resource allocation, as well as other concerns, such as spoofing and jamming. Fortunately, some of these challenges, specifically those having to do with the wireless channel, may be overcome with the help of AI [22]. Moreover, energy-related challenges may be overcome with energy harvesting and WPT [19].

7.2.2 *Satellite Network Layer*

The satellite network layer of 6G would consider two types of satellites: LEO and GEO. As mentioned in Chap. 1, LEO satellites would orbit at between 200 and 2000 km and would have lower propagation delay than GEO satellites, but they are non-stationary relative to the earth's surface [18]. GEO satellites have a fixed position, but the propagation delays might be prohibitive in 6G. This, alone, is sufficient to anticipate that 6G research will focus on incorporating LEO satellites only. Another benefit of using LEO and not GEO satellites is derived from the higher data rates that can be achieved when communicating to LEO satellites, owing to various factors, such as SNR.

It must be noted that the spaceborne network layer has a unique challenge, which is not shared with other layers. Once a satellite is deployed, hardware repairs are not possible anymore. As a result, satellites have to use well-tested hardware, as well as software. Moreover, they have to be self-sustainable when it comes to energy consumption, with light-harvested energy being the primary energy source. A spaceborne network layer, therefore, requires the use of proven technologies with a long track record of successful operation, whereas the use of cutting-edge innovative technologies is limited to the other network layers, owing to their physical accessibility [1].

Upload and download links to satellites have typically been established in the *S, C, X* and *Ku* bands (i.e. bands covering frequencies up to 20 GHz) in the past. However, the large bandwidth requirement of 6G networks means that they will require operation in, at least, the millimeter-wave frequency range (above 30 GHz). Strictly speaking, the millimeter-wave band is not the ideal band for satellite communication, because, as described in Chap. 3, path loss increases with carrier frequency [18]. The signal is also affected by the distance between the terrestrial node and the satellite, shadow fading and clutter, as well as by severe atmospheric absorption. The signal from LEO satellites, which appear in motion to the observer on earth, also suffers from Doppler shift [23]. Moreover, it is difficult to receive satellite signals from indoor network nodes, and even outdoor devices, such as IoT devices (sometimes satellites are mentioned as enablers of super IoT), need to be equipped with high-power transmitters and high-gain antennas [24].

The above limitations influence the Shannon capacity of the earth-satellite links. Thus, the Shannon capacity, or the maximum achievable data rate, of satellite links depends on the height (orbit) of the satellite, the frequency of the radio link operation, satellite antenna gain and attenuation factors, as noted in the previous paragraph [18]. It is interesting to note that, even with very directional antennas providing high gain, the Shannon capacity actually does not increase as the frequency increases beyond 70 GHz, owing to the increasingly harsh impact of atmospheric absorption in the higher millimeter-wave spectrum. The dependence of the link capacity on the satellite height, frequency of operation and the satellite antenna gain is shown in Fig. 7.7, illustrating the limitations on achievable data rates in ground-to-satellite communications.

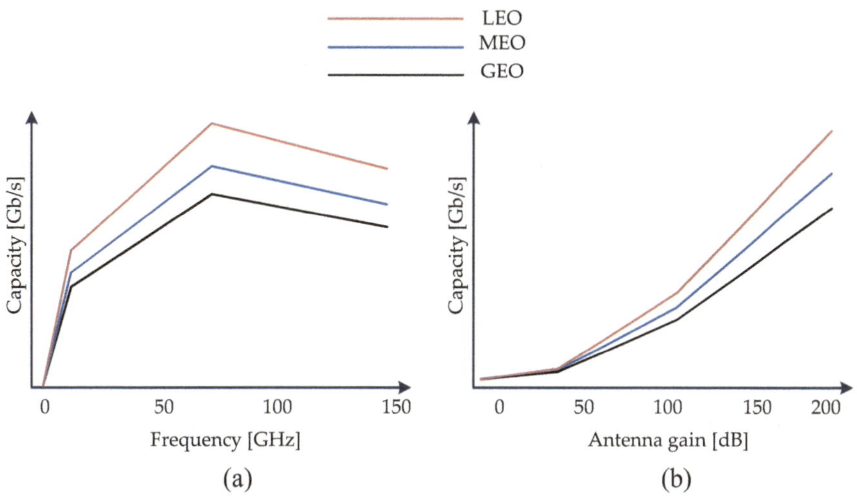

Fig. 7.7 Shannon capacity of satellite links as a function of **a** frequency and the satellite height (orbit) and **b** antenna gain and satellite height. Adapted from [18]

In November 2020, China became the first country to launch the experimental satellite capable of communicating at terahertz frequencies into orbit. The satellite was jointly developed by Chengdu Guoxing Aerospace Technology and Beijing MinoSpace Technology [25]. The aim of the launch was to investigate the feasibility of terahertz communications in space, making it the first 6G testbed located in space.

7.3 Underwater Communication Components of 6G Networks

In Chap. 5, underwater communications have been studied as a vertical of VLC communications. It was deduced that for underwater communication, VLC is by far the most promising alternative, because light propagates much better underwater than radio or acoustic signals. It was also stated that the demand for UOWC or underwater communication in general is increasing owing to various new applications, such as environmental monitoring and maintenance, underwater exploration, offshore oil field exploration, as well as port security, tactical surveillance, and warfare. What was not, however, mentioned in Chap. 5 is that the most feasible way to implement underwater communications practically is to build the UOWC network as part of future 6G network architecture. The feeder (backhaul) links to the underwater network would originate either from the relays built on the shore, or, more feasibly, from UAVs or satellites. The overall result is the integrated space–air–ground–underwater network (ISAGUN) [22], an extension of the space-air-ground network discussed in Sect. 7.2. The concept of ISAGUN is illustrated in Fig. 7.8.

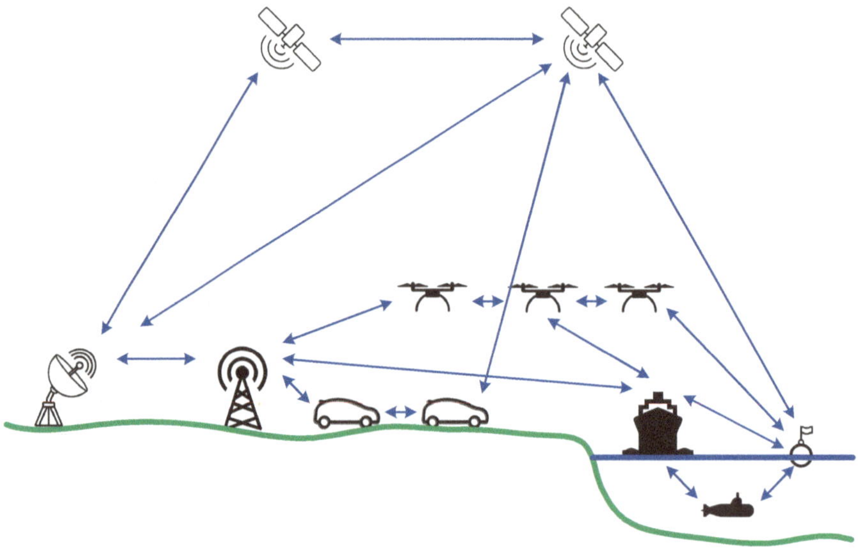

Fig. 7.8 6G pictured as an integrated space, terrestrial and underwater network

It should be noted that the underwater component of 6G will potentially be the most difficult one to realize in practice. In addition to the challenges introduced by absorption of light underwater, scattering and turbulence, there is also a question of how to build a reliable interface between the radio-waves propagating through air and light, which is needed for underwater communication. It thus remains more likely that 6G will be launched as an integrated space-air-ground network long before turning into ISAGUN.

7.4 Quantum Communications

In the quest for ever-faster communication for B5G networks, various challenges constantly keep emerging. One such recently identified challenge has to do with the need for extremely fast computing speeds to remain able to process the massive amounts of information that future wireless networks will need to handle. It is clear that the data-driven aspect of 6G requires a new computing paradigm. Most likely, the solution will be found in the concept of machine learning, which will be the topic of Chap. 8. However, there are suggestions that communications and computing that are required to process the transmitted and received information may also be enhanced by quantum computing and related technologies. The parallelism offered by the fundamental concepts of quantum mechanics, as well as recently demonstrated results in the quantum computing field that indicate the clear potential to outperform the conventional computing systems [20], may certainly move in the direction of

6G networks. In addition, quantum cryptography for a very high level of network and data security may be another related application that could be useful in the 6G network context.

One should be careful here, however, and differentiate between the concepts of quantum communications and quantum computing-assisted communications. Quantum communication systems are communication systems that are purely based on quantum mechanics concepts. On the other hand, quantum computing-assisted communications are traditional communications that make use of speed-ups in the form of quantum computing. There is also a third, related category, which is quantum computing assisted machine learning (QML), used in communications. QML exploits quantum computing to accelerate intelligent data analysis methods. Like that on machine learning, the discussion on the QML-assisted communication concept will be deferred until Chap. 8.

Before quantum communications and quantum-assisted communications are looked at in some detail, it makes sense to first look at the basic principles of quantum computing.

7.4.1 Principles of Quantum Computing

Quantum computing is proposed as a means of achieving significantly higher computing speeds, which is becoming increasingly important as the amount of collected data increases in 5G and B5G networks, as well as with the increase in the importance of the security of these data [26–28]. Quantum computing may prove to be an extremely powerful tool in applications that require intense mathematical computation. The proof-of-concept of quantum computing was achieved in the mathematical problem of prime number factorization. Various nanotechnologies can be used to build quantum computers, as will be described shortly.

Quantum computing relies on the principle that Boolean logic can be extended to include a more general set of transformations between the input and the output than just two states, the traditionally used binary states 0 and 1 [29, 30]. Quantum computations are possible within the laws of quantum physics; in the quantum system, traditional Boolean bits are replaced with qubits. In this naming convention, the prefix "qu" refers to their quantum state.

The information in the qubit is carried in the form of a vector in a two-dimensional space of complex numbers. This two-dimensional space represents a continuum of all possible states, which is actually a superposition of states 0 and 1. Note that irrespective of that, upon observation the qubit state will always be found to be exactly 0 or 1. If a and b are arbitrary complex numbers, the qubit wave equation, describing an unobserved qubit, can be formulated as

$$|\psi\rangle = a|0\rangle + b|1\rangle. \tag{7.12}$$

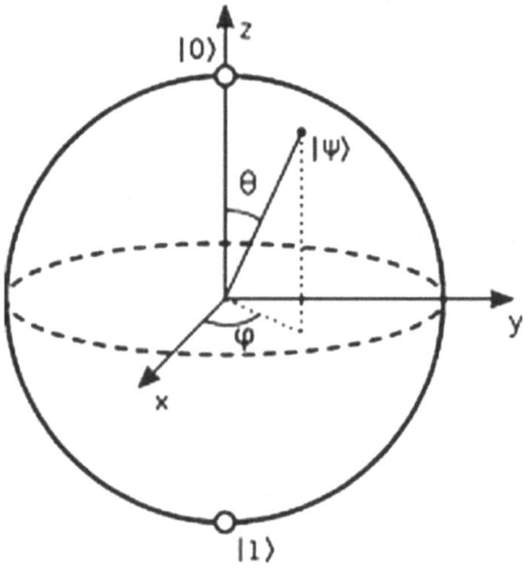

Fig. 7.9 Graphical representation of a qubit (Bloch sphere). Reproduced with permission from [30]

Graphical representation of a qubit is shown in Fig. 7.9. This 3D representation of qubits is called the Bloch sphere. The modulus squared of a and b represents the probability that when the qubit is observed, it will be found in the corresponding state (0 or 1).

The logic gates of the quantum logic system are defined by rotational matrices. While in Boolean logic, the number of logical operations is discrete (for example. two for one input, one output gate, as in the case of the NOT gate), quantum logic required for a continuous set of matrices to be defined. Fortunately, the NOT operation of the quantum NOT gate is represented similarly to the representation of the Boolean NOT gate:

$$\text{NOT}: \begin{pmatrix} 0 & 1 \\ 1 & 0 \end{pmatrix}. \tag{7.13}$$

Complex representation of qubits, however, requires the definition of the square of the NOT gate:

$$\sqrt{\text{NOT}}: \begin{pmatrix} \cos\pi/4 & \sin\pi/4 \\ -\sin\pi/4 & \cos\pi/4 \end{pmatrix}, \tag{7.14}$$

as well as the fourth root of the NOT gate:

$$\sqrt[4]{\text{NOT}}: \begin{pmatrix} \cos\pi/8 & \sin\pi/8 \\ -\sin\pi/8 & \cos\pi/8 \end{pmatrix}. \tag{7.15}$$

The gate in Eq. (7.15) is the main operator of the quantum logic, and is denoted by Q.

The NOT operation or the Q gate on its own is not sufficient and a gate that takes two inputs is also needed. The Q gate and the two-operator gate can subsequently be extended to all required operations, forming a universal gate set. A similar process happens in Boolean logic, where the NAND gate is added to the NOT gate, which allows for all other gates to be defined. The second gate for quantum computations is called a controlled NOT (cNOT) gate, and it acts as a vector on the combination on two qubit states. The two-state wave function becomes a superposition of four basis vectors:

$$|\psi\rangle = a|00\rangle + b|01\rangle + c|10 + d|11\rangle. \tag{7.16}$$

Similar to a and b, coefficients c and d are also complex numbers. In matrix notation, the cNOT gate can be represented as

$$cNOT : \begin{pmatrix} 1\,0 & 0\,0 \\ 0\,1 & 0\,0 \\ 0\,0 & 0\,1 \\ 0\,0 & 1\,0 \end{pmatrix}. \tag{7.17}$$

Q and cNOT now form a base of quantum computation algorithms.

Although understanding of the quantum computation theory has improved over the last three decades, real-world implementation of quantum computers is not trivial. In 2009, DiVincenzo [29] made an extensive list of research ideas and ongoing efforts on how to go about this in the laboratory environment. Some options include representing qubits as:

- nuclear spins in solute organic molecules, semiconductor surfaces (e.g. silicon or carbon);
- individual donors or acceptors in bulk semiconductors;
- atomic energy levels of individual ions in an ion trap or neutral atoms in optical traps or optical lattices;
- photons in optical cavities or optical networks;
- electron spins of vacancy complexes in diamond;
- electrons suspended above liquid helium;
- electrons trapped in semiconductor quantum dots;
- electron orbital states in semiconductor quantum dots;
- excitons in semiconductor quantum dots;
- rare-earth impurity levels in oxide crystals;
- collective excitations of superconducting condensates in Josephson junction circuits;
- Cooper-pair charge of superconducting islands; and
- quasiparticle states in the fractional quantum Hall effect.

Some of these implementations seem too abstract to be understood in the context of the book, and further discussion of these principles is not warranted. However, the list above may be narrowed, because according to the same author, implementation involving superconducting circuits, semiconductor quantum dots and ion traps are the only implementations that seem plausible at present. These implementations require low (cryogenic) temperatures for their operation, however, and a lot of work is still needed to make quantum processors at room temperatures [19].

The word "futuristic" in the title of this chapter is perhaps most applicable to quantum computing and quantum communication, seeing that this field is yet to produce some useful results. Predictions of the National Academies of Science, Engineering and Medicines indicate that a gate-based quantum computer, involving an order of 100 qubits, should become available in the early 2020s and would serve the purpose of demonstrating quantum supremacy over traditional logic [31]. Meanwhile, IBM emerged as the leader in research into the use of quantum computers, offering a platform called IBM Quantum Experience[2] via which the general public can get access to a set of IBM's prototype quantum processors. These processors use only several tens of gates and there are currently no predictions stating that high gate count commercial quantum computers (by IBM or otherwise) will come into existence before the mid- to late 2020s. On the other hand, given that 6G deployment is expected only to start in the early 2030 s, it is not quite implausible to expect that some commercial quantum technologies may come into existence some time during the lifecycle of 6G.

7.4.2 Quantum-Computing-Assisted Communications and Quantum Communications

Quantum-computing-assisted communications would be the first way in which quantum concepts are introduced into future wireless networks. Quantum computing is applicable in any situation where data need to be processed in a fashion that makes the processing otherwise extremely hard computationally [32, 33]. If this is translated into the domain of wireless communications, it means that the existing protocols can be enhanced with more efficient algorithms running on quantum processors [20]. This is particularly important when a network requires computationally efficient solutions to classical signal-processing problems. For example, channel capacity improvement, channel estimation and multi-user detection can all benefit from quantum-computing-assisted communications. Alternatively, signal processing could introduce processing steps that would be impossible to include with classical processing, such as introducing quantum cryptography for enhanced security.

On the other hand, pure quantum communication is more revolutionary than quantum-assisted communications. The aim of quantum communications is to utilize the quantum nature of information for designing novel communication protocols for 6G. Quantum communications could provide absolute randomness and security, while carrying much more information and significantly enhancing the transmission

[2]https://quantum-computing.ibm.com/.

quality. At the same time, all of this would come at speeds that are unheard of in classical systems. Another attractive feature of quantum communication is that it is suitable for communication over long distances [19]. However, there are at least two challenges of quantum communications that must be mentioned. The first is that the architecture of all the networks available at present (including the internet) is not geared for quantum communications. The network would require quantum switches, routers and repeaters, research into which is currently still being formulated. Yet, quantum repeaters are critical devices for building a long-distance global network [34]. The second is that the capacity of quantum capacity channels is far from being understood. This is because there are different ways of delivering quantum information.

Nawaz et al. [20] identify several potential enablers of quantum communications and related critical issues:

- *Quantum entanglement.* Quantum entanglement may allow for encoding both quantum and classical information together, for data transmission over a noisy quantum channel. The inherent security feature of quantum entanglement is that the information cannot be cloned or accessed without tampering with it [19].
- *Quantum-dot cellular automata (QCA).* QCA refers to a nano-scale computing mechanism that serves as a basis for binary computation that overcomes the limitation of CMOS technology scaling, by using cells or quantum dots to store and transfer information.
- *Quantum error correction codes (QECCs).* QECCs may serve as a way to overcome the harmful quantum perturbations that exist in quantum technology.
- *Quantum key distribution.* Quantum key distribution will enable quantum cryptography.
- *Quantum decision theory (QDT).* QDT is a promising approach to replace the classical decision-making process in communications.
- *Quantum game theory (QGT).* Like QDT, QGC has the potential to replace classical game theory. Classical game theory is at present used for rational decision-making. Game theory draws its name from the fact that it resembles players competing in a game.
- *Quantum-proof randomness extractors.* In information theory and cryptography, randomness is a fundamental aspect. Its objective is to transform the sources of correlated and biased bits into nearly uniform bits. It has been speculated that quantum-proof randomness will be required for applications such as quantum cryptography.

The possible applications of quantum principles are numerous [19, 20]. They can be applied to anything from underwater communication and terrestrial networks to aerial and satellite networks. Furthermore, their role in security of applications is probably unprecedented. Lastly, there are many 6G verticals that could specifically benefit from quantum principles. Some examples include:

- quantum-assisted multi-user detection;
- quantum-aided multi-user transmission;

- quantum-assisted indoor localization (for both millimeter-wave/terahertz variants);
- quantum-assisted joint routing and load balancing; as well as
- quantum-assisted channel estimation and detection.

What is not mentioned here is that a number of verticals will become possible if QML is also considered; this will, as already stated, only be considered in Chap. 8.

7.5 Molecular Communications

All the communications discussed up to now in this book were on a macro-scale. Communication between devices implanted inside human bodies, on a micro-scale, will, however, require yet another new communication paradigm—molecular communications [34]. Molecular communication may become another futuristic concept associated with 6G networks. Essentially, it involves communication using biochemical signals. Biochemical signals would typically travel in the form of small particles of a few nanometers to a few micrometers in size. These may be lipid vesicles and particles, which usually propagate in an aqueous or a gaseous medium. What is more, the concept of molecular communications can also be applied to the macroscale.

Certain advantages could be ascribed to molecular communications, when compared to traditional electromagnetic communications. First, the prospect of building nanoscale electromagnetic communication devices is at present not encouraging. This is where molecular communications could step in. Secondly, on a macroscale, these molecular communications could be used in some harsh environments where radio waves suffer from high path loss, e.g. in underground tunnels and gas pipelines.

Like quantum communications, research into molecular communications is in its infancy.

7.6 Nanophotonics

This section investigates the novel concept of nanophotonics, which is a branch of optoelectronics that involves replacing wire interconnects inside nanoelectronic circuits with optical links. These links may be on a single chip or between different chips. The main advantage of nanophotonics circuits is the speed at which the signals travel between circuits, thus allowing speeding up of circuit execution. This and many other advantages, described briefly, may be beneficial in computationally demanding circuits, such as AI and machine learning circuits, especially if these circuits are located on edge devices.

Several benefits can be ascribed to optical interconnects [28, 35, 36]. First, data rates in excess of 10 Gb/s can be established over a multitude of individual links. Second, this is achieved without signal interference (crosstalk), which is usually present if metal wires are packed close together. Third, in densely packed circuits, using metal interconnects also requires shielding, which is not necessary when optical interconnects are used. Thus, optical interconnects in ICs increase achievable interconnect density. Fourth, in optical interconnects, voltages and currents are not present (there is no need to charge the communication line), and therefore, there are no paths for the energy to leak out and optical circuits are, as a result, much more power-efficient. Finally, other benefits come from signal timing: optical interconnects are able to retain very precise timing in clocks and signals, and synchronization may not be needed. It is predicted that it will be possible to realize all these advantages at about the same cost tags as traditional circuits, once this technology is capable of entering mass production.

Optical interconnects in a typical IC would be realized using guided optics. Optical waves are guided via the principle of internal reflection, which results in the effect that light is confined inside the optical waveguide. Optoelectronic circuits also require an optical source and a detector (miniature versions of those described in Chap. 5). The substrate of optoelectronic circuits should ideally be able to support all three of these basic components and should also be commercially inexpensive. At this point, silicon can viably only be used to manufacture waveguides and photodetectors (photodiodes), whereas generation of light is typically reserved for III-V materials such as InP and GaAs. Thus, at present, it is most feasible to implement optoelectronic circuits as devices that incorporate different materials for each light emission, light guiding and light detection. As this approach is costly, it is to be expected that the full potential of optoelectronics for nanoscale devices will only be unlocked when challenges of light generation in silicon are overcome.

7.7 Blockchain Technologies

The last topic that will be discussed in this chapter is the role of blockchain technologies in 6G networks, especially from the viewpoint of security and privacy [19]. This is important, because 6G will require a holistic approach to secure the large amount of data on a diverse set of platforms, while complying with script privacy and security requirements.

Blockchain technology is one type of distributed ledger technology (DST). In the past decade, blockchain technologies have gained momentum and have been embraced by the industry and research communities alike across the globe [37]. Some of the characteristics of blockchain technology that make it appealing include:

- decentralization by eliminating the need for central trusted third parties and intermediaries;
- transparency and anonymity;

- provenance and non-repudiation of the transactions made;
- immutability and tamper-proofing of the distributed ledger's content;
- improving resiliency and resistance to attacks such as distributed denial of service; and
- a decrease in transaction processing delay and processing fee.

Whereas most people are probably more familiar with the role of blockchain in cryptocurrency, the blockchain features indicate than this technology may emerge as one of the key enabling technologies for 6G. Before describing the benefits that blockchain technologies may have for 6G networks, the basic principles of blockchain operation will be discussed.

7.7.1 Basic Principles of Blockchain Technologies

The main principle of blockchain is to build trust between networked applications, eliminating the need for trusted intermediaries [38]. Blockchain achieves this by building a distributed database, called the ledger, which collects the state changes of all participants as data blocks. These blocks are managed by the participants themselves. Blocks form a chronological chain, where the next block (b_{x+1}) is linked to the previous block (b_x) in a chronological order by their hash values. Blocks can always only be appended, without a possibility for removal. The blockchain is illustrated in Fig. 7.10. Blocks in the chain always remain in the same order, meaning that immutability, transparency and traceability are inherently ensured. In other (non-blockchain) DSTs, data can be handled differently, meaning that the whole blockchain does not necessarily always have to be retained.

Blockchain is considered the state-of-the-art trust technology. Blockchains are managed by a number of participants, which prevents falsification. Prior to being

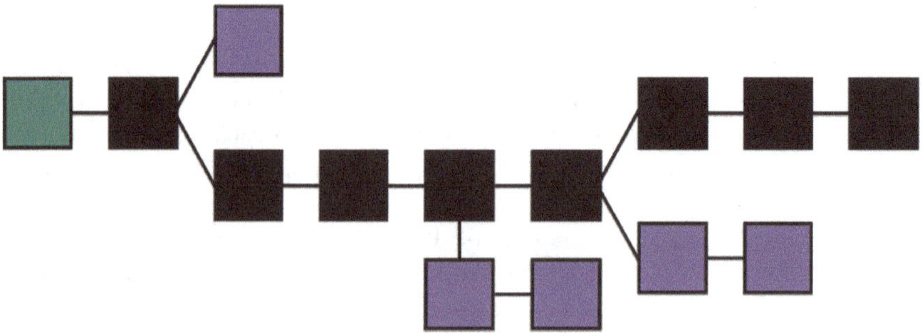

Fig. 7.10 Illustration of the block chain. Dark (black) blocks are the main chain, the green block is block b_{x+1} that is being added to the leftmost black block (b_x), purple blocks are unused (orphan blocks). Reproduced under Attribution 3.0 Unported license from https://commons.wikimedia.org/wiki/File:Blockchain.svg#/media/File:Blockchain_landscape.svg

added to the blockchain, a new data block is verified and agreed upon by a majority of the participants. During the verification process, those participants will require the witnessing of the previous data. Furthermore, blockchains are, by design, decentralized networks, which means that single points of failure (or attack) are avoided. This is the reason why many 6G verticals, including IoT and IoE, Industry 4.0 and 5.0, smart cities, smart healthcare, data storage and analytics, AI, V2V and airborne vehicle communications, as well as traditional communications, can all potentially benefit from the introduction of blockchain technology in 6G architecture [37].

7.7.2 Opportunities for Blockchain Technologies in 6G

According to Nguyen et al. [38], there are a large number of use cases for all mentioned verticals, that stand to benefit from the trust and security the blockchain introduces:

- *Edge computing.* Because edge cloud computations may evolve in 6G networks, blockchain can be used to secure sensitive information and ensure trustworthiness between the user devices and the edge servers.
- *Spectrum sharing.* Spectrum sharing means that other users can access ("lease") a portion of the spectrum that is dedicated to another user. Blockchain can ensure security of lease records.
- *D2D content caching.* Blockchain can increase security when data are bounced via other networked devices in the D2D chain.
- *Energy trading.* In SWIPT, some information may unintentionally leak during the transmission of power as part of the power beam. It is thus important that the information in a SWIFT network is secured, which is where the blockchain once again comes into the picture.
- *Federated learning.* In this case, blockchain can provide a decentralization framework.
- *Network architecture.* In open-source wireless networks, such as 6G, spectrum owners, infrastructure owners, and internet service provider owners freely exchange their resources. The security of each of those resources may be achieved with blockchain.
- *Network virtualization.* In the management of virtual network slices, blockchain can help with non-repudiation and immutability.
- *Interference management.* Lastly, blockchain can be used for optimal interference management.

7.7.3 Challenges Associated with Blockchain Technologies

Blockchain, like any other new and emerging technology, comes with a set of challenges. Three main groups of challenges of using blockchain, in the context of 6G, include security risks, scalability difficulty and the challenges associated with integration with quantum computing [38].

Where security is concerned, there are at least three types of attacks that could be made against the blockchain:

- Majority vulnerability: Because blockchain builds trust without third parties needing consensus, if the attacker gains control of over 50% of participants, theoretically he/she could control the blockchain.
- Transaction privacy leakage: Because blockchain relies in part on transparent transactions, user privacy may be an issue. This can, however, be resolved by using cryptographic algorithms.
- Double spending. This refers to attacks on the blockchain where a user performs two distinct equal-sized transactions.

Scalability has to do with the amount of computing power, storage space and bandwidth blockchain needs in order to ensure the integrity of the ledger. It also requires forced delays and a high number of messages being passed back and forth, which additionally introduce unwanted latency into the system. However, recall, from the earlier discussions on this topic, that high throughput, low latency, and synchronous data exchange are some of the most important KPIs of 6G. The communications and computational overhead that blockchain could introduce may prove prohibitive for 6G, unless the blockchain can be restructured into multiple abstraction layers, which would then be handled separately and more effectively.

Lastly, there is the challenge of getting the blockchain and quantum computing to work together, seeing that quantum computation can also address the issue of security. The answer to this challenge may be in a new concept, called the quantum blockchain.

7.8 Concluding Remarks

This chapter discussed the advanced, if not futuristic, technologies that may become a reality in future network generations. Whereas some of the technologies are already being deployed in the era of 5G, some of these technologies may only become reality sometime in the future, perhaps in the 6G era, or even beyond the 6G era, if at all [9].

However, there is still one group of futuristic concepts that has not received much attention in this book, despite being introduced in Chap. 1. This is the use of intelligence in 6G design, specifically AI, machine learning and EDGE computing. Thus, the discussion on network intelligence will form the topic of the next chapter, Chap. 8.

References

1. Rajatheva N, Atzeni I, Bjornson E, Bourdoux A, Buzzi S, Dore J-B et al (2020) White paper on broadband connectivity in 6G. arXiv:200414247 [eess] [Internet]. 2020 [cited 2020 Sept 16]. http://arXiv.org/abs/2004.14247
2. Björnson E, Sanguinetti L, Wymeersch H, Hoydis J, Marzetta TL (2019) Massive MIMO is a reality—What is next?: Five promising research directions for antenna arrays. Digital Signal Proc 94:3–20
3. Obeed M, Salhab AM, Alouini M-S, Zummo SA (2019) On optimizing VLC networks for downlink multi-user transmission: a survey. IEEE Commun Surv Tutor 21:2947–2976
4. Chen C, Zhong W-D, Yang H, Du P (2018) On the performance of MIMO-NOMA-based visible light communication systems. IEEE Photon Technol Lett 30:307–310
5. von Butovitsch P, Astely D, Friberg C, Furuskär A, Göransson B, Hogan B et al (2020) Advanced antenna systems for 5G networks (white paper). Ericsson
6. Yang P, Xiao Y, Xiao M, Li S (2019) 6G wireless communications: vision and potential techniques. IEEE Netw 33:70–75
7. Chen Y, Zhu P, He G, Yan X, Baligh H, Wu J (2020) From connected people, connected things, to connected intelligence. 2020 2nd 6G wireless summit (6G SUMMIT), pp 1–7
8. Huang C, Hu S, Alexandropoulos GC, Zappone A, Yuen C, Zhang R et al (2020) Holographic MIMO surfaces for 6G wireless networks: opportunities, challenges, and trends. IEEE Wirel Commun 1–8
9. Saad W, Bennis M, Chen M (2019) A vision of 6G wireless systems: applications, trends, technologies, and open research problems. IEEE Netw 1–9
10. Wu Q, Zhang R (2020) Towards smart and reconfigurable environment: intelligent reflecting surface aided wireless network. IEEE Commun Magaz 58:106–112
11. Bariah L, Mohjazi L, Muhaidat S, Sofotasios PC, Kurt GK, Yanikomeroglu H et al (2020) A prospective look: key enabling technologies, applications and open research topics in 6G networks. arXiv:200406049 [eess] [Internet]. 2020 [cited 2020 Sept 14]. http://arXiv.org/abs/2004.06049
12. Björnson E, Özdogan Ö, Larsson EG (2020) Intelligent reflecting surface versus decode-and-forward: how large surfaces are needed to beat relaying? IEEE Wirel Commun Lett 9:244–248
13. Liaskos C, Nie S, Tsioliaridou A, Pitsillides A, Ioannidis S, Akyildiz I (2018) A new wireless communication paradigm through software-controlled metasurfaces. IEEE Commun Magaz 56:162–169
14. Yildirim I, Uyrus A, Basar E, Akyildiz IF (2020) Propagation modeling and analysis of reconfigurable intelligent surfaces for indoor and outdoor applications in 6G wireless systems. arXiv: 191207350 [cs, eess, math] [Internet]. 2019 [cited 2020 April 21]. http://arXiv.org/abs/1912.07350
15. Basar E (2020) Reconfigurable intelligent surface-based index modulation: a new beyond MIMO Paradigm for 6G. IEEE Trans Commun 1:1
16. Björnson E, Sanguinetti L (2020) Power scaling laws and near-field behaviors of massive MIMO and intelligent reflecting surfaces. IEEE Open J Commun Soc 1:1306–1324
17. Özdogan Ö, Björnson E, Larsson EG (2020) Intelligent reflecting surfaces: physics, propagation, and pathloss modeling. IEEE Wirel Commun Lett 9:581–585
18. Giordani M, Zorzi M (2019) Non-terrestrial communication in the 6G era: challenges and opportunities. arXiv:191210226
19. Tariq F, Khandaker M, Wong K-K, Imran M, Bennis M, Debbah M (2020) A speculative study on 6G. IEEE Wirel Commun 27:118–125
20. Nawaz SJ, Sharma SK, Wyne S, Patwary MN, Asaduzzaman MD (2019) Quantum machine learning for 6G communication networks: state-of-the-art and vision for the future. IEEE Access 7:46317–46350
21. Wong K-K, Shojaeifard A, Tong K-F, Zhang Y (2020) Fluid antenna systems. arXiv:200511561 [cs, eess, math] [Internet]. 2020 [cited 2020 Oct 21]. http://arXiv.org/abs/2005.11561

22. Yang H, Alphones A, Xiong Z, Niyato D, Zhao J, Wu K (2020) Artificial intelligence-enabled intelligent 6G networks. IEEE Netw Early Access 1–9
23. Charbit G, Lin D, Medles K, Li L, Fu I (2020) Space-terrestrial radio network integration for IoT. 2020 2nd 6G Wireless Summit (6G SUMMIT), p 1–5
24. Zhang L, Liang Y-C, Niyato D (2019) 6G visions: mobile ultra-broadband, super internet-of-things, and artificial intelligence. China Commun 16:1–14
25. Makichuk D (2020) China leapfrogs world with first 6G experimental satellite [Internet]. Asia Times. 2020 [cited 2020 Nov 9]. . https://asiatimes.com/2020/11/china-leapfrogs-world-with-first-6g-experimental-satellite/
26. Rahman MS, Hossam-E-Haider M (2019) Quantum IoT: a quantum approach in IoT security maintenance. In: 2019 international conference on robotics,electrical and signal processing techniques (ICREST), p 269–272
27. Tegio, A (2019) Quantum computing and the IoT [Internet]. AZoQuantum.com. 2019 [cited 2019 Nov 30]. https://www.azoquantum.com/Article.aspx?ArticleID=101
28. Božanić M, Sinha S (2020) Millimeter-wave integrated circuits: methodologies for research, design and innovation. Springer Nature
29. DiVincenzo DP (2009) Quantum computing. In: Huff HR (ed) Into the nano era: Moore's Law beyond planar silicon CMOS [Internet], pp 297–313. Springer, Berlin, Heidelberg. [cited 2019 Nov 25]. https://doi.org/10.1007/978-3-540-74559-4_12
30. Bergou JA, Hillery M (2013) Introduction to the theory of quantum information processing. Springer Science & Business Media, Berlin
31. Grumbling E, Horowitz M (eds) (2019) Quantum computing: progress and prospects. National Academies Press, Washington, D.C
32. García-Álvarez L, Casanova J, Mezzacapo A, Egusquiza IL, Lamata L, Romero G et al (2015) Fermion-fermion scattering in quantum field theory with superconducting circuits. Phys Rev Lett 114:070502
33. Houck AA, Türeci HE, Koch J (2012) On-chip quantum simulation with superconducting circuits. Nature Phys 8:292–299
34. Huang T, Yang W, Wu J, Ma J, Zhang X, Zhang D (2019) A survey on green 6G network: architecture and technologies. IEEE Access 7:175758–175768
35. Tong XC (2013) Advanced materials for integrated optical waveguides. Springer Science & Business Media, Cham
36. Miller DAB (2009) Device requirements for optical interconnects to silicon chips. Proc IEEE 97:1166–1185
37. Hewa T, Gür G, Kalla A, Ylianttila M, Bracken A, Liyanage M (2020) The role of blockchain in 6G: challenges, opportunities and research directions. 2020 2nd 6G Wireless summit (6G SUMMIT), pp 1–5
38. Nguyen T, Tran N, Loven L, Partala J, Kechadi M, Pirttikangas S (2020) Privacy-aware blockchain innovation for 6G: challenges and opportunities. 2020 2nd 6G Wireless summit (6G SUMMIT), pp 1–5

Chapter 8
6G: The Intelligent Network

Abstract Perhaps one of the most disruptive technologies that will be mature by the time 6G is ready for deployment will be AI. Pervasive AI for communications and special cases of machine learning and edge intelligence are the three components that will turn 6G networks into intelligent networks. Each of these concepts will be discussed in this chapter in some detail. This chapter will also reconsider the topic of the ethics of AI.

Perhaps one of the most disruptive technologies that will be mature by the time 6G is ready for deployment will be AI.

Compared to 5G and earlier wireless network generations, 6G networks will undergo unprecedented transformation, and it will be characterized by a high degree of heterogeneity [1]. Multiple aspects, such as network infrastructure, radio access technologies, computing and storage resources, as well as application types, among others, will become difficult to manage using the simplistic, static and rigid approaches used in wireless networks today [2]. This problem will become aggravated by the fact that the amount of data generated in 6G networks will keep becoming larger and larger. It is evident that the new network will have to be organized as an intelligent network, where AI, at minimum, will handle the use of communications, computing, control, and storage resources in both the network edge and the network core in real time. Stated differently, this pervasive AI will be at the heart of the 6G network [3, 4], allowing the network to become self-sustaining, adaptive and capable of meeting the data and dynamic reconfigurability demands of future devices and services [5]. While some concepts of AI are already taking shape in 5G, in 6G networks, AI will grow from AI as a network enhancement to AI as the network foundation [6].

Machine learning, a form of AI that enables a system to learn from large amounts of data, will be the key AI concept that will empower 6G communications [7]. In particular, deep learning, reinforcement learning and federated learning are pictured to be the key functions giving intelligence to 6G. Machine learning becomes convenient in situations where no exact mathematical model of the system is available,

M. Božanić and S. Sinha, *Mobile Communication Networks: 5G and a Vision of 6G*, Lecture Notes in Electrical Engineering 751,
https://doi.org/10.1007/978-3-030-69273-5_8

but a sufficiently large amount of training data is available, the system is slow-varying along time, and the numerical analysis is acceptable [5]. For the most part, present and past wireless network generations are indeed mathematical model-based [2], deploying model-driven algorithmic design with data-driven approaches [8]. However, it will become close to impossible to expand such models into the context of 6G networks and thus, the opportunity for machine learning in 6G networks opens in full.

Furthermore, the devices generating and consuming the data are commonly located at the edge of the networks, in other words, near the network end users, or near systems that require monitoring, surveillance and control [9]. Analysis of the large amount of generated data requires these data to be transferred to the core of the network, which introduces additional data transfer demands as well as unwanted latency. The solution to this issue may be sought in reorganizing the network in such a manner that computing, or at least a big part of it, is actually performed at the edge of the network itself. This leads to a new concept called edge computing, a computing paradigm where the computing happens at the edge of the network (in the edge cloud) instead of the central location (in the core cloud). Similarly, some of the network intelligence may also be located in the edge, leading to the concept of edge AI. Edge AI, edge computing and some other edge concepts can collectively be referred to as edge intelligence.

Pervasive AI for communications, as well as special cases of machine learning and edge intelligence, are the three components that will turn 6G networks into intelligent networks. Each of these concepts will be discussed in this chapter in some detail. This chapter will also reconsider the topic of ethics of AI, briefly discussed in Chap. 1.

8.1 Pervasive AI for Wireless Communications

The 6G network is envisioned as a network where AI is spread in all spheres or layers in the network, from network orchestration and management to coding and signal-processing in the physical layer, manipulation of smart structures, to data mining at network and device level for service-based context-aware communications, among others [3]. The pervasive nature of the AI presence in the 6G architecture makes 6G the first fully intelligent radio (IR). Previous network generations could be regarded as software-defined radio (SDR) networks, which had evolved into cognitive radio (CR) networks by the time 5G emerged [1]. Key differences between each of these approaches to network intelligence are highlighted in Table 8.1. The comparison includes differences in frequency selection and spectrum sharing, a comparison of how hardware is accessed and upgraded, differences in how physical transmitter and receiver hardware are organized, differences in approaches to multiple access, differences in protocols that can be used, as well as a comparison of verticals that are made possible in each case. As expected, IR introduces extreme flexibility into

Table 8.1 Differences between SDR, CR and IR. Adapted from [1]

Characteristic	SDR	CR	IR
Frequency band	Fixed	Adaptive to environment	Adaptive to environment and hardware
Spectrum sharing	Fixed	Opportunistic	Controlled by AI
Multiple access	Predetermined	Sensing based	Machine-logic-based
Hardware capability	Pre-claimed	Pre-claimed	Real-time estimated
Hardware upgradability	No	No	Yes
Transmitter/receiver module	Modulation/coding/detection/estimation	Modulation/coding/detection/estimation	DNN
Protocols	Fixed	Fixed	Self-upgradeable
Mainstream apps	Voice, data	Same as SDR, plus multimedia	Same as CR, plus AI and in-network computation

the network architecture, which is seen as a leading advantage for future wireless networks.

The simplest way to visualize where AI contributes to all aspects of 6G network architecture is to introduce a physical network model consisting of four layers, as done by Yang et al. [10]. The four layers, ranked from the most primitive to the most advanced, are:

- Intelligent sensing layer;
- Data mining and analytics layer;
- Intelligent control layer; and
- Smart application layer.

These four layers can control almost all the functions of the wireless network, from data collection to service provisioning, as illustrated in Fig. 8.1. These network layers, and the functions they can perform, are discussed in more detail in the subsections that follow.

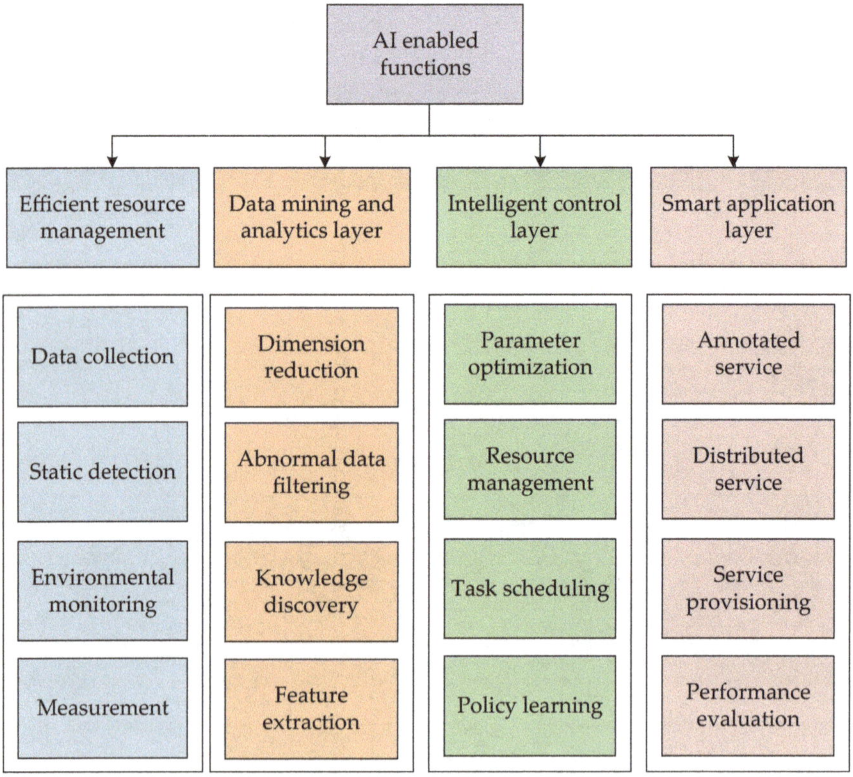

Fig. 8.1 Architecture of intelligent 6G as per [10]

8.1.1 Intelligent Sensing Layer

In this layer, data are collected either by various sensors (e.g. those networked in IoT, for example for environmental monitoring, or those operating as part of the Industry 4.0 umbrella), or by a large amount of connected user devices (such as smartphones, smart wearables and cameras) as well as by vehicles, drones, and other connected machines [10]. This is the most primitive layer in the four-layer structure of 6G and is in essence the physical layer of the network. The role of AI in this layer is to enable sensing and detection of a large amount of dynamic, diverse, and scalable data by directly interfacing with the physical environment. In this layer, AI can assist with RF utilization identification, spectrum sensing, imaging, interference detection, and other tasks. Sensing happens in real time; thus, it must remain robust to changing environmental and physical conditions, and it should be robust in scenarios when the users are in motion. In addition to the required data, this layer can also collect contextual data that may later be useful in making certain decisions, including the information that describes the state and health of the hardware deployed in this layer. Other context factors may include the environment, traffic patterns, mobility patterns and physical location [6].

One of the key functions in this layer is spectrum sensing. This finding can improve spectrum usage efficiency and contribute to resolving spectrum scarcity problems, allowing for cognitive spectrum use [6]. AI technologies can identify spectrum characteristics, and intelligently establish suitable training models to sense spectrum working status. This is the first step in end-to-end optimization of the full channel of the physical layer, from the transmitter to the receiver [1]. Other scenarios where this layer makes a key contribution is in the case of autonomous driving and in the navigation of UAVs, as well as in associated handling of information transfer to and from these vehicles and among these vehicles themselves [10]. The problem in this case is the constant high-speed mobility, meaning that the environment changes at a rapid pace and that low-latency real-time sensing becomes paramount.

Data collected in the intelligent sensing layer is sent to the layer above, the data-mining and analytics layer.

8.1.2 Data-Mining and Analytics Layer

Data collected from the massive amounts of devices physically connected to the network are processed in the data-mining layer [10]. In physical terms, the storage and computational capabilities of the 6G network are located in this layer.

The data are generally heterogeneous, and can also be nonlinear and high-dimensional, which means that the processing of these data is a challenging task. On the other hand, it is not feasible to transmit massive amounts of data, because this can become costly in terms of available resources. Thus, the role of AI is to transform the higher-dimensional data into a lower-dimensional subspace, which has the overall result of reducing the total computing time as well as the amount of data transmitted and stored. In addition, data analytics allows understanding of the essential characteristics of wireless networks and achieving more clear and in-depth knowledge of the behavior of 6G networks. In this layer, valuable patterns or rules can be discovered as more data become available, enabling the provision of suitable solutions for tasks such as resource management, protocol adaptation, data traffic control, architecture slicing, signal processing and minimizing the end-to-end link delay [1, 3, 10].

Because of the massive amount of data that will be collected in 6G networks, data mining becomes a big data analytics problem. Four types of big data analytics have been discussed in Chap 1 (from [1]):

- Descriptive analytics refers to mining historic data for gaining insight into network performance, channel conditions and traffic.
- Diagnostic analytics, among other uses, is necessary to detect faults in the network and improve network reliability.
- Predictive analytics refers to using the data to predict future events, including traffic patterns and resource availability.
- Prescriptive analytics is used to suggest the decision options for resource allocation, network slicing and virtualization, edge computation and other uses.

Once analyzed, the collected data can be passed onto the intelligent control layer.

8.1.3 Intelligent Control Layer

The intelligent control layer is the layer in which network control is situated [10]. It includes the functions of learning, optimization and decision-making. The decisions made in this layer are converted into actions that trigger the functions in lower layers (power control, spectrum access, routing and resource management, network association, etc.), thus forming a closed loop [1]. In other words, this layer is the brain of the network, and it is in this layer where machine learning and AI-based decision-making happens. The core network cloud, as well as the edge cloud and edge intelligence, are all located in this layer.

Generally speaking, learning can be defined as the process of utilizing or modifying existing knowledge and experience to improve one's behavior, whereas decision-making is an important cognitive task that enables massive agents to intelligently reason, plan, and choose the most suitable decisions to meet high-quality service requirements. In the wireless network context, AI or machine learning techniques are the best learning techniques for network design; thus AI/machine learning is the most feasible option for network optimization for 6G wireless networks. In short, by deploying machine learning, 6G networks can learn to achieve self-configuration, self-optimization, self-organization and self-healing. In essence, this approach allows 6G networks to become "zero-touch" operation-based networks [11].

The importance of machine learning in 6G networks cannot be stressed enough. For this reason, machine learning, including various specific machine learning techniques such as reinforcement and deep learning, which are regarded as the most powerful machine learning techniques [3, 7], will be discussed separately, in Sect. 8.2. Similarly, the increasingly important concept of edge intelligence will be examined separately, in Sect. 8.3.

The three layers discussed up to now (the intelligent sensing layer, data mining and analytics layer, intelligent control layer) support the existence of the smart application layer, which is closest to the end user.

8.1.4 Smart Application Layer

The highest layer in the proposed four-layer intelligent organization of 6G networks is the smart application layer. The application layer is responsible for delivering application-specific services, supporting different verticals of 6G. These services include but are not limited to old and new types of communication, data transfer, sensing, imaging and radar sensing, localization, IoT and IoE, Industry 4.0 and Industry 5.0, VR, AR and XR, robotics and autonomous systems, as well as tactile internet, as will be illustrated in the closing chapter of this book in Fig. 10.1.

Many services are not unique to 6G and may be supported without the help of AI. However, the majority of listed verticals can benefit to various degrees from machine learning and AI decision-making. For example, IoT-related applications such as automated services, smart cities, smart industry, smart transportation, a smart grid, and smart health are managed more efficiently by AI than by pre-coded algorithms.

The smart application layer is also involved in evaluation of the performance of various services [10]. Evaluation considerations include QoS, quality of experience (QoE), quality of collected data, and quality of learned knowledge, among others, as well spectrum utilization efficiency, computational efficiency, energy efficiency and storage efficiency. Based on these evaluations, the network can seek to improve its performance parameters constantly and sustainably.

The four AI layers into which the architecture of the 6G network is organized interact with one another in real time. It is worth reiterating that AI algorithms will be deployed and trained at all layers of the network, encompassing network management, core, physical radio base stations and user equipment [11]. Machine learning is at the center of this architecture, keeping it fused together, and it is thus fitting to discuss it next.

8.2 Machine Learning

Machine learning is generally understood as a subbranch of AI, as illustrated in Fig. 8.2a. Although some sources argue that machine learning can be used to develop AI, not all machine learning applications can be considered AI [12], as illustrated in Fig. 8.2b. In machine learning, machines perform and improve their operations by exploiting operational knowledge and experience gained in the form of data [5]. Machine learning plays a crucial role in systems in which modeling cannot be presented by a mathematical equation [2]. Thus, as already suggested, it is envisioned that machine learning will assist in several spheres of 6G network operation. It will enable real-time data analysis and automated zero touch operation and control of the network, propagating through all layers of 6G architecture.

The general process of conventional machine learning is shown in Fig. 8.3 [13]. Conventional machine learning methods start by assuming an inductive bias and are based on the availability of training data and time for using these data to train the system for a configuration of interest. The inductive bias includes a model class (for example, a given architecture neural network) and a predetermined learning procedure, as well an optimization algorithm. Looking mathematically, the typical goal of conventional learning is to minimize the out-of-sample, population, or generalization loss [13]

$$L_k(\varphi) = E_{x \sim P_k}[l(x, \varphi)], \tag{8.1}$$

where $E_{x \sim P_k}$ is the expectation that is taken over the true distribution P_k of data point x for task k, and function $l(x, \varphi)$measures the loss for data point x when using

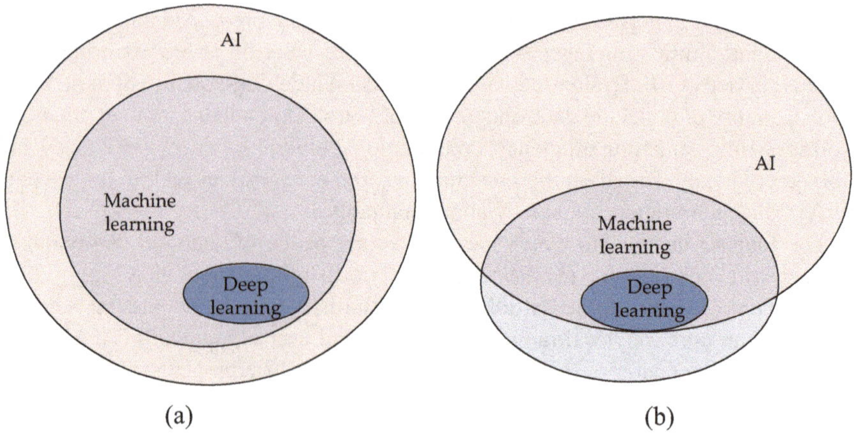

Fig. 8.2 Two definitions of machine learning: **a** Machine learning as part of AI; **b** Machine learning as a separate discipline, with some machine learning applications placed outside the scope of AI. Deep learning is considered a dominant approach of machine learning irrespective of the definition of machine learning

Fig. 8.3 Process of conventional learning. Adapted from [13]

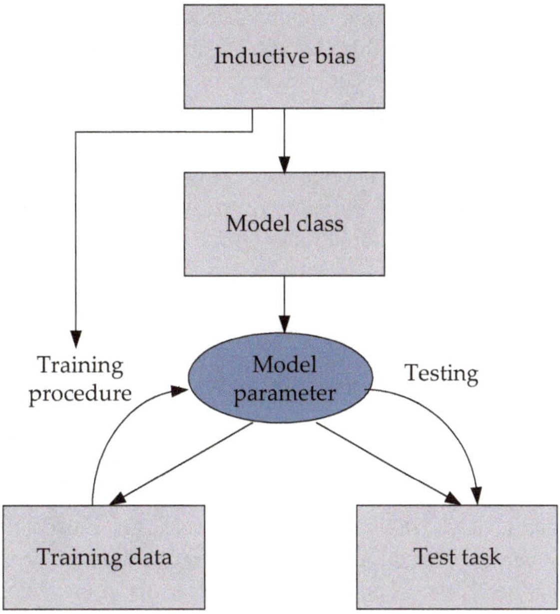

the model characterized by a parameter vector in the given model class. Since the distribution P_k is usually assumed to be unknown to the learner in the learning problem, most learning procedures minimize some function of the training loss, which is an empirical approximation of the population loss.

On the practical level, three machine learning paradigms are usually isolated [2]:

(1) *Supervised learning:* In supervised leaning, the model is taught by presenting input samples and their associated outputs.
(2) *Unsupervised learning:* In unsupervised learning, there are no output labels and the model can only learn to classify samples of the input.
(3) *Reinforcement learning:* In this learning technique, an agent interacts with an environment and learns to map any input to an action, similar to how the learning process in a human brain uses the trial and error approach [7].

In recent years, deep learning has emerged as the preferential learning method for communication networks. Deep learning resembles the perception process in a brain enabled by a DNN [7]. In machine learning, the DNN is a part of an artificial neural network [5]. Machine learning can be performed in a supervised, unsupervised, reinforcement or hybrid fashion. Federated learning has also been suggested for 6G networks, with the aim of protecting the privacy of mobile devices. Federated learning is a distributed machine learning algorithm that enables mobile devices to collaboratively learn a shared machine learning model, while avoiding data exchange among mobile devices [2]. Other learning methods recently proposed include meta-learning [13] and QML [5]. Genetic programming may also be used in certain scenarios, typically for optimal antenna selection, power control and symbol detection [5]. It should be noted that the use of blockchains has also been suggested as a vehicle for making the decision process of machine learning methods more understandable and coherent. This is because the blockchain structure enables all the underlying elements on which the decisions are made to be traced [14].

With the fundamental concept of machine learning covered, the most important machine learning techniques, viz. deep learning, federated learning, reinforcement learning, meta-learning and QML, will be discussed in more detail in the sections that follow.

8.2.1 Deep Learning

Increased adoption of the deep learning approach has predominantly been triggered by an increase in computational power and access to large datasets [2]. Deep learning has advanced rapidly over the last decade and is used in several applications, including image classification, computer vision, social networks and security [6]. These application areas are generally regarded as the areas where significant amounts of data are readily available for training, which is paramount for deep learning.

Deep learning is based on DNNs. Typically, the DNN consists of the input layer, several hidden layers and an output layer, as shown in Fig. 8.4 [7], with the neuron as the smallest element, responsible for processing and passing the information between the different layers. The input layer provides the information observed by the learning agent, the hidden layer extracts the features of the input information, and lastly, the output layer makes decisions (classifications). In order for a neural network to perform induction, inductive bias, consisting of a set of explicit or implicit assumptions, is required [15, 16]. Among other factors, inductive bias depends on the neural network topology, the activation function and the network training regime. In general, activation functions are nonlinear functions. Some examples of activation functions include the sigmoid function, tanh function, and the rectified linear units (ReLU) function. Plots of these functions are shown in Fig. 8.5 [7].

In essence, the neuron network functions in the following way: If the input vector of a neuron is

$$\mathbf{x} = (x_0, x_1, x_2, \ldots x_n)^T \tag{8.2}$$

and the corresponding weight vector is

$$\mathbf{w} = (w_0, w_1, w_2, \ldots w_n)^T, \tag{8.3}$$

then the output of the neuron is

$$y = \sigma\left(\mathbf{w}^T \mathbf{x}\right), \tag{8.4}$$

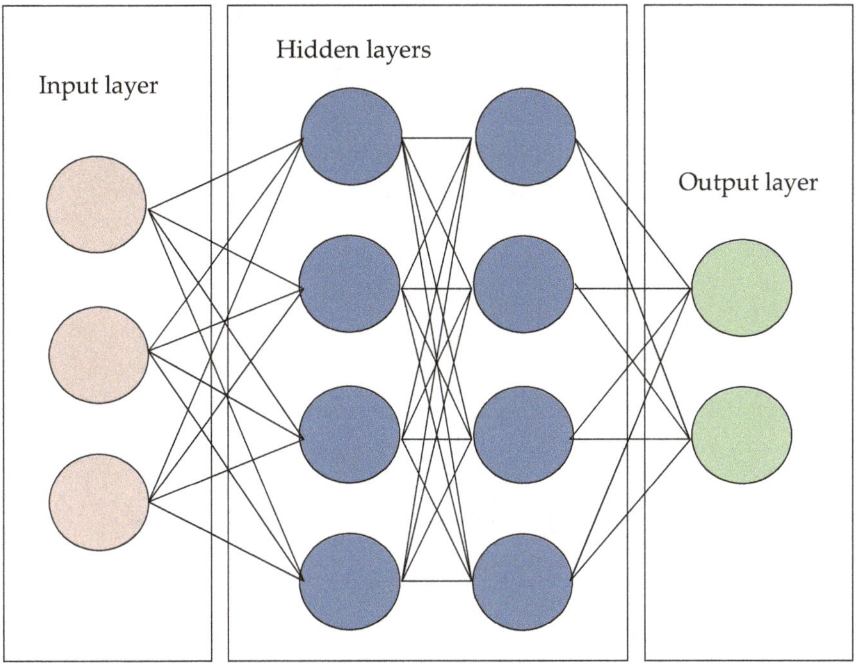

Fig. 8.4 Framework of the DNN. Adapted from [7]

where $\sigma(\cdot)$ is the activation function defined above. The weight vector of each neuron plays a critical role in the corresponding output and thus directly influences the performance of the DNN.

Before using the deep learning algorithm, it needs to be trained off-line. Labeled data are used to train the weights of the whole DNN by minimizing a specific loss function, such as mean-squared error and categorical cross-entropy functions. The DNN training data can come from various sources, for example manual designs, experimentation or computer simulations based on practical situations. One of the widely used training algorithms of the DNN is called stochastic gradient descent through back-propagation.

The DNN shown in Fig. 8.4 is just one possible DNN architecture, called the multi-layer perception network. Other types of neural networks are possible, including recurrent neural network (RNN), convolutional neural network (CNN), deep belief network (DBN), autoencoder (AE) and general adversarial networks (GAN) [17]. The architectures of these DNNs are shown in Fig. 8.6 and are briefly described in Table 8.2. One of the pressing challenges in using deep learning for wireless communication lies in finding the optimal design of the DNN for different scenarios the network may come across. Finding the perfect model for the communication networks has thus been the burning topic of research in machine learning for future wireless networks in recent years.

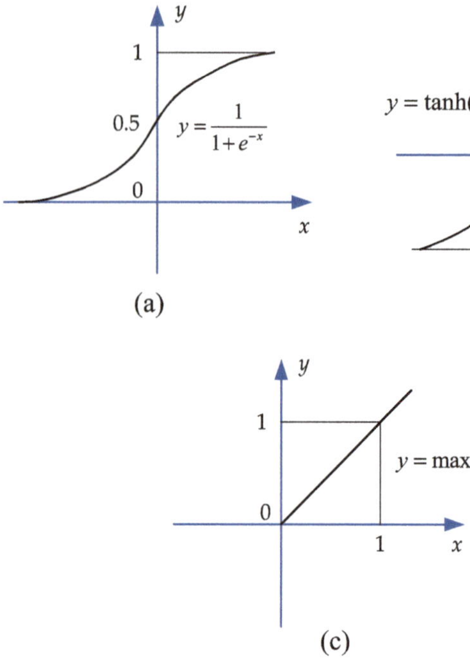

Fig. 8.5 Commonly used activation functions of a neuron: **a** Sigmoid, **b** Tanh and **c** ReLU

8.2.2 *Federated Learning*

Federated learning is a learning approach in which multiple clients collaborate with the data center to solve a machine learning problem [18]. This process can be pictured as shown in Fig. 8.7. As already mentioned, federated learning enables mobile user devices to collaboratively learn a shared machine learning model, while avoiding data exchange among mobile devices. In the case of federated learning, each user device and the data center have their own machine learning models, called the local and global federated learning mode respectively [2].

The training in federated learning is performed as follows:

1. Each user device uses its collected data to train its local federated learning model and sends the trained model to the data center.
2. The data center integrates the local learning models to generate the global learning model and broadcasts it back to all user devices.
3. Steps 1 and 2 are repeated until the federated learning model that minimizes the federated learning model loss function is found.

The main disadvantage of federated learning is that the training data need to be sent over wireless links. However, the main advantage comes from the fact that federated learning can play a major role in incorporating end-to-end operations between devices

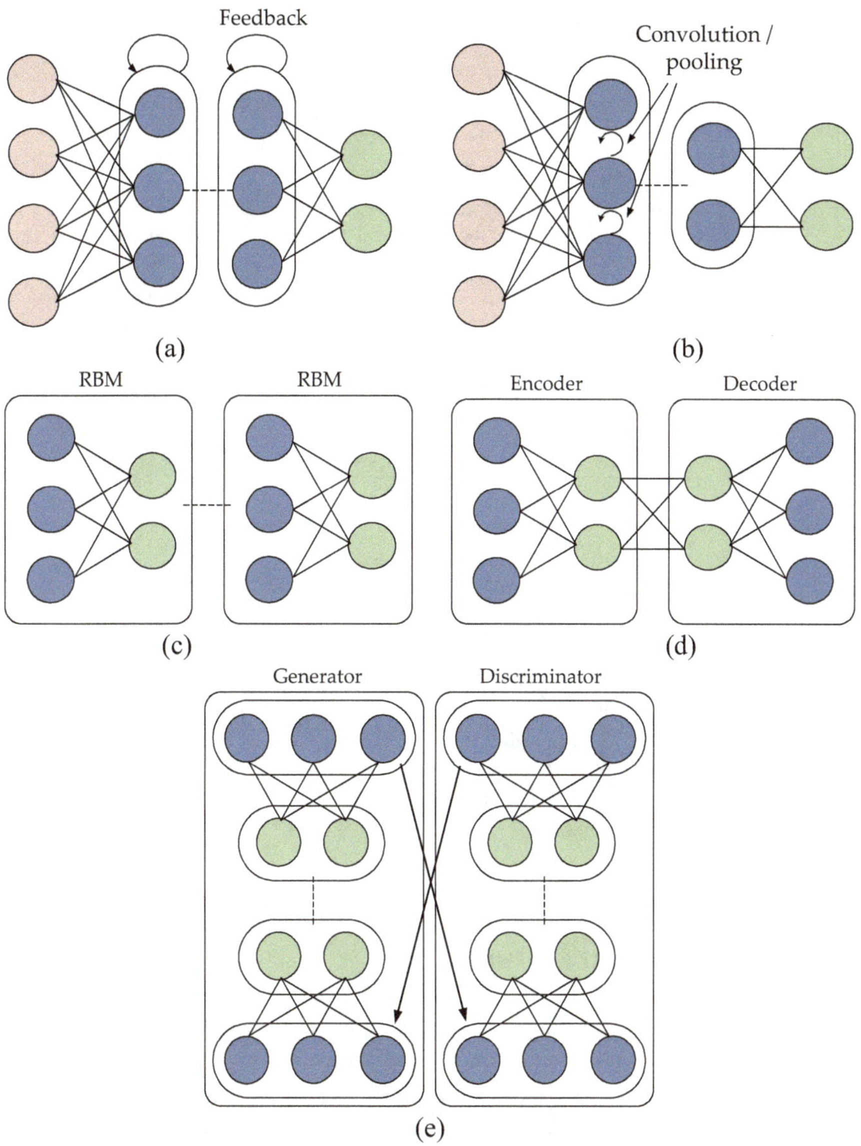

Fig. 8.6 Alternative neural network types: **a** RNN, **b** CNN, **c** DBN, **d** AE and **e** GAN

and network infrastructure into 6G [6]. Furthermore, federated learning is one of the learning techniques that can directly benefit from the introduction of blockchain. This further improves the reliability and scalability of federated learning. Blockchained federated learning is sometimes referred to as BlockFL [17].

Table 8.2 Differences between different DNN topologies. From [17]

DNN type	Description
RNN	The output of a layer has recursions (feedback) within the same layer
CNN	Two pre-processing layers, the convolutional layer and pooling layer, are inserted
DBN	Stack of pre-trained reduced Boltzmann machines
AE	A stack of two feedforward neural networks, copying the input to its output in an unsupervised learning way; the first neural network handles the encoding and the other one handles the decoding
GAN	A stack of two neural networks, a generator and a discriminator, "playing" a zero-sum game: the generator produces fake samples to fool the discriminator and the discriminator tries to identify the fake samples

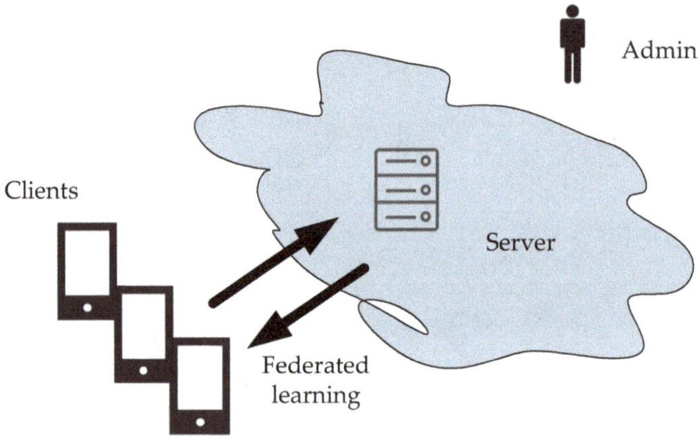

Fig. 8.7 Federated learning. Adapted from [18]

8.2.3 Reinforcement Learning

In reinforcement learning, an agent interacts with an environment, thus learning how to take action [2, 7]. The process, illustrated in Fig. 8.8, also involves receiving a numerical reward every time upon choosing the action, before moving to the next state. The aim of reinforcement learning is to maximize the cumulative reward. An example of a problem that can be formulated as a reinforcement learning problem is network resource allocation.

Reinforcement learning is structured as a recursive learning approach. This means that if the learning agent has experienced the best action for a specific state of the environment, it can use this experience to improve the long-term reward by selecting the best action. In many instances, experience is lacking, and in those cases, the agent needs to search for (explore) the best action to perform, which slows the learning down.

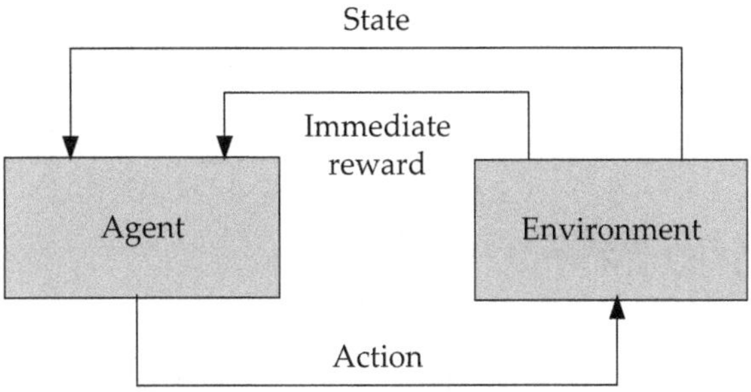

Fig. 8.8 Reinforcement learning framework

One of the main advantages of reinforcement learning is that it focuses on a long-term reward, which includes both an immediate reward and future rewards, effectively making intelligent predictions on the future radio environment. On the other hand, conventional learning algorithms only attempt to optimize immediate reward. Its main drawback is its high reliance on training. In addition, for a large state-action space, the performance of reinforcement learning may drop, since many state-action pairs may not have been explored by the reinforcement learning agent at the given time.

8.2.4 Meta-Learning

Meta-learning is a proposed method that introduces a way to automate the selection of the inductive bias. For this reason, meta-learning is also called "learning to learn" [13]. Meta-learning has the potential to reduce both training data and training time requirements.

In meta-learning, inductive bias selection is performed by leveraging data or active observations from tasks that are known to be related to a future, a priori unknown, system configuration of interest. The key strong point of meta-learning is the fast adaptation to a new task using task-specific training examples. A block diagram, illustrating meta-learning, is shown in Fig. 8.9. This figure shows that the learning algorithm of meta-learning is almost identical to that of conventional learning. However, from this figure it can also been seen that the inductive bias, which is fixed in traditional learning, is in this case selected dynamically. Meta-learning algorithms differ mostly in the way the inductive bias is optimized. One of the more popular meta-training algorithms is called model-agnostic meta learning (MAML). The details of MAML are discussed in [13] and will not be replicated here.

In future wireless networks, meta-learning could be used both on the side of the transmitter and on the side of the receiver. For example, on the receiver side, supervised learning of a demodulator or a decoder could deploy meta-learning. In another scenario, end-to-end learning of an encoder and a decoder of a communication link (in the transmitter and the receiver respectively) could be performed in unsupervised fashion.

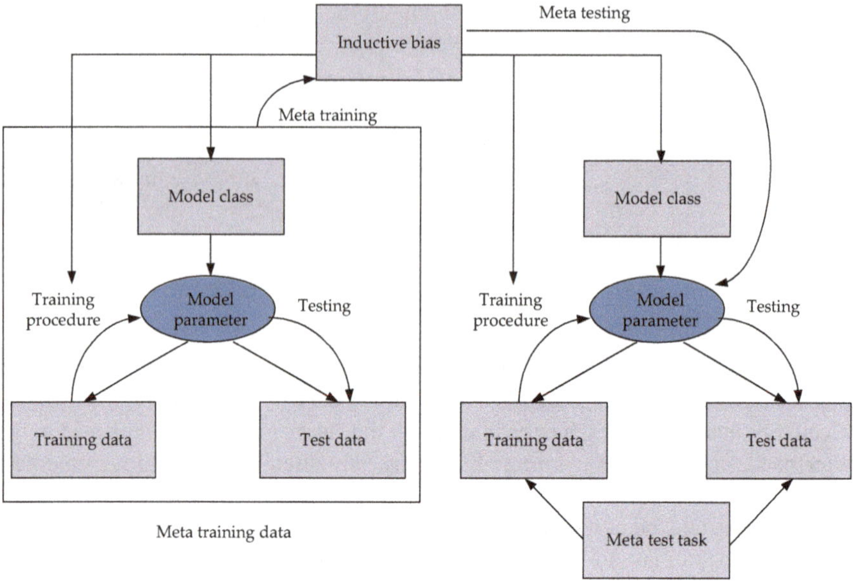

Fig. 8.9 Meta-learning. From [13]

8.2.5 *Quantum Machine Learning*

QML uses the principles of quantum computing, discussed in the previous chapter, to speed up the process of machine learning in various applications.

QML has the potential to run several orders of magnitude faster than conventional machine learning [5]. At the same time, QML can enhance security and privacy in communication networks, or alternatively, enable applications that would be prohibitive if conventional machine learning were deployed because of the massive computation requirement. Some tasks that can notably benefit from QML include quantum pattern recognition, quantum classification, quantum process tomography and regression, as well as boosting quantum computing and adiabatic quantum computing. Like regular machine learning, its quantum variation could also proceed in the supervised, unsupervised or reinforcement manner. Similarly, it is expected that quantum deep learning would be the equivalent of the popular deep learning described in Sect. 8.2.1.

Quantum machine learning suffers from the same set of drawbacks that quantum computing in general encounters: lack of commercial redistributable quantum computing hardware. This problem may still take a while to be solved, as discussed earlier in Sect. 7.4.

8.2.6 *Taxonomy of Machine Learning Applications*

The discussion of machine learning will be wrapped up with a brief review of machine learning applications.

As already discussed, machine learning will penetrate all architecture layers of future wireless networks, starting with 6G. In addition to realizing functions that have not been realized properly without machine learning, it will enable some functions that would not be possible at all without machine learning [2, 8, 19]. Machine learning will contribute equally on the physical, network control and application layers. On the physical layer, machine learning has the potential to assist with channel coding, synchronization, positioning, channel estimation and detection, beamforming and physical layer optimization. On the network control layer, machine learning could assist with resource allocation, power management, traffic control and security. On the application layer, it could assist with network performance management, UAV control, opportunistic data transfer for vehicular networks, data transfer for smart city networks, as well as for sensing, imaging and localization, to mention just a few.

Nawaz et al. [5] have created a comprehensive taxonomy of machine learning applications. In their paper, potential machine learning applications were classified according to different protocol layers, learning types (supervised, unsupervised and reinforcement) and data availability on the side of the network and on the side of the user device. The taxonomy of Nawaz et al. was adapted for this book and is shown in ual edge and the central cloud Table 8.3.

What is more, new applications of machine learning in the context of wireless applications keep emerging as research into machine learning for communications is progressing. Applications such as blockage detection [20], compressed sensing [21] and prediction of achievable data rates in mobile and vehicular networks [22], for example, have not been captured in ual edge and the central cloud Table 8.3.

8.3 Edge Intelligence

The edge consists of computing and network resources located on the path between physical data sources and network clouds [23]. This may, for example, be the radio access network part of a mobile network. Edge intelligence is considered one of the missing elements in all wireless networks used at present, including 5G networks. Even though the 5G network is already using some concepts relating to AI and machine learning, the architecture of the network is such that the various services (verticals) cannot explore the benefit of AI to its full potential [9]. Consider, for example, a scenario of a self-driving vehicle that is connected to a network that tries to explore the machine learning capability of the network that is located in its core (cloud) to plot the course of its further movement or to make a decision on whether it needs to apply its brakes [17]. This vehicle needs to communicate with the network with an incredibly short latency. In addition, it may not be the only vehicle, nor the

Table 8.3 Taxonomy of machine learning applications across different layers of a future communication network. Adapted and expanded from [5]

	Learning type			Available data	
Network layer	Supervised	Unsupervised and partially supervised	Reinforcement	Side of the network	Side of the user device
Physical	Channel equalization, decoding, and prediction, path loss and shadowing prediction, localization, sparse coding, filtering, adaptive signal processing, beamforming	Modulation, interference cancellation, mobility prediction, spectrum sensing, radio resources optimization, localization, security, transmission optimization, nodes clustering, duplexing configuration, multiple access, beam switching	Link preservation, channel tracking, on-demand beamforming, secure transmission, energy harvesting, transmit power selection, nodes selection, channel access management, modulation mode selection, coverage optimization, anti-jamming radio identification	Baseband signals, channel models, channel state information, spatio-temporal statistics, received power record, etc.	Baseband signals, temporal statistics, channel models, received power record, etc.
Network and other layers	Caching, traffic classification, network anomalies identification, throughput optimization and adaptation, latency minimization, optimization of other KPIs	Multi-objective routing, traffic control, network state prediction, source encoding and decoding, network parameters prediction, intrusion detection, fault detection, anomaly detection, etc.	Multi-objective routing, packet scheduling, access control, adaptive rate control, network security, capacity and latency demand prediction, traffic prediction and classification, network slicing	Traffic load, service demands, random access, latency, user type	Mobility tracks, traffic statistics, outage statistics, etc.

(continued)

Table 8.3 (continued)

	Learning type			Available data	
Network layer	Supervised	Unsupervised and partially supervised	Reinforcement	Side of the network	Side of the user device
Application	Smart health care, smart home, smart city, smart grid, query processing, data mining, crime detection, etc.	Data processing, data ranking, data analysis (spatial, temporal, etc.) data flow prediction, dimension reduction, malware detection and classification, network anomaly prediction, demographics features extraction and prediction, fraud detection	Proactive caching, data offloading, error prediction, traffic rate determination and allocation, data rate selection for segments	Media traffic demands, user behavior, services ranking, resources ranking, computational load, KPI records, etc.	Services utilization frequency, user behavior, local app data (health, location, screen time, media, etc.), subscription record, etc.

only smart terminal, user device or IoT sensor located at the edge of the network that is trying to send data for processing in the core at that given moment. In fact, these and many other devices are expected to generate several hundreds of ZB (10^{21} B) of data at the network edge that need to be transferred and processed simultaneously by the time 6G is deployed. It is clear that this scenario has little prospect of being successful in reality.

In an alternative solution, the self-driving vehicle may be allowed to only push the data to the edge of the network, where part of the service-specific computations, data storage and decisions can happen instead. This leads to a much shorter data exchange turnaround time and less data traffic, thus enabling the vehicle in this example to react swiftly.

Computing that does not require data processing in the cloud but rather utilizes the computing capacity located at the edge only is called edge computing and is illustrated in Fig. 8.10. Furthermore, if computing decisions involve the use of AI or machine learning, the concept can be referred to as edge intelligence or "AI at the edge" [24] instead. Fog computing is another concept in edge computing, but this time, computing capacity at the edge is expanded to include computers located between the actual edge and the central cloud.

As the levels of AI at the edge increase, it becomes possible to bring some AI features to each node, as well as on clusters of nodes. Clusters and nodes can learn

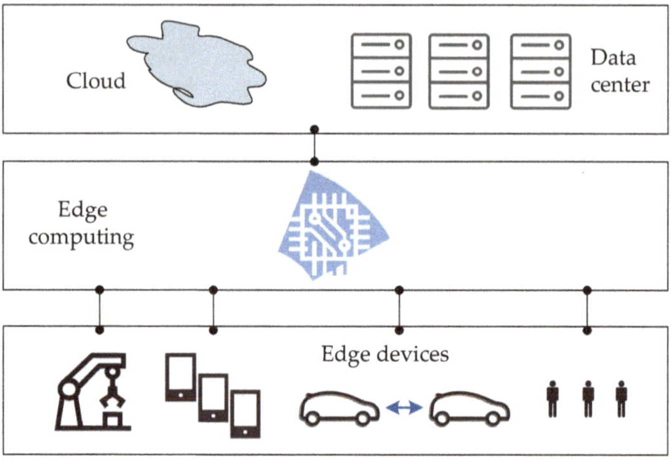

Fig. 8.10 Illustration of the edge computing concept

progressively and possibly share what they learn with other similar edge nodes, thus providing new services or optimizing old ones, in addition to benefits such as increasing the data-processing throughput, decreasing the latency of inference and increasing the reliability of the network, which are considered to be the trigger for considering edge intelligence in the first place. The addition of a number of advanced services and QoS functionalities is an appealing proposal and various sources predict that 6G network designers will consider highly distributed AI, where the intelligence is located not only on the central cloud, but also on the edge computing resources [4, 5, 9, 17, 23–26].

The sections that follow describe the basic aspects of edge intelligence as envisioned for 6G.

8.3.1 Levels of Edge Intelligence and Edge Intelligence Classification

Naturally, the task of designing the network that supports distributed AI is not without complexity. Edge devices, cloud software, gateways and other tools need to be designed in such a manner as to enable edge intelligence. This requires so-called "liquid" software, software that "flows" from one device to another to be designed for 6G, among other challenges that need to be overcome before edge intelligence can become a reality [9]. In addition, computationally efficient dedicated hardware capable of locally training and running AI and machine learning algorithms on edge devices in an online and decentralized manner is required. This deviates from the classical offline and centralized concepts of AI and machine learning and training [17, 24]. Bringing the intelligence to the edge is, as a result, likely to be a task that will be carried out in steps, resulting in different stages of edge intelligence autonomy and capability.

Zhou et al. [26] categorize edge intelligence into six levels, based on data path length and the amount of data that needs to be offloaded to the cloud. The categorization was extended in the white paper on edge intelligence written by the 6G flagship program members [9] to include an additional, seventh, level. In this seven-level model, the edge can either be viewed as a set of single, autonomous, intelligent nodes or as a cluster or a collection of federated/integrated edge nodes. The 6G flagship group also added differentiation in terms of a different degree of autonomy in the operation of the edge nodes. The updated 6G flagship model is shown in Fig. 8.11.

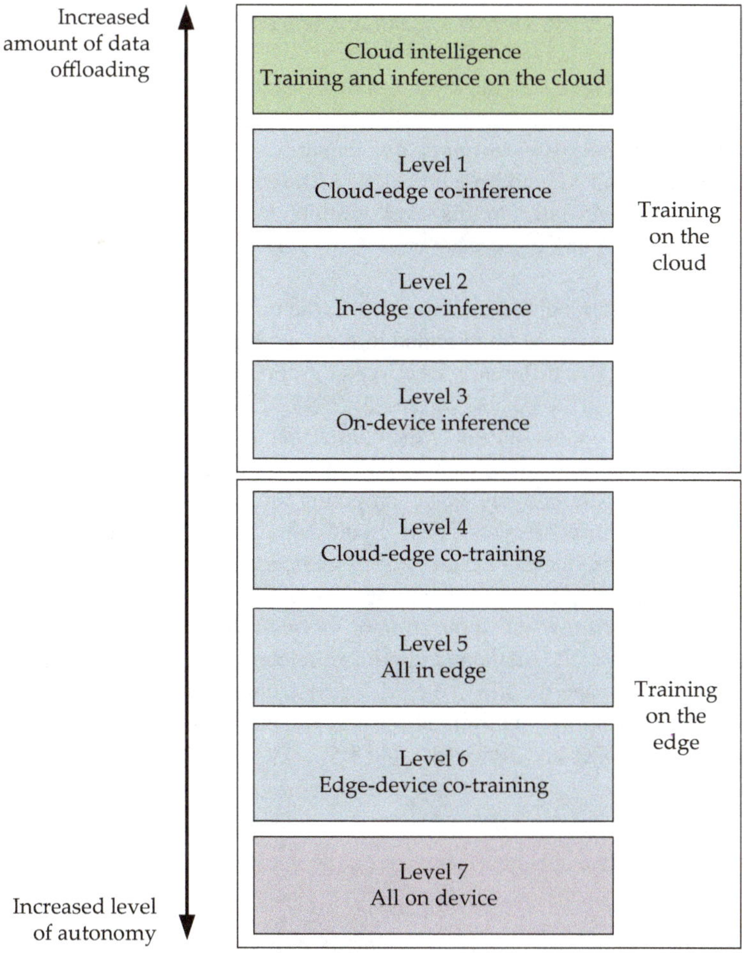

Fig. 8.11 Level rating for edge intelligence

Various levels of edge intelligence can be described as follows:

- *Level 1 (cloud-edge co-inference and cloud training)*: On this level, training of the AI model happens in the cloud, but inferencing of the AI model is done in an edge-cloud cooperation manner. Furthermore, data are partially offloaded to the cloud.
- *Level 2 (in-edge co-inference and cloud training)*: In this level, training of the AI model happens in the cloud but inferencing the AI model is done in an in-edge manner. The model inference is carried out within the network edge, by either fully or partially offloading the data to the edge nodes or nearby devices in an independent or coordinated manner.
- *Level 3 (on-device inference and cloud training)*: On this level, training of the AI model happens in the cloud but inferencing the AI model happens in a fully local on-device manner. No data are offloaded.
- *Level 4 (cloud-edge co-training and inference):* On this level, training and inferencing of the AI model are both done in the edge-cloud cooperation manner.
- *Level 5 (all in-edge):* On this level, training and inferencing of the AI model are both done in the in-edge manner.
- *Level 6 (edge-device co-training and inference):* On this level, training and inferencing of the AI model are both done in the edge-device cooperation manner.
- *Level 7 (all on-device):* On this level, training and inferencing of the AI model both occur in the on-device manner.

Figure 8.11 also shows cloud intelligence, i.e. the complete absence of intelligence in the edge, where both the training and inferencing happen fully in the cloud.

Several sources also differentiate between the concepts of AI for edge computing and edge computing for AI (AI on edge) [9, 17, 25]. AI for edge represents the aim of providing a better solution to the constrained optimization problems in edge computing with the help of AI technologies, i.e. AI is used for enhancing the edge with intelligence. Alternatively, edge computing on AI aims to provide a platform to carry out the entire lifecycle of AI models on edge. This interplay of AI and edge computing introduces a clear set of benefits, as shown in Fig. 8.12 [25]. This also allows for further classification of edge computing to communication, platform control, security, privacy, and application or service-specific aspects.

Some examples of IC methods to optimize telecommunication infrastructure in the 6G era and manage the life-cycle AI for edge computing are shown Table 8.4. Similarly, some examples of how the edge acts as a platform for application-oriented distributed AI services are shown in Table 8.5.

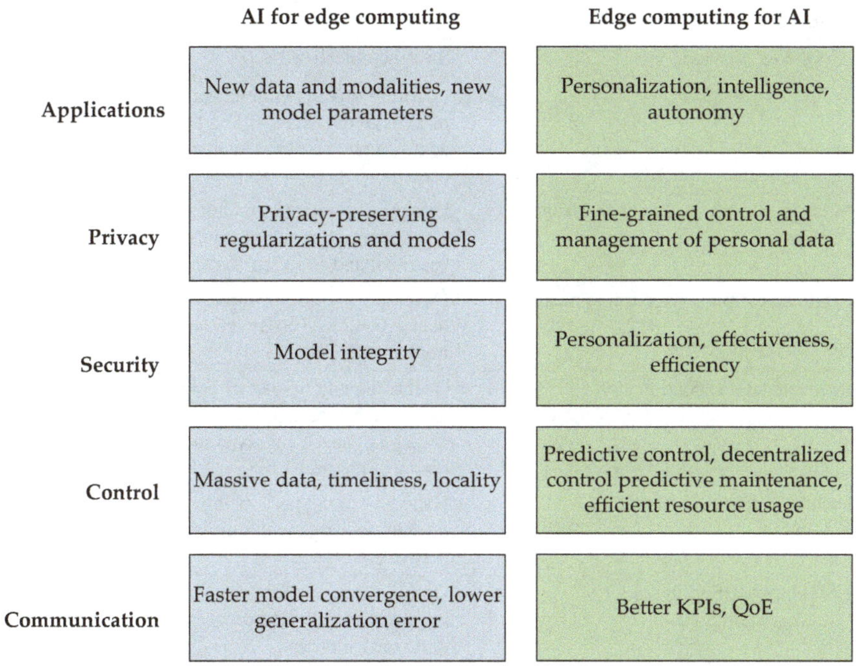

Fig. 8.12 Differentiating between AI for edge computing and edge computing for AI. From [25]

8.3.2 Challenges and Key Enablers of Edge Intelligence

Seeing that the research into edge intelligence is very much in its infancy, it does not come as a surprise that a multitude of challenges will need to be overcome before edge intelligence can be successfully deployed. These challenges are not only related to those involving lack of or limited hardware infrastructure or the liquid software mentioned in the previous section, but also include challenges related to [9]:

- Edge platform orchestration;
- Data and network management:
- Intersection of edge and device intelligence;
- Real-time requirements and online learning;
- Developing distributedly trained algorithms;
- Security and privacy: and
- End user aspects.

Table 8.6 [9] lists the detail of key challenges grouped as in the list above. Similarly, Table 8.7 shows the key enablers that could potentially help in resolving these sets of challenges over time. Also note that according to Park et al. [17], federated learning is by far the optimal (best-fitting) machine learning principle for machine learning at the edge.

Table 8.4 Objectives of AI for edge computing services [9]

AI for edge service	Specific objective
Wireless networking	Learning-driven communication across the whole process of data acquisition. Includes multiple access, radio resource management and signal encoding
Millimeter-wave backhaul and fronthaul systems	Dynamic optimization, fault detection and resource management for millimeter-wave backhaul and fronthaul systems
Communication service implementation	Development, optimization and run-time determination of communication service implementation
Dynamic task allocation	Offloading and onloading computational tasks and data between participating devices, edge nodes, and cloud; smart and dynamic allocation and reallocation of tasks
Liquid computing handover	Handover of the tasks being shared between devices and edge nodes as devices move in the network
Location-based optimization	Optimization of network coverage based on the information collected progressively on the local radio environment
Predictive QoS	The edge nodes exploit the knowledge available in the environment and the devices in that environment and predict their behavior to improve QoS
Energy management	Utilizing AI to improve energy efficiency as the degree of autonomy increases

Table 8.5 Objectives for edge computing for AI services [9]

Edge for AI service	Specific objective
Novel application areas	Autonomous and driving-assisted vehicles, autonomous drones, traffic control, smart factories, smart farms, smart homes, smart health, smart cities, etc.
Data intelligence	Data collection, aggregation, fusion, processing, distribution and services; the ability to learn, infer and control from data
Cooperative intelligence	Jointly solving AI problems on a range of heterogenous platforms
Real-time requirements	Localized AI functions with constrained computation resources and strict real-time requirements
Computing as a service	Intelligent computing capabilities available whenever user needs them
Advanced IoT models	Services that can adapt based on the available IoT devices and services on the network, where connected smart objects operate as an intelligent virtualized computational environment that is deployed across cloud, edge and mixed layers vertically and horizontally

Table 8.6 Challenges of edge intelligence. From [9]

Aspect	Key challenges
Edge computing infrastructure	• Support for dynamically changing resources and configurations • Reliable feature development • Device mobility
Edge platform orchestration	• Resource allocation • Resource distribution • Dynamic allocation and distribution
Data and network management	• Fusion of heterogenous data types • Understanding and definition of data and model provenance and lifecycle • Fitting of models into a given context • Enabling incremental learning methods • Inconsistencies in heterogenous sensory data • Modeling and understanding of the network behavior and performance • Introducing common interoperability practices and standards • Introducing centralized system management and operation • Limiting control latencies
Intersection of edge and device intelligence	• Introducing common interoperability practices and standards • Introducing centralized system management
Software development for edge	• Enabling dynamic configurations • Security • Flexible deployment • Debugging capabilities in development time
Real time requirements and online learning	• Adapting along the network dynamics • Need for short response times
Distributedly trained algorithms	• Reducing the cost of coordination among edge devices • Lowering generalized errors in trained models
Security and privacy	• Guarantee the implementation of security and privacy strategies according to user needs and requirements • Guarantee the recognition of abnormal behavior according to user requirements and the operator criteria
End-user aspects	• Understanding user context

Table 8.7 Associated enablers of edge intelligence. From [9]

Aspect	Key enablers
Edge computing infrastructure	• Container technologies • Virtualization • Isomorphic software architectures • Handover protocols and techniques
Edge platform orchestration	• Virtualization optimization • Data intelligence algorithms • Data analysis algorithms • Multilevel and fully distributed dynamics optimization algorithms
Data and network management	• Extension of the sensor and data fusion algorithm to support proper network and data management • Data and network fault detect and identification • Knowledge extraction and incremental learning algorithms • Knowledge-sharing solutions • Lightweight AI solutions to increase autonomy • Agents with cognitive capabilities • AI techniques for collaboration and cooperation (swarm intelligence, game theory, genetic algorithms)
Intersection of edge and device intelligence	• Lightweight AI solutions to increase autonomy • Agents with cognitive capabilities
Software development for edge	• Container technologies • DevOps practices • Virtualization • Liquid software that can flow from one node to another
Real time requirements and online learning	• Transfer learning, knowledge distillation and reinforcement learning • Codesign of communication, control, and machine learning
Distributedly trained algorithms	• Federated learning, knowledge distillation, transfer learning • Model pruning coded and quantized machine learning • Meta distributions, extreme value theory, risk management framework
Security and privacy	• Allow the application of the security strategy on the physical level. Use physical level solutions to increase security • Allow the proper management of personal information ownership at all stages
End-user aspects	• Digital twins and software agents with cognitive capabilities

8.3.3 Prospective Use Cases of Edge Intelligence in 6G

If the challenges identified in the previous section can be overcome, then there would be a multitude of use cases of edge intelligence in 6G. As the edge intelligence becomes more flexible, more and more services could be enabled by edge AI. In this section, some of the most important prospective use cases are identified [5, 9, 17].

The use case of autonomous driving has already been discussed as an example to introduce edge intelligence. In addition to driver assistance as well as data processing for fully autonomous driving, edge intelligence could be detrimental in so-called platoon driving—coordinated moving of a group of autonomous vehicles. Low-latency communications and decisions made together by all the vehicles in the platoon could be enabled by edge AI. Specifically, edge AI would allow for dynamic spectrum access that should allow for fast and reliable processing of data generated within platoons, as well as any other useful data that could be collected by all users of the road. Platoon driving would, in turn, result in more efficient road use, exploiting techniques such as adaptive cruise control. Note that the benefits of autonomous driving and autonomous platoon driving could be extended to drones and drone swarms, respectively, allowing for so-called massive UAV control.

Edge AI may be detrimental in protecting the privacy of smart places, such as smart homes, offices, farms, hospitals and factories. Keeping the data at the edge instead of transferring it to the central cloud implies that not only the smart places would react to the sensor (e.g. image camera) stimuli faster, but that all the privacy and security data can be confined to the smart place, thus decreasing the possibility of data security breaches and corruption.

Environmental sensing could also generate massive amounts of data that would need to be processed in real time. In this case, edge AI could assist in identifying optimal locations for sensor deployment, in their calibration as well as in providing solutions for processing the raw data in a manner that reduces the network load and parallelizes the computation.

Similarly, edge AI may prove detrimental in mobile multi-sensory XR, tactile internet, as well as in various concepts of CRAS. Many use cases mentioned in this section correspond to verticals of IoT, IoE, Industry 4.0 and 5.0 and others that will be discussed in much greater detail in Chap. 10. Of course, various communication paradigms, new and existing (voice, video, data, and holographic communications), also discussed in Chap. 10, will, likewise, benefit from what edge AI has to offer. Lastly, concepts such as content streaming would benefit from intelligent proactive caching at the edge.

8.3.4 Roadmap for Edge Intelligence in 6G

The white paper on edge intelligence compiled by the 6G flagship group [9] also discusses the roadmap for edge intelligence in 6G. This roadmap considers the evolution and deployment of a new generation of edge intelligence systems, applications and services that will take place during the next ten years (2020–2030), as 5G networks mature and 6G networks are developed. This is shown in Fig. 8.13 and justified by the prediction of the completion of different technological steps that will provide new devices, technologies and applications, as well as by the fact that some of the challenges will be overcome over time.

8.4 Ethics in Artificial Intelligence

The last topic that will be considered in this chapter is the topic of ethics in AI that was introduced in Chap. 1.

As mentioned there, the user might not always be satisfied with the decision made automatically by AI algorithms. Thus, a set of prohibitive clauses, which would ensure that intelligent services are only provided within some permissible boundaries, called the code of ethics, is required [27, 28]. It is equally important for laws and regulations to be established to address data ethics, as well as ethics of data ownership and storge in the context of 6G, while striking a proper balance between risk and benefit [1].

For a technology that crosses all international and cultural boundaries, developing a code of ethics, laws and regulations becomes difficult, and this development should be undertaken in parallel with the evolution of 5G and 6G networks [29]. Any regulatory framework established in this way will need to include cooperation across domains and across borders [30]. Furthermore, a scenario that must be avoided is one in which the codes of ethics are influenced by a small number of people, potentially consisting only of powerful groups or individuals. Rather, in order to drive such a process ethically on its own, it is fundamental to involve citizens in a participative process [30].

8.5 Concluding Remarks

This chapter expanded on the involvement of AI in the design of future wireless networks. 6G networks, specifically, will be using a diverse range of technologies, and will have to satisfy the needs of a massive number of heterogeneous users, both humans and machines. Controlling a network that remains human-centric in part, and in part becomes machine-centric, is not trivial, and requires a paradigm shift. In the case of 6G, the paradigm shift is to replace decision-making based on deterministic models with decision-making that uses a machine learning approach instead, and to

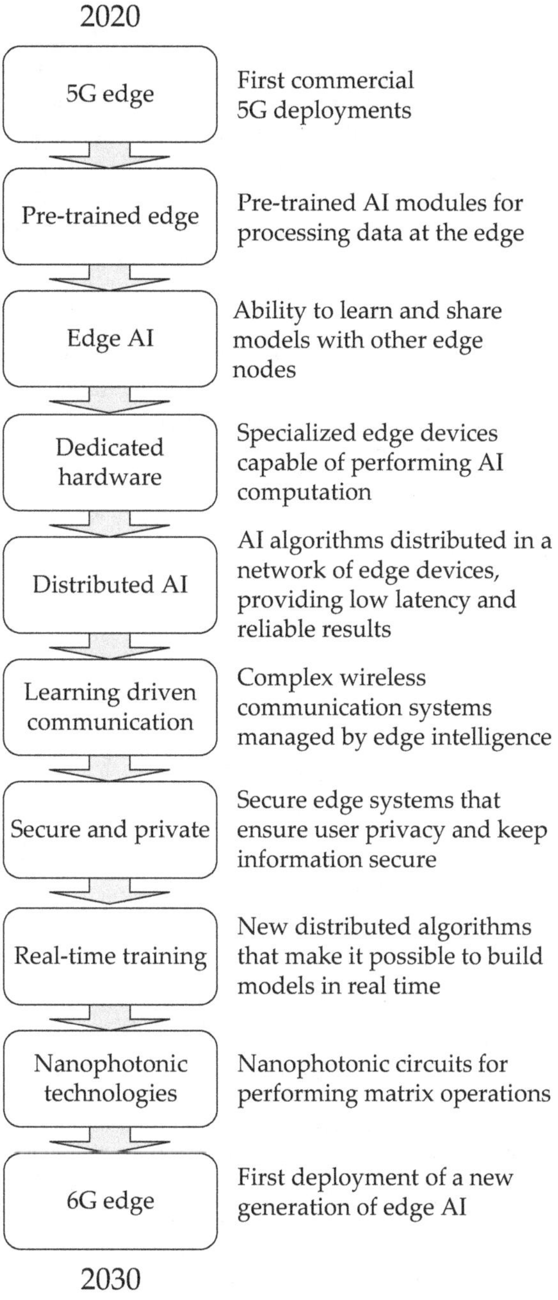

Fig. 8.13 Roadmap for achieving edge intelligence by 2030

introduce intelligence in the form of AI in general into the network. Separately, the machine learning techniques used in other fields are not necessarily the best fit for 6G networks and should rather be adapted before 2030 (the year of the envisioned launch of the first 6G networks). One of the directions 6G network researchers are pursuing is moving the location of the AI to the edge of networks, instead of keeping all intelligence at the core of the network. This has the net effect of decreasing network response time (latency), while simultaneously reducing the data load, by allowing the data to remain at the network edge when not needed at the central location. The introduction and expansion of AI, in addition, raises ethical concerns, which will also have to be addressed before this technology would be allowed to see the light of day.

With the discussion of AI and machine learning, the discussions of all the technologies associated with 6G networks are concluded, and what remains to be addressed are many services or verticals that are envisioned in the 6G network era, which have been mentioned in this and several previous chapters. This will be done in Chap. 10; however, Chap. 9 will take a step back and digress into a discussion of potential health issues, if any, that could be associated with present and future wireless network technologies.

References

1. Letaief KB, Chen W, Shi Y, Zhang J, Zhang Y-JA (2019) The roadmap to 6G: AI empowered wireless networks. IEEE Commun Magaz 57:84–90
2. Ali S, Saad W, Rajatheva N, Chang K, Steinbach D, Sliwa B et al (2020) 6G White paper on machine learning in wireless communication networks. arXiv:200413875 [cs, eess, math] [Internet]. 2020 [cited 2020 Nov 9]. http://arXiv.org/abs/2004.13875
3. Tariq F, Khandaker M, Wong K-K, Imran M, Bennis M, Debbah M (2019) A speculative study on 6G. IEEE Wirel Commun 27:118–125
4. Calvanese Strinati E, Barbarossa S, Gonzalez-Jimenez JL, Ktenas D, Cassiau N, Maret L et al (2019) 6G: The next frontier: from holographic messaging to artificial intelligence using subterahertz and visible light communication. IEEE Veh Technol Magaz 14:42–50
5. Nawaz SJ, Sharma SK, Wyne S, Patwary MN, Asaduzzaman MD (2019) Quantum machine learning for 6g communication networks: state-of-the-art and vision for the future. IEEE Access 7:46317–46350
6. Viswanathan H, Mogensen PE (2020) Communications in the 6G era. IEEE Access 8:57063–57074
7. Zhang L, Liang Y-C, Niyato D (2019) 6G Visions: Mobile ultra-broadband, super internet-of-things, and artificial intelligence. China Communications. 16:1–14
8. Rajatheva N, Atzeni I, Bjornson E, Bourdoux A, Buzzi S, Dore J-B et al (2020) White paper on broadband connectivity in 6G. arXiv:200414247 [eess] [Internet]. 2020 [cited 2020 Sept 16]. http://arXiv.org/abs/2004.14247
9. Peltonen E, Bennis M, Capobianco M, Debbah M, Ding A, Gil-Castiñeira F et al (2020) 6G White paper on edge intelligence. arXiv:200414850 [cs] [Internet]. 2020 [cited 2020 Nov 9]. http://arXiv.org/abs/2004.14850
10. Yang H, Alphones A, Xiong Z, Niyato D, Zhao J, Wu K (2020) Artificial intelligence-enabled intelligent 6G networks. IEEE Netw Early Access 1–9
11. Wikström G, Peisa J, Rugeland P, Johansson N, Parkvall S, Girnyk M et al (2020) Challenges and technologies for 6G. 2020 2nd 6G Wireless Summit (6G SUMMIT), pp 1–5

12. Raschka S (2020) Chapter 1: Introduction to machine learning and deep learning [Internet]. Dr. Sebastian Raschka. 2020 [cited 2020 Nov 18]. https://sebastianraschka.com/blog/2020/intro-to-dl-ch01.html
13. Simeone O, Park S, Kang J (2020) From learning to meta-learning: reduced training overhead and complexity for communication systems. 2020 2nd 6G Wireless Summit (6G SUMMIT), pp 1–5
14. Hewa T, Gür G, Kalla A, Ylianttila M, Bracken A, Liyanage M (2020) The role of blockchain in 6G: challenges, opportunities and research directions. 2020 2nd 6G Wireless Summit (6G SUMMIT), pp 1–5
15. Hüllermeier E, Fober T, Mernberger M (2020) Inductive bias. In: Dubitzky W, Wolkenhauer O, Cho K-H, Yokota H (eds) Encyclopedia of systems biology [Internet], pp 1018–1018. Springer, New York, NY. [cited 2020 Nov 18]. https://doi.org/10.1007/978-1-4419-9863-7_927
16. Ziyin L, Hartwig T, Ueda M (2020) Neural networks fail to learn periodic functions and how to fix it. arXiv:200608195 [cs, stat] [Internet]. 2020 [cited 2020 Nov 18]. http://arXiv.org/abs/2006.08195
17. Park J, Samarakoon S, Bennis M, Debbah M (2019) Wireless Network Intelligence at the Edge. Proc IEEE 107:2204–2239
18. Kairouz P, McMahan HB, Avent B, Bellet A, Bennis M, Bhagoji AN et al (2020) Advances and open problems in federated learning. arXiv:191204977 [cs, stat] [Internet]. 2019 [cited 2020 Nov 23]. http://arXiv.org/abs/1912.04977
19. Sarieddeen H, Saeed N, Al-Naffouri TY, Alouini M-S (2020) Next generation terahertz communications: a rendezvous of sensing, imaging, and localization. IEEE Commun Magaz 58:69–75
20. Korpi D, Yli-Opas P, Jaramillo MR, Uusitalo MA (2020) Visual detection-based blockage prediction for beyond 5G wireless systems. 2020 2nd 6G Wireless Summit (6G SUMMIT), pp 1–5
21. Leinonen M, Codreanu M (2020) Quantized compressed sensing via deep neural networks. 2020 2nd 6G Wireless Summit (6G SUMMIT), pp 1–5
22. Sliwa B, Falkenberg R, Wietfeld C (2020) Towards cooperative data rate prediction for future mobile and vehicular 6G networks. 2020 2nd 6G Wireless Summit (6G SUMMIT), pp 1–5
23. Reijonen J, Opsenica M, Kauppinen T, Komu M, Kjällman J, Mecklin T et al (2020) Benchmarking Q-learning methods for intelligent network orchestration in the edge. 2020 2nd 6G Wireless Summit (6G SUMMIT), pp 1–5
24. Tomkos I, Klonidis D, Pikasis E, Theodoridis S (2020) Toward the 6G network era: opportunities and challenges. IT Profess 22:34–38
25. Lovén L, Leppänen T, Peltonen E, Partala J, Harjula E, Porambage P et al (2019) EdgeAI: a vision for distributed, edgenative artificial intelligence in future 6G networks. The 1st 6G Wireless Summit. Levi, pp 1–2
26. Zhou Z, Chen X, Li E, Zeng L, Luo K, Zhang J (2019) Edge intelligence: paving the last mile of artificial intelligence with edge computing. Proc IEEE 107:1738–1762
27. Boddington P (2017) Towards a code of ethics for artificial intelligence. Springer
28. Etzioni A, Etzioni O (2017) Incorporating ethics into artificial intelligence. J Ethics 21:403–418
29. David K, Berndt H (2018) 6G vision and requirements: is there any need for beyond 5G? IEEE Veh Technol Magaz 13:72–80
30. Mucchi L, Jayousi S, Caputo S, Paoletti E, Zoppi P, Geli S et al (2020) How 6G technology can change the future wireless healthcare. 2020 2nd 6G Wireless Summit (6G SUMMIT), pp 1–6

Chapter 9
5G and 6G Networks: Should There Be a Health Concern?

Abstract This chapter discusses the health issues associated with underlying technologies of 5G and 6G. It is mostly dedicated to the investigation of the short-term and long-term influence of different types of electromagnetic radiation on various human health aspects. Both the physics of wave propagation and medical research studies are discussed. Potential health issues associated with VLC are also discussed in this chapter.

Around April 2020, in the midst of what was later dubbed the "first wave" of the SARS-CoV-2 ("COVID-19") virus outbreak, some half of the world's population had limited freedom of movement as a way of enforcing physical distancing, which was initially deemed necessary by various governments to curb the spread of the disease [1]. As a result, many people turned to technology to organize their lives, thus using the internet as a way of working from home, remaining in touch with friends and family or staying entertained. Even the elderly, who have traditionally been slow to adopt new technologies, have found themselves "cornered" into using technology. The media reported that internet traffic was about 25% higher than normal in countries that were in lockdown. As a consequence, the existing networks became congested and lack of bandwidth became evident. Many service providers resorted to making higher-speed internet packages available to subscribers without additional costs, and many, in order to do that, resorted to using infrastructure that had been decommissioned or had even not been commissioned yet. For example, in South Africa, cellular network providers were allowed to open up the 5G spectrum temporarily, ahead of planned deployment of the 5G network [2]. As a result, in a span of about a month, a large number of 5G network towers were erected in South Africa's main hubs such as Pretoria and Johannesburg (see Fig. 9.1).

At about the same time, in South Africa and globally, an increased level of concern about the health-related safety of 5G wireless networks was observed. Resistance to 5G deployment with health concerns as a basis was already evident globally several years prior to 2020, but these concerns seem to have been aggravated in the turbulent period of 2020, possibly fueled by the sudden expansion of social

M. Božanić and S. Sinha, *Mobile Communication Networks: 5G and a Vision of 6G*, Lecture Notes in Electrical Engineering 751,
https://doi.org/10.1007/978-3-030-69273-5_9

Fig. 9.1 Photo of a 5G tower in a suburb of Pretoria, South Africa, taken in May 2020. At the bottom right, another 5G tower can be spotted, illustrating the required tower density for successful network deployment

networking use.[1] Citizens' response even triggered some countries to delay or slow down 5G deployment instead of speeding it up, for example Slovenia, which stopped the deployment of 5G in March 2020, until the health effects of the network had been evaluated [3].

These various negative responses to the planned or sudden 5G deployment beg a question: How (if at all) harmful is 5G and should humanity be concerned about this deployment? And if 5G is harmful, will 6G be even more harmful? This chapter attempts to find a scientific basis to answer these questions.

The first step is to rephrase these layman's questions into a scientific problem. Clearly, network users are worried about the electromagnetic field (EMF) radiation that is emitted by the base towers (which are much easier to spot nowadays because of the dense deployment required by 5G), as well as about the radiation emitted by the increasing number of user devices (cellphones, tablets) and sensors (IoT) with wireless connectivity (which is typically always enabled). It is often argued that adding so many additional devices to the global network increases electromagnetic

[1]Some sources went to the extent of claiming that the COVID-19 outbreak was related to the deployment of 5G. The authors will not honor that claim with a response in this text.

"pollution" of the environment [4]. The second step is to identify on which levels 5G and 6G are different from the historical network deployments, such as Wi-Fi and past network generations up to 4G, the uncertainty of which could be (another) possible reason for resistance among lay users, specifically to 5G. Clearly, this is because it has been advertised that 5G is going to use millimeter-waves instead of (or in addition to) lower frequency waves. Then, physical laws and principles governing electromagnetic radiation can be analyzed, and it can be investigated how they correspond with findings of experimental health studies conducted to date, while keeping in mind the recommended electromagnetic radiation levels proposed by various sources and organizations. When it comes to experimental evidence, note that the aim of this chapter is not to analyze the data presented by various health studies carried out to date in detail; this has been left to medical researchers and practitioners. Rather, the authors have consulted several review papers, which collectively cover hundreds of individual studies that investigate the possible linkage of electromagnetic radiation with health risks, as well as some regular papers, typically speculating on the possible or probable risk.

Before this subject is reviewed in full throughout several sections in this chapter, it is noted that people's "resistance" to new, specifically wireless technologies, based on health concerns, is a not a recent development. The relationship between health and wireless networks has been discussed for decades, practically with the release of every new wireless technology. This phenomenon is discussed next.

9.1 Health Concerns and Resistance to Past Wireless Networks

As discussed in Chap. 2, cellular telephony was introduced some 40 years ago with 1G. Since then, a new generation of the network has been released every ten years. With every release, or every network deployment, the general population has been concerned with the health consequences of the new release. The 5G deployment is thus certainly not unique in this regard.

As a matter of fact, the World Health Organization started the International EMF project as early as 1996 (in the 2G/GSM era) with the goal to assess the scientific evidence of possible EMF health effects. In the rationale of the project, it was stated that the project was initiated in response to "to public concern over health effects of EMF exposure" [5]. This indicates that the health concerns go back at least 30 years, which is at least three-quarters of the way back into the history of cellular networks.

Over the years, a so-called "base-station myth" has emerged: cellphone towers are regarded as dangerous by the public, resulting in initiatives and protests against the erection of cellular towers. In addition to quoting health concerns, cellphone masts have been regarded as ugly, degrading, criminal, ruining, eyesores, unsightly or devastating, according to a study published by Drake [6] in 2006, which coincides with the 3G era. As will be discussed later in this chapter, physics dictates that the radiation from cellular towers is much less concerning than that of the user devices, thus the use of the term "myth" above. In the 4G era, it was well understood, even by the greater public, that the only possible harm could come from user devices and not at all from the cellphone towers (for the moment, the focus is on the word "possible"). That is only because cellphones and tablets are kept very close to the body for prolonged periods of time, thus allowing the body to absorb the near field electromagnetic radiation (whereas radiation from the tower is heavily attenuated far-field radiation).

At the beginning of the 5G era, the cellphone tower has come back into focus. Small cell deployment, which is required to provide coverage when operating at the higher frequencies of the electromagnetic spectrum, calls for cell towers to be packed in much more densely than 2G, 3G and 4G towers. Some countries (for example, again South Africa [7]), could even potentially force house owners to allow cell towers to be installed in their backyards. Needless to say, this regulation has opened a new wave of controversy. However, the question that remains, appearance aside, is even whether this move is a reason for concern. The remainder of this chapter continues to try to find the answer.

9.2 5G: Is There a Health Risk?

At this point in this chapter, the focus shifts to 5G, and the discussion on 6G is deferred until later (Sect. 9.3). To answer the question in the title of this section, the first step is to isolate the differences between 4G and 5G, which would introduce renewed concern about the harm of cellular networks to human health over and above the concerns the public may have had with 4G and other older technologies.

Recall that there are, in fact, two main differences. First, to support better coverage and faster wireless data transfer, more base stations are needed, which have more antennas emitting electromagnetic radiation in a small area simultaneously. Second, 5G and/or future network generations will use a wider range of frequencies to deliver the required bandwidth, with maximum frequency shifting from 6 to 86 GHz and beyond. With 4G transmitting frequencies of only up to 6 GHz, 5G poses the additional concern of the effect of frequencies that are orders of magnitude higher than those used before.

9.2.1 Sources of 5G Electromagnetic Radiation

Before moving on to a more detailed discussion of the effects of EMF on health, the reader is owed an explanation of why user devices (e.g. mobile phones) are potentially more harmful than base stations.

As reviewed in Sect. 9.1, over the last 20 years, a myth has emerged that the radiation emitted from a base station is harmful to people living close to the base station. This is exactly that—a myth: although the amount of power emitted at the base station is orders of magnitude higher than that of a cellphone, the power at any point away from the antenna decreases quadratically with the increase in distance. Once again, recall the Friis formula, which describes the amount of power received at a distance r from the transmitting antenna:

$$P_R = \frac{P_T G_T G_R \lambda^2}{(4\pi r)^2}. \tag{9.1}$$

It can be assumed that human body (or skin) is a receiving antenna [8], therefore, this equation may be applicable to a human standing at a distance r from the antenna. Equation (9.1) states that for every two meters one moves away from the antenna, the signal will decrease four times. At a distance of 10 m from the antenna, the signal is one 100th of the signal at a distance of 1 m from the antenna. Hence, unless the antenna is installed in the person's living room, the electromagnetic radiation received by the person standing in the network cell is so small that it is practically insignificant.

This part of the analysis is common to all electromagnetic waves emitted by the 5G antenna, as well as to the electromagnetic radiation emitted by older networks. However, a typical 5G base station will emit both the lower frequency signals (signals below 6 GHz) and the higher (millimeter-wave) frequency signals (signals above 6 GHz) simultaneously. It seems fitting to now analyze the difference in the broadcast frequency as well. This can be done by noting that wavelength λ also features in Eq. (9.1), this time in the numerator. As the wavelength decreases with the increase in frequency, the attenuation $(4\pi r)^2/\lambda^2$ of the channel increases.

This concept has already been analyzed in Chap. 3 in Table 3.2. It is worthwhile to convert the decibels in Table 3.2 into ratios, which is done in Table 9.1. With this done, one can work out the power received at different distances from a typical network antenna. Recall from Chap. 6 (Table 6.2), that the typical power emitted from a 4G antenna is 40 W and that of a 5G antenna is 240 W [9]. (This is the total power for the base station, but for the sake of argument, it will be assumed that all the power is radiated directly to the observer some distance away.) Then, the absolute maximum power levels of received radiation at various distances from the base station for applicable distance/frequency combinations in Table 9.1 can be worked out, as applicable to 4G and 5G. This is done in Table 9.2, with the assumption that the transmit antenna has about 10 dBi antenna gain [10]. This table shows that

Table 9.1 Attenuation for different combinations of frequency and distance

Distance/frequency	3 GHz	30 GHz	60 GHz	300 GHz	600 GHz
3.16 m	6.31×10^{-6}	6.31×10^{-8}	1.58×10^{-8}	6.31×10^{-10}	1.58×10^{-10}
10 m	6.31×10^{-7}	6.31×10^{-9}	1.58×10^{-9}	6.31×10^{-11}	1.58×10^{-11}
100 m	6.31×10^{-9}	6.31×10^{-11}	1.58×10^{-11}	6.31×10^{-13}	1.58×10^{-13}
1 km	6.31×10^{-11}	6.31×10^{-13}	1.58×10^{-13}	6.31×10^{-15}	1.58×10^{-15}

Table 9.2 Received power for various distances from the base station. It is assumed that the 4G base station emits 40 W, the 5G base station emits 240 W and the antenna gain is 10 dBi for all configurations

Distance/frequency	Large base tower		Street poles (5G and beyond)	
	3 GHz (4G)	3 GHz (5G)	30 GHz (5G)	60 GHz (5G)
3.16 m	2.52 mW	15.1 mW	151 μW	13.79 μW
10 m	252 μW	1.51 mW	15.1 μW	1.379 μW
100 m	2.52 μW	15.1 μW	151 nW	37.9 nW
1 km	25.2 nW	151 nW	1.51 nW	Less than 1 nW

the received power at a distance of 10 m is 1.5 mW or less for the 3 GHz signal, and that it drops to just over 1 μW at the same distance at 60 GHz. At a distance of 100 m, the signal is in the nanowatt range for millimeter-wave transmission and in the order of several microwatts at 3 GHz. Lastly, at a distance of 1 km, there is virtually no power at 60 GHz. Hence, even the power perceived by the regular RF signal as close as 10 m is too small to do much harm to the network user (this will be discussed later), and the millimeter-wave power is practically miniscule. If one considers that most cellphone towers are at least 10 m high, a person will never be exposed to power higher than about 1 mW, because he/she would never stand closer than this to the actual transmitter, as shown in Fig. 9.2a. It may be slightly different for antennas mounted on street poles, but these are typically only applicable to millimeter-wave deployments. Typically, street poles would not be shorter than 5 m, so even if an average 1.84 m tall person is standing right at the bottom of the street pole, the closest point of his/her body would be exposed to only several hundreds of microwatts of millimeter-wave radiation, as illustrated in Fig. 9.2b.

Lastly, recall that all the estimates in Table 9.2 are pessimistic estimates, which were done assuming that all the power coming from the base station is directed directly at the person in question. In reality, this is not the case, and the true potential exposure is much lower. Furthermore, these calculations do not take any (additional) atmospheric attenuation, such as the attenuation due to oxygen or water vapor, or attenuation due to blockage and scattering, into consideration.

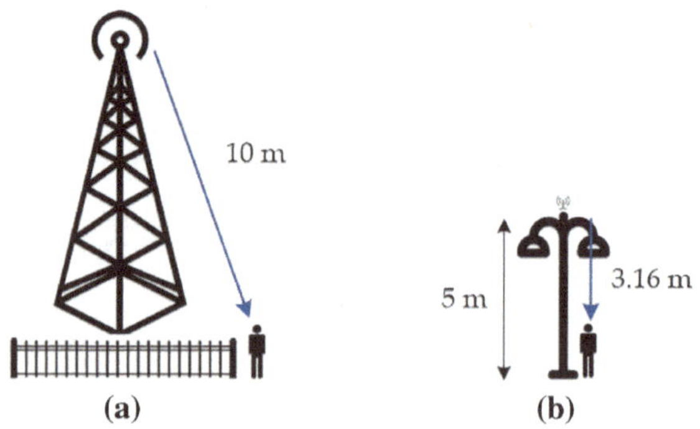

Fig. 9.2 The minimum distance from the transmitter when a person is standing at the bottom of a large base tower (**a**) and at the bottom of a lamppost (**b**)

It is interesting to take a look at the signal attenuation from another angle. If one considers a scenario when 1 W of power is transmitted by a wireless access point and that the typical power perceived by the user device is not more than 1 μW, this means that 99.9999% of the signal is lost elsewhere [11]. To a professional RF engineer, this might sound like stating the obvious, but it may serve as a very good argument when discussing the principle of attenuation with the general public.

Now the case of dense deployment can be considered. Two scenarios can be considered—the higher density of regular base towers as well as the density of street poles. In the first scenario, assume that the person is standing in the center of a triangle made by three towers spaced about 175 m apart, as shown in Fig. 9.3. At this point the person would be equally irradiated by each of the three towers. Since he/she receives three times the radiation registered at a distance of 100 m from the tower (the height of the tower is ignored in this case), the amount of power received will be $3 \times 15.1\ \mu W = 45.3\ \mu W$ for the higher perceived power associated with the 3 GHz transmission. Although this is still three times more than what he/she would receive if only one tower was in the vicinity, the power is nevertheless some 33 times less than if the same person should just be standing at the bottom of the first tower. Therefore, the density of the tower deployment has virtually zero influence on the amount of electromagnetic pollution.

Now consider a case of street pole deployment of millimeter-wave transmitters. Assuming that street poles 5 m high are placed 5 m apart, at the center point between the two street poles the person's head will be 4 m away from the transmitter on each pole, as shown in Fig. 9.4. Perceived radiation at this point will be about 95 μW for the single signal at 30 GHz, resulting in total received power of 190 μW if transmitters on both lampposts are considered. This is still only about 25% higher

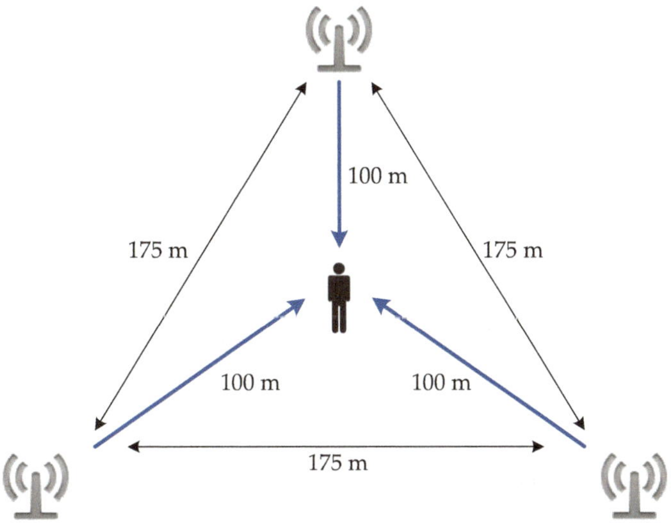

Fig. 9.3 Scenario in which a person is standing equidistant from three cellular base stations

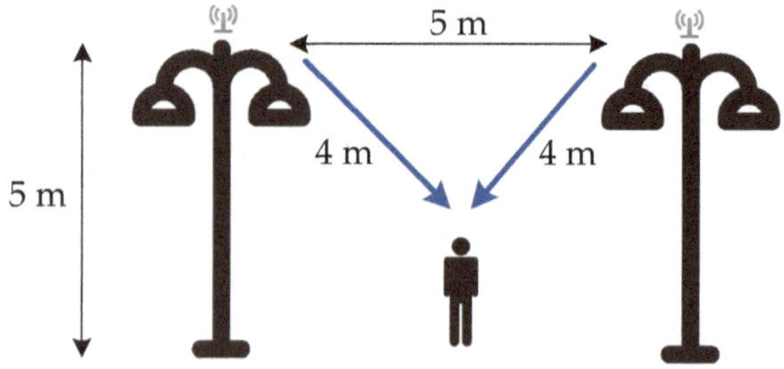

Fig. 9.4 Person located at the exact midpoint between two lampposts spaced 5 m apart and 5 m tall

than the power received if the person should be standing at the bottom of a single lamppost. If the distance between lampposts increases to roughly 6 m and beyond this point, the power received at the midpoint will be lower than at the bottom of a single lamppost. Therefore, dense deployments, even using lampposts, do not significantly increase maximum exposure to electromagnetic radiation, even if the lampposts are placed very close together, and the amount becomes irrelevant as spacing between lampposts increases.

From this discussion, it should be clear that the potential threat to human health does not come from the side of the base station. Potentially more harmful is the radiation emitted by cellphones, tablets, laptops and other devices, held by users close to their faces or other parts of the body when these devices actively transmit—during a phone call or during data transfer. Health concerns, with some merit, have also been raised about wireless networks in homes offices and schools, automotive radar and other technologies such as smart meters and sensors wired into IoT [12]. Even though the radiation emitted by these devices is much lower than that of the base station, the problem may potentially arise from the proximity of, for example, the phone to the face or another part of the human body. Even when the phone transmits only several tens of milliwatts, a large portion of that radiation could potentially be absorbed by the body. Everybody who has been on a long phone call must have noticed that the device became hot during the call—thus, at least one potential problem that could be identified is the heating of human tissue. However, the question that remains is whether the type of radiation that is emitted by devices operating in the 5G regime itself is harmful in any way. As already mentioned, 5G will be allowed to operate at frequencies up to 86 GHz, which are classified as millimeter-wave frequencies.

9.2.2 Ionizing and Non-ionizing Radiation

Like light, electromagnetic radiation can be described in terms of travelling waves, but it also has a particle-like nature; the particles are called photons [13]. Each photon has a certain energy, which increases with frequency, according to the formula:

$$E = hf, \tag{9.2}$$

where $f = c/\lambda$ is the frequency of the general electromagnetic wave emitted and h is Planck's constant, equal to 6.626×10^{-34} J·s or 4.135×10^{-15} eV·s. Recall that the same formula applies to bandgap energy, as described in Chap 5, hence the similarity between Eqs. (9.2) and (5.3). At 6 GHz, photon energy is 24.8 μeV. For millimeter-wave radiation, this energy increases, ranging from 124 μeV at 30 GHz to 1.24 meV at 300 GHz. These energies are non-ionizing, because the photon energy is insufficient to remove an electron from an atom or a molecule. For this to happen, about 12 eV is needed. Even the energy from the sun, including most ultraviolet light, which is higher than the visible light in the electromagnetic spectrum, remains below this threshold, despite have a frequency four orders of magnitude higher than millimeter-waves. Gamma-rays and X-rays have energies that are ionizing, but their frequency spectrum lies above that of visible and ultraviolet light, as shown in Fig. 9.5. This figure was adapted from Chap. 1 to include the classification of electromagnetic radiation into ionizing and non-ionizing classes as well.

Four main differences between ionizing and non-ionizing radiation are shown in Table 9.3. Other than the obvious difference in frequency and the photon energy described above, ionizing radiation can break chemical bonds, whereas non-ionizing radiation cannot. This means that ionizing radiation can have both thermal and non-thermal effects on the human body if this is exposed, whereas non-ionizing radiation potentially only has thermal effects. From this analysis, it appears that electromagnetic radiation in both the RF and the millimeter-wave regimes cannot cause any notable harm. It should be added here that one notable exception is microwaves used in microwave ovens, because the frequency of these waves is tuned exactly to resonate with water molecules (2.45 GHz); however, they are used in enclosed environments

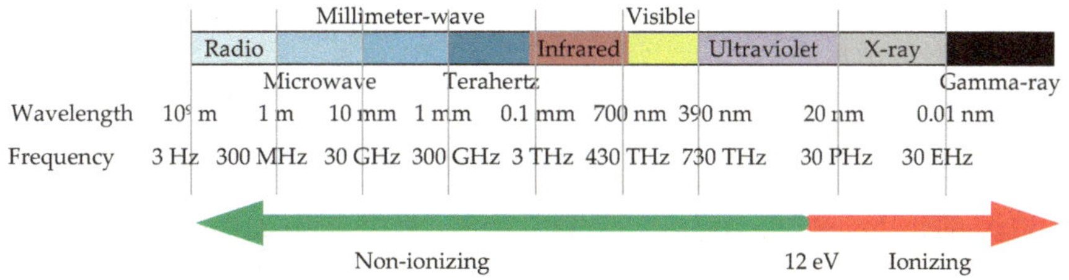

Fig. 9.5 Frequency spectrum as described in Chap. 1, with differentiation between non-ionizing and ionizing radiation added

Table 9.3 Differences between ionizing and non-ionizing radiation

Ionizing	Non-ionizing
Higher frequency (>30 PHz)	Lower frequency (< 30 PHz)
Higher energy (>12 eV)	Lower energy (<12 eV)
Breaks chemical bonds	Does not break chemical bonds
Thermal and non-thermal effects	Thermal effects

with levels regulated such that the energy does not escape. Furthermore, the fact that even the energy from the sun (or the energy of light in general), to which humans are universally exposed, has higher photon energy than electromagnetic radiation is possibly the best proof that electromagnetic radiation is something that people should not be concerned about. In fact, some argue [14] that the term "radiation" itself is misleading in this context, as it is normally associated with something that is universally harmful (such as nuclear radiation).

Nevertheless, there must be another angle to this: it is important to understand how much electromagnetic radiation is absorbed at any time (instantaneous exposure) and over time (long-term exposure), as it is important to know how strong sunlight is and how long one spends in direct sunlight. It is for this reason that dosimetric quantity metrics for measuring electromagnetic exposure were introduced a few decades ago.

9.2.3 Measuring Electromagnetic Radiation Exposure

Exposure to electromagnetic radiation is normally measured using three quantities:

- Specific absorption rate (SAR);
- Plane-wave equivalent power density; and
- State-state and/or transient temperature.

The SAR measures the RF power absorbed in living body tissue and has units of W/kg [13]. It thus represents a power level per body mass. SAR can be expressed as

$$SAR = \frac{P_{diss}}{m} = \frac{\sigma |E|^2}{\rho}, \tag{9.3}$$

where P_{diss} is the radiation power dissipated in the tissue (in W), m is the tissue mass under exposure (in kg), σ is the conductivity of the tissue (in S/m), ρ is the tissue mass density (kg/m^3) and E is the root mean square (RMS) value of the electric field strength dissipated in the tissue (in V/m). The SAR at the surface of the exposed tissue will have a different value from the SAR deep within the tissue. SAR is usually used to specify the safety recommendations for limiting close proximity to the body [4].

The plane-wave equivalent power density is the power per surface area metric and is expressed in W/m^2. In the far field of the antenna, it can be expressed either in terms of the RMS value of the electric field or the magnetic field incident of the surface of the tissue:

$$PD = \frac{|E_i|^2}{\eta} = \eta |H_i|^2. \tag{9.4}$$

In this equation, η is the wave impedance (377 Ω in free space). If the gain of the transmit antenna G_t and its transmit power P_t are known, then the power density at the distance d from the antenna can be worked out from

$$PD = \frac{G_t P_t}{4\pi d^2}. \tag{9.5}$$

For the near field of the antenna, which is more applicable to exposure to user device radiation, the calculation gets complicated because E, H and even η are functions of the position and have different values from those in the case of the far field. Unlike SAR, power density does not require prior knowledge of tissue absorption characteristics. As such, power density is not as useful as SAR for assessing the safety in scenarios involving close proximity and high skin depths.

The steady state and/or transient temperature is useful in evaluating the effect of medium and high-power radiation, which causes rapid heating of the skin. The steady-state temperature refers to the temperature to which the skin is heated, and the transient temperature is essentially the temperature difference due to sudden exposure to radiation. Humans can detect temperature changes as small as 0.1 °C, whereas temperature changes of 1 °C or even more are considered safe. The important thing to note is that the heating due to RF radiation goes deeper into the tissue than that of millimeter-wave radiation. At millimeter-wave frequencies, most of the heat is absorbed in the first millimeter of the skin.

The next step is to determine how these metrics have been used to control the limits to electromagnetic exposure in practice.

9.2.4 Regulation of Electromagnetic Radiation Exposure

The regulation of EMF exposure is done in terms of SAR and power density, whereas heating is generally not considered as a strict metric, but rather as the consequence of exposure. Electromagnetic exposure is regulated by bodies such as the FCC in the USA and the International Commission on Non-Ionizing Radiation Protection (ICNIRP) in most European countries [4, 13, 15]. Some other countries, such as Russia and China, elect to provide their own, more conservative regulations.

Table 9.4 shows the power density limits adopted by ICNIRP, FCC and the governing bodies for China, Russia, Switzerland and Italy, as well as the year of adoption of these limits and the frequency range to which the figures apply. These limits are applicable to far field radiation, and thus dictate the location of base stations. There is an obvious difference in the figures used in China, Russia, Switzerland, and Italy from those recommended by the ICNIRP and FCC. The rationale for this differs from country to country, but in principle, the lower figures allow the governing bodies to err on the side of caution, thus minimizing the as yet unknown risks. The lower figure of 0.1 W/m^2 practically stops base stations from being deployed inside or in very close proximity to apartment or office buildings.

Table 9.4 Power density exposure limits to RF radiation in various countries (Adapted from [13])

Country of governing body	Power density restriction for general public (W/m^2)	Year adopted	Frequency range (GHz)
ICNIRP	10	1998	2–300
FCC	10	1996	1.5–100
China	0.1	1987	0.3–300
Russia	0.1	2003	0.3–300
Switzerland	0.1	2000	1.8–300
Italy	0.1	2003	0.0001–300

The regulation of radiation exposure for user devices is much more complex, at least for two reasons: first, they typically operate very close to human bodies, and secondly, factors such as the shape of the parts of the body (such as the head) need to be taken into account. This is aggravated by the fact that different radiation frequencies result in different EMF penetration depth. This calls for using a combination of the SAR and power density metrics when specifying exposure limits. The split is best illustrated in Table 9.5, which shows how the FCC evaluates the radiation exposure criteria for frequencies below and above 6 GHz. Note that the FCC does not consider the SAR measurement to be applicable at all to millimeter-wave frequencies; however, it does consider the way the power density metric is obtained, depending on whether the device is placed closer or further away than 5 cm from the body. Numerical modeling, such as that using the final element method, rather than direct measurement, is applicable for short distances. SAR is only used as a metric for RF (< GHz) and for distances below 20 cm, while the power density is applicable to distances above 20 cm or frequencies above 6 GHz. Note that for the most part, ICNIRP guidelines are very similar to FCC guidelines; however, specifically in this case, the frequency differentiation between RF and the millimeter-wave domain is done at 10 GHz instead of 6 GHz.

Scientific literature, as reviewed in [13], considers that the 4 W/kg whole-body SAR level is the level at which RF energy begins to induce undesirable biological effects on humans. The FCC and ICNIRP use large safety factors to determine the maximum allowed and whole-body averaged SAR figures for general public and occupational exposure. The FCC allows 0.08 W/kg as the whole-body average for

Table 9.5 Exposure compliance criteria for different frequencies and different distances, as used by FCC

Frequency	Distance between the emitter and the human body	Criterion
Less than 6 GHz	Less than 20 cm	SAR
Less than 6 GHz	More than 20 cm	Power density, direct measurement
More than 6 GHz	Less than 5 cm	Power density, indirect measurement
More than 6 GHz	More than 5 cm	Power density, direct measurement

public exposure and 0.4 W/kg for occupational exposure. The maximum allowed level of exposure at any point of the human body increases by a factor of 20–1.6 W/kg. At the same time, the ICNIRP allows the maximum level of exposure at any point to go as high as 2 W/kg, which, in principle, is the same general ballpark figure.

The FCC and ICNIRP power density guidelines depend on the frequency of operation. The FCC guidelines are the stricter of the two. The FCC allows a power density of 10 W/m^2 for radiation up to 400 MHz only for occupational exposure. From 400 MHz to 2 GHz, there is a ramp-up in the allowed exposure, and above 2 GHz and all the way into the millimeter-wave range, the 10 W/m^2 figure can be used for the general public and the 50 W/m^2 figure becomes applicable to occupational workers.

As a last step towards determining whether 5G networks are a health risk to humans, one must consider how practical it is for user device manufacturers to consider and obey the regulations stipulated in this section, and, if this is achieved, whether there is a definite way to answer the question in the heading of Sect. 9.2.

9.2.5 *Effect of Electromagnetic Radiation on the Human Body*

In this section, it may be helpful to differentiate between instantaneous exposure and long-term exposure to electromagnetic radiation. Whereas the evidence presented in this section so far indicates that as long as network operators and user device manufacturers follow the radiation exposure guidelines, the physics of radio-wave propagation and absorption dictates that there is no risk to humans other than some heating of body tissue, typically that of the skin and of the eyes [13], medical researchers argue that long-term effects may still be unexplored and unknown; the evidence collected to date generally does not exclude the probability of long-term risks. The long-term effects may be due to heating itself, but also to some other mechanisms that will be listed shortly.

Effects Due to Heating

Heating effects on the human body are generally much better understood than non-heating effects.

As a result of low penetration of electromagnetic radiation, heating is typically associated with the eyes and skin as the two predominantly affected organs. The eyes are particularly vulnerable because they are located on the surface of the body with no skin protection, and in close proximity to the ears and mouth, meaning that they are subject to a relatively large amount of exposure when the phone headset is pressed against one's face [4, 13]. The main issue with the eyes is that heat is not easily dispersed, which in the rest of the body is achieved by blood flow. The problem with the effect of the EMF on the eyes has been aggravated in recent years with VR devices that are put on people's heads and used in entertainment and gaming, as well as for scientific exploration and remote work, causing exposure to near-field radiation for prolonged periods of time. There are also some indications that ocular injuries, such are cataracts and corneal damage, may arise from long-term exposure.

The heating of the skin is potentially worse in the case of millimeter-wave radiation than in the case of microwave radiation [4, 13]. This is because millimeter-wave radiation reaches a lower skin depth, so it does not reach deep into the tissue where it is easily dispersed across the body volume (recall that this is the reason why SAR is usually not used as a measure of electromagnetic exposure for millimeter-wave radiation). Instead, at least 90% of the millimeter-wave energy is absorbed in the deeper epidermis and dermis. This may lead to localized elevations near the skin surface and care needs to be taken to limit the amount of radiation so that extensive heating does not happen.

In principle, Wu et al. [13] conclude that at a power density of 10 W/m^2, no adverse effects are expected in the eyes or skin.

Non-heating Effects

When long-term, non-heating effects are concerned, the range of potential health issues considered to date is vast. Russel et al. [4] list adverse effects on deoxyribonucleic acid (DNA) integrity, cellular membranes, gene expression, protein synthesis, neuronal function, the blood-brain barrier, melatonin production, sperm damage and immune dysfunction as possible effects associated with electromagnetic radiation. Kostoff et al. [12] expand the list much more meticulously to include:

- *Carcinogenicity (brain tumors/glioma, breast cancer, acoustic neuromas, leukemia, parotid gland tumors);*
- *genotoxicity (DNA damage, DNA repair inhibition, chromatin structure), mutagenicity, teratogenicity, mutagenicity, teratogenicity;*
- *neurodegenerative diseases (Alzheimer's Disease, Amyotrophic Lateral Sclerosis);*
- *neurobehavioral problems, autism, reproductive problems, pregnancy outcomes, excessive reactive oxygen species/oxidative stress, inflammation, apoptosis, blood-brain barrier disruption, pineal gland/melatonin production, sleep disturbance, headache, irritability, fatigue, concentration difficulties, depression, dizziness, tinnitus, burning and flushed skin, digestive disturbance, tremor, cardiac irregularities;*
- *adverse impacts on the neural, circulatory, immune, endocrine, and skeletal systems.*

as several categories of potential health issues of RF radiation, even those at levels below the FCC guidelines.

The biggest issue with claims like those above is the lack of evidence supporting direct correlation between various diseases and electromagnetic radiation. A typical study to investigate the influence of radiation on health takes a long time, and the presence of a control group to compare the results against, which in the scenario of many potential risks above is not practical. Furthermore, Miller et al. [16], among others, argue that it is challenging to do these kinds of studies because the technology is changing so rapidly that by the time some meaningful insight is obtained, a new technology with its own potential risks has been released and the old technology may have been abandoned or changed. Moreover, these authors argue that lack of epidemiological evidence does not necessarily indicate the absence of an effect, but could indicate inability to find a relation, given the experimental conditions and study duration. On the other hand, Kostoff et al. [12] argue that industry-funded research is far more likely to show no effects than research that secures funding from non-industry sources. In the context of health, it is also essential to consider additional evidence from in vivo and in vitro studies [17], but studies that involve animals such as rodents generally do not lead to accurate results, because electromagnetic radiation exposure is dependent on the exposed tissue area and mass.

Several recent review articles attempted building an overall picture of the available research data that seek to make this connection. The detail of the review coverage, as well as the study findings, have been tabulated in Table 9.6 [17–20]. It does not come as a surprise that the results are inconclusive. Also note that these studies, with the exception of [20], although recent, typically cover the material collected in the 3G and 4G eras, which is due to the already discussed fact that these studies need to be conducted over an extended period. The long-term studies that would be relevant to technologies introduced by 5G would only become available many years from now, and any results available at present may be deceiving.

Special Case: What about Cancer?

In the context of solely the relation of electromagnetic radiation and cancer, the International Agency for Research on Cancer (IARC) gave electromagnetic radiation a 2B Classification in 2011 [21]. 2B in this classification means that the agent in question is possibly carcinogenic, as shown in Table 9.7. To put this into perspective, coffee is also put into the category of possibly carcinogenic agents, together with many elements that can consumed in food or drinks. Some researchers have also been calling for electromagnetic radiation to be re-classified into group 2A, which would make it "probably carcinogenic" (for example, [19] included in Table 9.6), but for this to happen, much more concrete evidence of the carcinogenic effect of radiation is needed. The present classification, as it stands, is probably a good indication that electromagnetic radiation is safe.

Electromagnetic Radiation and Children

Another major concern mentioned in the literature is the influence of electromagnetic radiation on the development and growth of children. There is definite evidence that children are more susceptible to EMF radiation [16, 22, 23]. A group led by Fernandez [22, 23] has determined that children experience two to three times more RF exposure than adults from radiation that has the same incident power level. This is due to their smaller size, the mass of their heads and thickness of their skulls, because the head typically remains the most exposed part of the body, even in children.

If one looks from the volumetric perspective, the same volume exposed to radiation in the case of a child's head, is in fact a much larger portion of the complete head.

Table 9.6 Several recent review articles that attempt to study the body of existing research into the long-term effects of electromagnetic radiation

Authors, year of publication and reference	Material reviewed	Conclusion
Halgamuge et al. [17]	Data from 300 peer-reviewed scientific publications (1990–2015) describing 1 127 experimental observations of non-thermal effects in cell-based in vitro models	Cumulative effect (SAR × exposure time) study leading to inconclusive results. Roughly half of the studies found no cellular changes, while the other half showed some response
Keykhosravi et al. [18]	Data from articles indexed by PubMed, Scopus, Springer, Cochrane and Google Scholar (1995–2013) for evidence on the correlation of electromagnetic radiation and skin diseases	Overall evaluations showed that the level of evidence associated with the effects of radiation from mobile phones and tablets on the skin, and skin-related diseases (including skin cancer) is weak and has no statistical significance. Authors describe the associated health risk as "mildly increased"
Morgan et al. [19]	Data from INTERPHONE, CERENEAT and some other studies aiming to find the correlation between brain cancer and glioma with electromagnetic exposure	The risk increases with hours of device use; radiation could have a stronger impact on developing brains. Brain cancer remains a rare disease and extremely large samples are required to reach a definite conclusion. This study also recommends that electromagnetic radiation be classified as a class 2A carcinogen
Simkó and Mattsson [20]	Data from 94 publications performing in vivo or in vitro investigations on the effect of millimeter-wave radiation on human health	Contradictory information from the in vivo and in vitro investigations. No conclusive evidence on any of the non-thermal effects. The number of available studies too small for definite conclusions

Table 9.7 Classification of carcinogenic hazards by IARC [21]

Group	Description
1	Carcinogenic to humans
2A	Probably carcinogenic to humans
2B	Possibly carcinogenic to humans
3	Not classifiable as to its carcinogenicity to humans

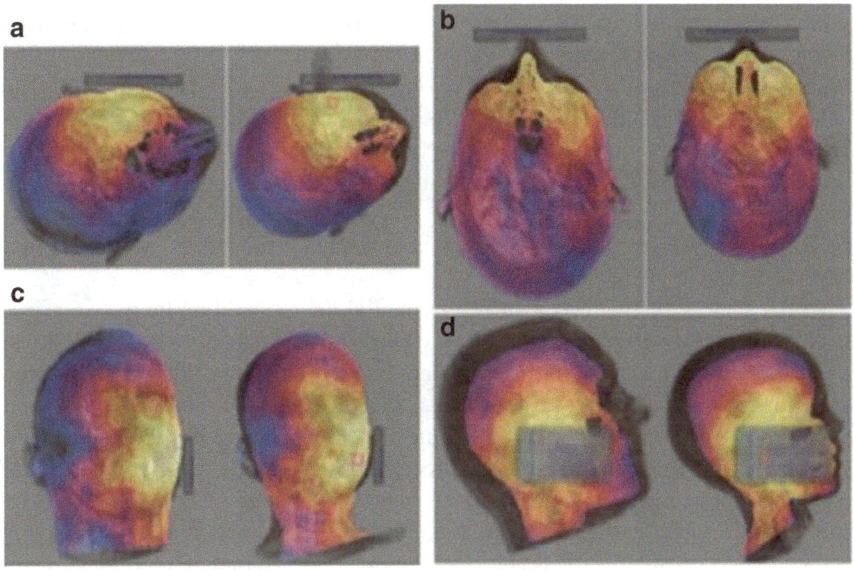

Fig. 9.6 SAR in cross-sectional views of heads of adult male (34 years old) and child (6 years old), with phone used in talk mode pressed against their faces. Views A, B, C and D represent different angles of observation. Yellow indicates the largest exposure and purple the lack thereof. Reproduced with permission from [22]

Thus, the radiation can correspondingly penetrate deeper into the head. This is shown in Fig. 9.6. Some of the harmful effects of higher doses of radiation include the higher level of heating of children's eyes, as well as the influence of radiation on their early childhood brain development.

Three main recommendations can be made here. Firstly, beyond doubt, younger children should not use cellphones or any other devices that can be placed close to their body or head while transmitting. Rather, children can be allowed devices that run in the so-called "airplane mode", or they should use devices that establish the connection to the internet via cable. Secondly, regulations could be amended to forbid the sale of phones that attract children, as was done in Belgium in 2014 [24]. And thirdly, metrics for electromagnetic radiation exposure (SAR, power density) should be adjusted for the age of the child [16], while ensuring compliance by technology providers. Essentially, the last point raises a question on how to improve the certification of mobile phones and other RF-emitting devices in future.

9.2.6 Conclusion: Is 5G Harmful?

Again, the main concern with EMF exposure could be the heating of the face—the eyes and the skin—and other parts of the human body because of the absorption of the energy emitted by the headset. Fortunately, human bodies are reflective and do not absorb a lot of radiation. Thus, it is clear that short-term effects are not problematic, as long as the general radiation guidelines are followed. This holds true for lower—RF—frequencies as well as for millimeter-wave radiation. Even airlines, which traditionally prohibited the use of cellphones on board aircraft, on the pretext that the electromagnetic radiation could interfere with the aircraft systems, have been cleared to allow cellphone use during all stages of the flight [25], which enforces the point that the electromagnetic waves are weak several meters away from their source.

At the same time, there is a small chance that there could be long-term effects of prolonged EMF exposure, such as a chance of cancer development; however, scientific studies conducted to date could not find any close correlation of this nature yet. Thus, more long-term research studies are needed. It seems unlikely that overall EMF exposure at present and in the future will be more harmful than spending a few minutes in the sun or having some processed meat, coffee or a glass of wine (all known to cause cancer in humans). Nevertheless, it is never a bad idea to err on the side of caution, and to take certain precautions to prevent overexposure, especially when it comes to children. This is for at least two reasons: first, it takes long for full recognition of a health risk [15], and second, the risks may increase as the technology becomes widespread [4]. Some of the recommendations on steps that can be taken to minimize exposure, taken from various studies, can be seen in Table 9.8 and can be practiced by all individuals.

A case may also be made for reviewing the official RF exposure guidelines and regulations in future, especially when it is considered that present regulations date

Table 9.8 Recommendations on steps to minimize exposure from cellphone radiation

Recommendation	Description
1	Avoid carrying the mobile phone close to one's body all the time (e.g. in a pocket)
2	Avoid holding the mobile phone pressed against one's face during a call
3	Use the loudspeaker or a hands-free kit whenever possible
4	When use of a hands-free kit or a loudspeaker is not possible, limit the duration of the call
5	Use airplane mode when not using the phone or disable the second sim card (e.g. work phone can be switched off outside work hours)
6	Do not use the phone when on the move (car, bus, train), unless in hands-free mode, because the movement requires the device to transmit at a higher power level
7	Keep the phone some distance away from the bed at night
8	Forbid children to use mobile phones up to a certain age; rather use devices with no connectivity or wired connectivity

Table 9.9 Received powers at various distances from imagined 5G and 6G base stations. It is assumed that both the 5G and 6G base stations emit at 240 W and with a transmitting antenna gain of 10 dBi

Distance/frequency	30 GHz	60 GHz	300 GHz	600 GHz
3.16 m	151 μW	13.79 μW	1.5. μW	379 nW
10 m	15.1 μW	1.379 μW	151 nW	37.9 nW
100 m	151 nW	37.9 nW	1.51 nW	Less than 1 nW
1 km	1.51 nW	Less than 1 nW	Less than 1 nW	Less than 1 nW

to about three decades ago. This could be done by specifying different recommendations for children, or amending regulations associated with the general public to be universally applicable to children or in some other manner.

An additional point that can be made concerns the fact that 5G networks are the first networks that introduce M2M communication. It is fortunate that machines (sensors, robots, etc.) can be exposed to much higher radiation levels than humans. Therefore, radiation limits and concerns that are associated with human radiation exposure are not applicable in all the contexts of 5G deployment, specifically not those that are involve machines. These devices ("machines"), depending on their deployment scenario, may emit much higher power levels than those allowed for cellphones, tablets, laptops, and other equipment, without any risk of bodily harm, thus opening up the possibility of other types of 5G deployment. The regulations for machine exposure could be set at power levels that are used for occupational exposure; this limit is only set because of the necessity for human maintenance.

9.3 6G: What is Different?

The previous section dealt with the presently burning issue of 5G and its effect on human health. It is now a good time to consider how different 6G is from 5G in terms of its impact on human health. To begin this analysis, it should be considered what the further differences are when it comes to 6G deployment. The following list identifies the key points for this discussion:

- 6G will add another range of frequencies to the frequencies that are already utilized for 5G. To be more specific, 6G is likely to use frequencies above 100 GHz (higher millimeter-wave frequencies and terahertz frequencies).
- Because of the increase in frequencies, and the increase in path loss, even smaller cells may need to be used, calling for possibly even denser network deployment.
- Even more man-made ambient radiation will consequently exist in areas covered by 6G, with a combination of frequencies below 6 GHz, between 6 and 100 GHz and above 100 GHz. With the ubiquitous coverage promised by 6G, it may be argued that there would be no place on earth to hide from EMF radiation exposure.

Fortunately, none of the differences considered above are showstoppers to the technology evolution. First, when the increase in frequency is concerned, one must again recall Fig. 9.5 and the fact that terahertz radiation is also non-ionizing, and is no more harmful than millimeter-wave radiation. If anything, the penetration of terahertz radiation into the human body is even shallower than that of millimeter-wave radiation and the exposure to radiation becomes even more localized. As a matter of fact, terahertz radiation has already been studied extensively for imaging and sensing, including for applications such as full-body scanning for security, medical scanning, or for high-precision radar, including that for gesture recognition [26]. The primary reason for considering THz radiation in imaging in the first place was that it was considered less harmful than the radiation generated by ionizing X-ray scanners, which are widely used today (albeit with great care, e.g. radiologists wear a steel cage or stand behind a wall when performing a scan on a patient).

Secondly, the density of cells for 6G will not necessarily be much higher than the density of 5G cells, because some of the path loss that is much higher (Table 9.1) can be compensated for by high-directivity antennas, an approach that was extensively discussed in Chap. 3. If this results in the transmitter not complying with the regulation of maximum power or power density guidelines, this attenuation compensation may still be executed on the receiver side. There are also some other factors that work in favor of terahertz transmission over distances comparable to millimeter-wave transmission distances, including lower atmospheric (O_2) attenuation than, for example, observed at 60 GHz.

Lastly, the high path loss means that at any reasonable distance from the access point, the physical power density of terahertz radiation becomes negligible. According to the analysis in Table 9.9, the perceived power barely exceeds 1 μW, and there is already virtually zero terahertz pollution 100 m away from the base station.

Additional concern brought about by terahertz radiation may be about the effects of short wavelengths on absorption parameters in various parts of the human body, including the skin [27]. This is because the sub-millimeter wavelengths are now of similar size as the smallest features in human tissue, and thus the assumption of a uniform tissue used in modeling of tissue absorption does not hold anymore. Research on the possible effects of this phenomenon continues. In fact, it may be even more challenging to find credible research that investigates the general effects of terahertz radiation on human bodies than on RF or millimeter-wave radiation. One recent review that can be considered is that completed by Cherkasova et al. [28], which, not surprisingly, concludes that "there currently exists no scientific consensus as to whether THz radiation has a damaging effect on biological objects of different complexity." This is attributed to the fact that currently there are no standardized experimental procedures, that the diversity of available THz sources is insufficient, and that there is a clear lack of tools that would enable accurate assessment of the initial functional state of the biological object under observation.

In view of everything above, for the time being at least, terahertz-wave transmission, and thus 6G, will not be worse when it comes to human health considerations than 5G is at present.

9.4 Visible Light Communications and Health

There is still one aspect that has not been considered in this chapter, namely the fact that in 6G networks, communications using visible light may complement traditional radio communication, as discussed in Chap. 5, with the purpose of achieving extremely high data rates.

Additional to the high-speed general networking benefit, VLC is usually quoted as a potential alternative means of high-speed communication for environments such as hospitals, chemical plants and aircraft, because light does not have a tendency to interfere with sensitive equipment [29]. In principle, several arguments can be made in favor of declaring VLC safe for human use in these and many other environments:

- *VLC transmitters require lower power to operate than RF systems.* This is due to the high efficiency of modern LEDs. As a result, tissue damage due to heating will not be an issue in VLC.
- *Light does not travel through walls and other obstacles.* Given that VLC communications are envisioned for indoor environments, all the power from the VLC transmitter is contained in a single room. Consequently, it is relatively easy to protect oneself from "light pollution" if needed, which is not necessarily the case for electromagnetic pollution.
- *VLC transmitters are often deployed as add-ons on existing lighting LEDs.* Because VLC receivers are able to distinguish the AC currents representing the recovered transmission signal from the DC current that is generated due to ambient light, a dark environment is not necessary for VLC transmission. This means that for all practical purposes, the additional light energy that is generated by signal transmission can be worked into the total room illumination requirements. Therefore, the light generated for information transfer does not require any additional safety requirements to be considered, other than those that are typically considered for illumination. This excludes flickering, which will be discussed shortly.
- *LEDs are dimmable.* For VLC transmitters that are not part of existing lighting, dimming control is typically used to regulate the brightness of the transmitter to stay within the acceptable guidelines for illumination intensity.

One concern that needs to be considered, however, is the unavoidable flickering that is a result of the modulation of the transmitted signal on the LED. Typically, this flickering is so fast that it is not observable, i.e. human eyes perceive it as continuous light. It is interesting to note that flickering of LEDs occurs even in LEDs solely used for lighting, which is due to driving voltage fluctuations [30]. In both cases (intentional and non-intentional flickering), it is possible to take steps to minimize the flickering effect on human eyes. For this purpose, in 2015, the IEEE developed a standard that lists the recommended practices for modulating current in high-brightness LEDs with the aim of mitigating health risks to viewers [31]. In practice, this is done by minimizing the potential of intensity fluctuations by virtue of modulation symbol design. It entails constraining the total optical power of each symbol to an equal value for each symbol that is used in the transmission

[32]. in addition, the transceiver system design also needs to consider periods of no transmission: unlike radio, a VLC transmitter should never stop transmitting entirely, but should rather transmit dummy information or light with constant intensity. This also means that sleep modes used in radio to conserve power (Sect. 6.3.6) are not practical in the VLC context; rather the focus in energy saving and conservation must be turned to harvesting the "wasted" light on the receiver side, as described in Chap. 5.

Lastly, it should be noted that VLC may not be practical for UL communication from the health and safety perspective [33]. The reason is that if the VLC transmitter is installed in the user device, the light that would constantly be emitted would create nuisance or could even potentially harm the eyes if the device is held close to the face.

9.5 Concluding Remarks

Discussions in this chapter indicate that at least for now, there is no reason to be alarmed by 5G deployment, nor should one be wary of 6G deployment ten years from now, as far as general public health is concerned. As it stands, 5G and 6G are not paving to the way for human and environmental destruction. Many other technologies used every day carry a substantially higher measurable risk [14]. There are also many activities people undertake willingly that carry a substantially greater risk, such as drinking alcohol or lying on the beach at noon. This holds true for potential risks from both electromagnetic radiation as well as light used in VLC.

It should not be denied that overexposure to electromagnetic radiation, predominantly from user devices placed close to the human body, does carry certain risks, but these are easily mitigated with individual behavior changes (Table 9.8). This is to err on the side of caution—most leaders in this field agree that more research is needed before any real association between human health and electromagnetic radiation can be confirmed. Meanwhile, the authors are of the opinion that the benefits of 5G networks and of 6G networks in future are currently irreplaceable by any other technology. If nothing, one should at least honor the lives saved by these technologies. What about roadside assistance and medical assistance sought on a remote road? It would not be possible without cellular coverage. What about all the car safety systems, such as automatic braking, which use V2X communication and millimeter-wave or terahertz radar to prevent crashes and the medical attention to be sought in the first place? Or children lost and found using user device geolocation features, such as those on their smart watches? All this would be impossible with these radio waves that are dubbed potentially harmful. This is even before we mention technologies such as smart healthcare and remote surgery that will become a reality in 6G with concepts such as VR, technologies like smart cities that aim at green living and energy efficiency, or other sustainable solutions for the developed and developing world. Thus, the authors of this book are of the opinion that with careful wireless technology use, network coverage planning, possible amendment of certain emission regulations and technology education, it is possible to focus on the advancement of humanity as a whole without compromising on the overall state of health. In parallel, engineers can work with health scientists on structuring more conclusive studies, while looking for approaches to mitigate the risks found, if any.

To reinforce the argument of the benefits of 6G for society as a whole further, Chap. 10 will conclude the book by reassessing the many verticals that will be enabled by 6G, which make this field of research exciting.

References

1. Sandford A (2020) Coronavirus: half of humanity now on lockdown as 90 countries call for confinement [Internet]. Euronews. https://www.euronews.com/2020/04/02/coronavirus-in-europe-spain-s-death-toll-hits-10-000-after-record-950-new-deaths-in-24-hou. Accessed 13 Aug 2020
2. SA's 5G spectrum is now wide open to help fight Covid-19—but only until November [Internet]. Business Insider. https://www.businessinsider.co.za/temporary-5g-spectrum-for-cellphone-operators-in-sa-during-the-covid-19-crisis-2020-4. Accessed Aug 13 2020
3. (. 2020) Slovenia stops the introduction of 5G technology [Internet]. Eng News 24 h. https://engnews24h.com/slovenia-stops-the-introduction-of-5g-technology/. Accessed 10 Apr 2020
4. Russell CL (2018) 5 G wireless telecommunications expansion: public health and environmental implications. Environ Res 165:484–95
5. WHO (2020) What is the international EMF project? [Internet]. World Health Organization. World Health Organization. http://www.who.int/peh-emf/project/EMF_Project/en/. Accessed 13 Aug 2020
6. Drake F (2006) Mobile phone masts: protesting the scientific evidence. Public Underst Sci 15:387–410 SAGE Publications Ltd
7. Waterworth T, Guy D (2020) 5G tower can be set up in your garden. https://www.iol.co.za/ios/news/5g-tower-can-be-set-up-in-your-garden-80dab300-d8a6-4c1e-84c7-dcc9dd0a8efe. Accessed 13 Aug 2020
8. Feldman Y, Puzenko A, Ben Ishai P, Caduff A, Agranat AJ (2008) Human skin as arrays of helical antennas in the millimeter and submillimeter wave range. Phys Rev Lett 100:128102. American Physical Society
9. I C-L, Han S, Bian S. Energy-efficient 5G for a greener future. Nat Electron 3:182–184. Nature Publishing Group
10. Arnold O, Richter F, Fettweis G, Blume O (2010) Power consumption modeling of different base station types in heterogeneous cellular networks. In: 2010 Future network mobile summit, pp 1–8
11. Gati A, Salem FE, Serrano AMG, Marquet D, Masson SL, Rivera T et al (2019) Key technologies to accelerate the ICT Green evolution—an operator's point of view. arXiv:190309627 [cs] [Internet]. http://arxiv.org/abs/1903.09627. Accessed 10 July 2020
12. Kostoff RN, Heroux P, Aschner M, Tsatsakis A (2020) Adverse health effects of 5G mobile networking technology under real-life conditions. Toxicol Lett 323:35–40
13. Wu T, Rappaport TS, Collins CM (2015) Safe for generations to come: considerations of safety for millimeter waves in wireless communications. IEEE Microwave Mag 16:65–84
14. Johnson D (2020) How worried should you be about the health risks of 5G? [Internet]. How-To Geek. https://www.howtogeek.com/423720/how-worried-should-you-be-about-the-health-risks-of-5g/. Accessed 13 Aug 2020
15. Lin JC (2016) Human Exposure to RF, microwave, and millimeter-wave electromagnetic radiation [health effects]. IEEE Microwave Mag 17:32–6

16. Miller AB, Sears ME, Morgan LL, Davis DL, Hardell L, Oremus M et al (2019) Risks to health and well-being from radio-frequency radiation emitted by cell phones and other wireless devices. Front Public Health [Internet]. Frontiers. https://www.frontiersin.org/articles/10.3389/fpubh.2019.00223/full. Accessed 7 Aug 2020
17. Halgamuge MN, Skafidas E, Davis D (2020) A meta-analysis of in vitro exposures to weak radiofrequency radiation exposure from mobile phones (1990–2015). Environ Res 184:109227
18. Keykhosravi A, Neamatshahi M, Mahmoodi R, Navipour E (2018) Radiation effects of mobile phones and tablets on the skin: a systematic review [Internet]. In: Advances in Medicine, Hindawi, p e9242718. https://www.hindawi.com/journals/amed/2018/9242718/. Accessed 7 Aug 2020
19. Morgan LL, Miller AB, Sasco A, Davis DL (2015) Mobile phone radiation causes brain tumors and should be classified as a probable human carcinogen (2A) (Review). Int J Oncol 46:1865–71 Spandidos Publications
20. Simkó M, Mattsson M-O (2019) 5G wireless communication and health effects—a pragmatic review based on available studies regarding 6 to 100 GHz. Int J Environ Res Publ Health. 16:3406. Multidisciplinary Digital Publishing Institute
21. Agents Classified by the IARC Monographs, Volumes 1–127—IARC Monographs on the Identification of Carcinogenic Hazards to Humans [Internet]. In: International agency for research on cancer. https://monographs.iarc.fr/agents-classified-by-the-iarc/. Accessed 11 Aug 2020
22. Fernández C, de Salles AA, Sears ME, Morris RD, Davis DL (2018) Absorption of wireless radiation in the child versus adult brain and eye from cell phone conversation or virtual reality. Environ Res 167:694–9
23. Fernández-Rodríguez CE, De Salles AAA, Davis DL (2015) Dosimetric simulations of brain absorption of mobile phone radiation-the relationship between psSAR and Age. IEEE Access 3:2425–30
24. Furniere A (2020) Belgium bans sale of mobile phones designed for children| Flanders Today [Internet]. Flanders Today. innovation/belgium-bans-sale-mobile-phones-designed-children. Accessed 25 Aug 2020
25. Airlines cleared for mobile phone use during flights [Internet]. BBC News. https://www.bbc.com/news/business-29381193. Accessed Aug 19 2020
26. Božanić M, Sinha S (2019) Emerging transistor technologies capable of terahertz amplification: a way to re-engineer terahertz radar sensors. Sensors 19:2454
27. Betzalel N, Feldman Y, Ishai PB (2017) The Modeling of the absorbance of Sub-THz radiation by human skin. IEEE Trans Terahertz Sci Technol 7:521–8
28. Cherkasova OP, Serdyukov DS, Ratushnyak AS, Nemova EF, Kozlov EN, Shidlovskii YV et al (2020) Effects of terahertz radiation on living cells: a review. Opt Spectrosc 128:855–866
29. Obeed M, Salhab AM, Alouini M-S, Zummo SA (2019) On optimizing VLC networks for downlink multi-user transmission: a survey. IEEE Commun Surv Tutor 21:2947–76
30. Orgulan A, Sukič P, Ribič J (2019) A procedure for mitigating the light flicker in office LED lighting caused by voltage fluctuations. Energies 12:3975. Multidisciplinary Digital Publishing Institute
31. IEEE (2015) 1789-2015 - IEEE recommended practices for modulating current in high-brightness LEDs for mitigating health risks to viewers [Internet]. IEEE Standards Association. https://standards.ieee.org/standard/1789-2015.html. Accessed 20 Aug 2020
32. Xu W, Wang J, Shen H, Zhang H (2015) Multi-colour LED specified bipolar colour shift keying scheme for visible light communications electronics letters. IET Digital Libr 52:133–5
33. Sarieddeen H, Saeed N, Al-Naffouri TY, Alouini M-S (2020) Next generation terahertz communications: a rendezvous of sensing, imaging, and localization. IEEE Commun Mag 58:69–75

Chapter 10
6G Networks: Fusion of Communications, Sensing, Imaging, Localization and Other Verticals

Abstract Throughout this book, 6G has been introduced as a network that might be the first to truly fuse the likes of communications, sensing, imaging, localization and many other verticals. Some of the verticals that are expected to be introduced in 6G, as well as those that might become available earlier, over the course of the 5G evolution that is expected in the next ten years, have been discussed sporadically in the previous nine chapters. It seems fitting, however, to dedicate this complete chapter to a more detailed discussion of each of the most important envisioned 6G services or applications. At the same time, this topic presents the most logical place to conclude this book.

Throughout this book, 6G has been introduced as a network that might be the first to truly fuse the likes of communications, sensing, imaging, localization and many other verticals [1, 2]. 6G will practically be a network of services, with communication being only one of many available services [3]. 5G networks were initially thought of in a similar way, and certainly take a step in this direction, but it appears that, at least in early releases, the emphasis is still predominantly on delivering fast and reliable communications, albeit no doubt of far better quality than previous wireless network generations. However, 5G does still come with limited verticals, and ten years is a very long period in terms of wireless technologies. Thus, it can be said without any doubt that one can expect to see notable changes in the (beyond) 5G architecture over the coming years, eventually paving the road for the ultimate 6G experience. At the same time, 6G enabling technologies will continue to evolve, allowing some 6G ideas that have not been able to see the light of day due to technology limitations to date, to appear in the fresh 6G release.

Some of the verticals that are expected to be introduced in 6G, as well as those that might become available earlier, over the course of the 5G evolution that is expected in the next ten years, have been discussed sporadically in the previous nine chapters. It seems fitting, however, to dedicate one full chapter to a more detailed discussion of each of the most important envisioned 6G services or applications. At the same time, this topic presents the most logical place to conclude this book.

M. Božanić and S. Sinha, *Mobile Communication Networks: 5G and a Vision of 6G*, Lecture Notes in Electrical Engineering 751,
https://doi.org/10.1007/978-3-030-69273-5_10

To be able to define the topics for discussion in this chapter, the introductory section will attempt to summarize the innovations that 5G and 6G are bringing to wireless communications.

10.1 Summary: Which Innovations Do 5G and 6G Bring to Wireless Communications?

The main verticals of 5G applicable to various economic sectors, including the agriculture, automotive, construction, energy, finance, health, manufacturing, media, public safety, retail and consumer and transport fields, were summarized in Chap. 2, as part of the historical discussion of wireless networks, including 5G. In addition to improved communication and media delivery efforts, 5G will introduce the first steps to automatic driving and traffic safety, factory automation, remote healthcare and indoor navigation. 5G is also considered the key enabler of concepts such as IoT and smart cities, while for the first time allowing the concepts of VR and AR to move from an enclosed room into the open, mobile environment, although with a number of limitations pertaining to QoE. The main verticals introduced by 5G are summarized in Table 10.1 for ease of future reference.

In Chap. 1, 6G was introduced as the network that will introduce even faster and more enhanced conventional communications, with the focus on improving the personal experience, while, at the same time, tying up all the loose ends that may have been introduced by 5G. In, 6G, additionally, new communication paradigms will be introduced, including holographic and tactile communication and communication between new devices, including wearables and implantable sensors. New services will include XR and tactile internet, robotics, autonomous systems including full autonomous driving and drones, accurate indoor positioning, IoE and possibly even Industry 5.0. Many of the verticals will only become possible with ubiquitous coverage that will be achieved with a mix of new technologies, including satellite communications and hovering base stations for airborne objects and VLC for communicating under water. The additional verticals of 6G are summarized in Table 10.2.

Figure 10.1 shows the verticals from both Tables 10.1 and 10.2 that will be discussed in the sections that follow. Since communications and data transfer (internet connectivity) will remain an integral, if not the most important, service of future wireless networks (be it 5G or 6G), these two verticals will be discussed first.

Table 10.1 Some of the verticals introduced by 5G, as identified in Chap. 2

Possible vertical	Short description
Autonomous vehicle control	First steps to automatic driving; better traffic safety; control signaling, exchange of video information; networking of vehicles for fleet driving
Emergency communication	Reliable network for emergency communications, such as earthquakes, floods, etc.
Communication while traveling at high speed	Retaining high data rates even if the user is moving at high speed, for example on board a high-speed train
Communication at large events	Concerts, sports games or any events where a large number of network users are concentrated in a small area in which normally, such a high traffic load does not occur
Factory automation	High level of manufacturing automation encompassed by the umbrella of Industry 4.0
IoT	Fast and low-latency connections to a large number of geographically spread devices, including devices used for surveillance, monitoring, environmental sensing, etc.
Media on demand	High-quality media (e.g. HD streaming) accessed at any time, irrespective of the physical geographical location
Remote healthcare	Virtual doctor consultations, virtual examinations and remote surgery in extreme situations
Indoor navigation	Using the 5G network for localization in large indoor areas, such as shopping malls and office buildings
Smart cities	Including concepts such as smart homes, smart offices, smart buildings, smart grid, smart traffic control
VR and AR	Enabling mobile access for VR and AR services that have traditionally been bound to a single location (a room with a wired connection)

Table 10.2 Additional services and verticals introduced by 6G, as identified in Chap. 1

Service	New or improved verticals/Short description
MBRLLC	VR, AR, XR, autonomous vehicles, autonomous drones
mURLLC	IoE, sensing, robotics
Human-centric services	Improving QoPE
MPS	Convergence of communication, computing, control, localization and sensing: CRAS, telemedicine, environmental imaging, mapping, improved localization, Industry 5.0

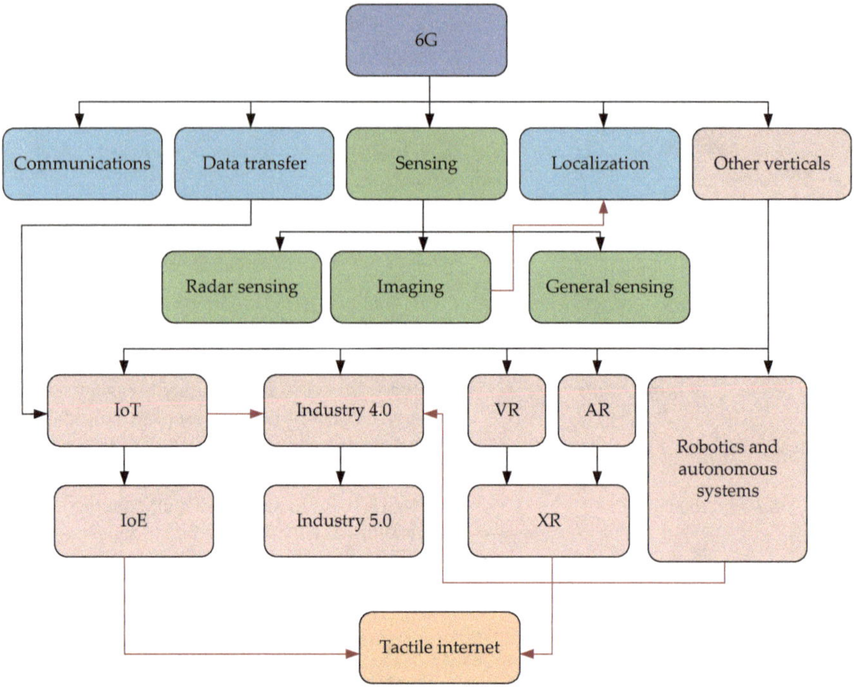

Fig. 10.1 6G verticals that will be discussed in this chapter, with most prominent relationships between different verticals indicated

10.2 Communications and Data Transfer

As already stated, in 6G, the conventional communications between human users, which generally also include data transfer, will be enhanced in many ways.

10.2.1 Communications Vision of 6G

6G promises to offer enhanced mobile broadband with ubiquitous connectivity in all three dimensions, even in remote or isolated areas. Furthermore, the connectivity will come at faster data rates and lower latency, as well as lower cost. The green vision of 6G will additionally work towards extending the battery life of user devices. With the increase in the number of devices requiring access to the internet, 6G will have to continue supporting the ever-increasing number of users in any given small area, without any major compromise in the QoS and irrespective of whether any of the users are stationary or moving.

When it comes to new communication paradigms, wearable devices, integrated interfaces and implantable nanodevices and nanosensors are expected to start complementing or even replacing traditional communication devices, which today come in the form of cellphones, tablets, laptops and desktop computers. Sensing, which will also become an integral part of 6G, may also enable control of some of these devices, in the form of technologies such as gesture recognition or haptic communications. As discussed in Chap. 9, future 6G devices should be declared safe for human use (which is achieved in wearables by using extremely low power levels), while also having reliable power supplies with prolonged battery lifetime, because of the inconvenience of charging such devices, as discussed in Chap. 6.

10.2.2 Services Associated with Communications

Many new communication possibilities will be introduced by the two underlying technologies—terahertz communication and VLC, which are both able to support extremely high data rates, albeit in a very small area, typically an enclosed room or small outdoor sections. Nevertheless, these technologies will enable various other sub-types of communication [4–8].

Localized High Bandwidth User Communications

In small indoor environments where distances between communicable devices are lower than 10–15 m, it would be possible to establish LOS connections via radio or even light waves for fast data transfer that would not involve cabling. These types of WLANs, sometimes called enhanced hotspots [9], would support extremely fast data rates, but depending on the positioning or routers, connections could occasionally be broken by moving people and other objects. In small rooms, however, this could be compensated for by receiving signals through indirect paths arising from reflections from walls. For outdoor coverage, femtocell deployment of 6G by utilizing street poles in urban environments could achieve a similar effect over distances of tens of meters.

Wireless Personal Area Networks

Another application of the fast data rate provided by 6G could be for wireless personal area networks (WPANs). These networks would be similar to WLANs, but the communication would be point-to-point communication between different peripherals of a single device or between different devices belonging to one user, again without cables. For example, a WPAN connection could involve a solid-state drive connected to a laptop, desktop or a server.

Fast Data Sharing

Fast data sharing is another important vertical related to modern communications. Some sources refer to this application as "kiosk downloading", whereas others use the

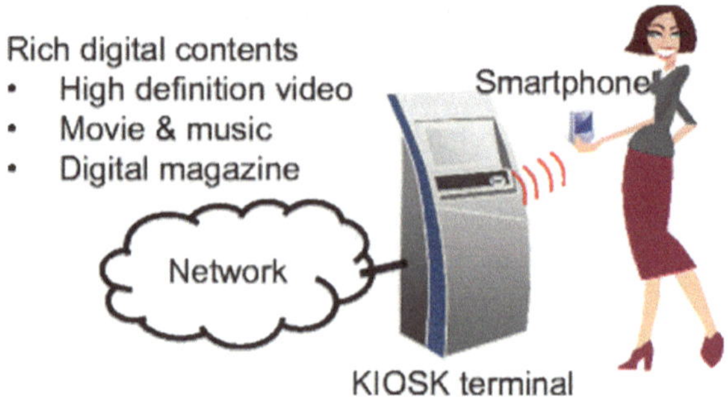

Fig. 10.2 Kiosk downloading concept. Reproduced with permission from [10]

term "information shower". The content would be shared by bringing two devices in proximity to each other, with the 6G network mediating the traffic transfer on its edge. With distances of only some centimeters, the feasible data rates increase considerably owing to the minimal amount of attenuation, and it would take only a few seconds to download full 1080 p or 4 K movies. This concept is illustrated in Fig. 10.2, where Ref. [10] envisioned the transfer between a mobile device (e.g. a cellphone) and a data terminal ("kiosk").

Communication Between Servers

Servers and cloud-based storage will also benefit from 6G. First, with the speed of available connections, most of the storage and computations could be offloaded to the cloud. In the case of local servers, it can be observed that a large portion of space taken up in data centers is occupied by cabling, which could be replaced by directional high-speed terahertz or millimeter-wave wireless links.

Wireless Backhauling

Wireless backhauling has also been discussed earlier in the book. In 5G and 6G, dedicated millimeter wave, terahertz or optical links will be used as one of several alternatives to cross the last mile between the backbone internet network and end users in rural environments in which the LOS can easily be established. The wide bandwidth that operation at high frequencies or in the OWC domain allows basically enables a scenario in which wireless technologies are used both to service the users and to provide backhaul access [9].

Depending on the exact technology deployed and the topology of the terrain, the distances that can be covered will vary from several hundreds of meters to a few kilometers. In many instances (for example in areas with poor infrastructure), these kinds of links may be the only option to provide high-data-rate services, such as multimedia streaming. It is thus generally understood that wireless networks such as 6G are by far the most cost-effective option for "connecting the unconnected" in developing countries, thereby assisting in approving social fairness issues reflected by the Gini index, and in turn contributing to reaching many of the SDGs [11]. As such, 6G is expected to succeed where previous technologies have failed.

Holographic Communications

Another major innovation in the world of communication may be establishing holographic communications. Practical applications of holographic communications include entertainment, education, training, telepresence, medical imaging, interactive design and scientific visualization [12]. In office environments, a particularly appealing concept is that of co-workers getting together and interacting without being physically present in a meeting [11]. The beginning of holographic communications (other than those imagined in sci-fi movies) dates back to 1990 when the first 3D holographic imagining system was demonstrated by the Media Laboratory at the Massachusetts Institute of Technology. The concept of holographic image creation and projection is shown in Fig. 10.3 (also see [13]); this principle was already discussed in more detail in Chap. 7 when the principle of holographic radio operation was introduced.

Holographic real-time imaging requires an enormous amount of information to be transferred, with 3D display bandwidths that are some 20,000 times larger than those of conventional 2D displays [12]. Thus, the initial holographic demonstrations were projecting still images, which only had the ability to refresh every several minutes. However, as processing power and data capability increase, holographic communications can become a reality. It is estimated that 6G data rates in excess of 1 Tb/s are sufficient for projecting a 60 cm hologram that has 10^{12} samples (1 terabyte of data) at a refresh rate of 1 s. With some extra bandwidth, not only 3D image projection, but also the transfer of other sensory (tactile) information, as described in Chap. 1, should be possible.

High Secrecy/Security Communications

In another general case, terahertz communications and VLC can be used in any scenario where security or secrecy of information is necessary. 6G may manage the establishment of the required link between two parties [14]. Because of the directivity of terahertz or light waves, it is not trivial to eavesdrop on such a link, or even to detect the presence of the secure link without prior knowledge. Thus, at least from the hardware perspective, terahertz and VLC communications are some of the most secure types of communication available at present.

10.3 Radar Sensing

Sensing in general, after communication, is possibly the second most important vertical of 6G. Radar sensing is a special case of sensing, performed actively (with the help of amplification), with imaging, discussed in Sect. 10.4, being considered

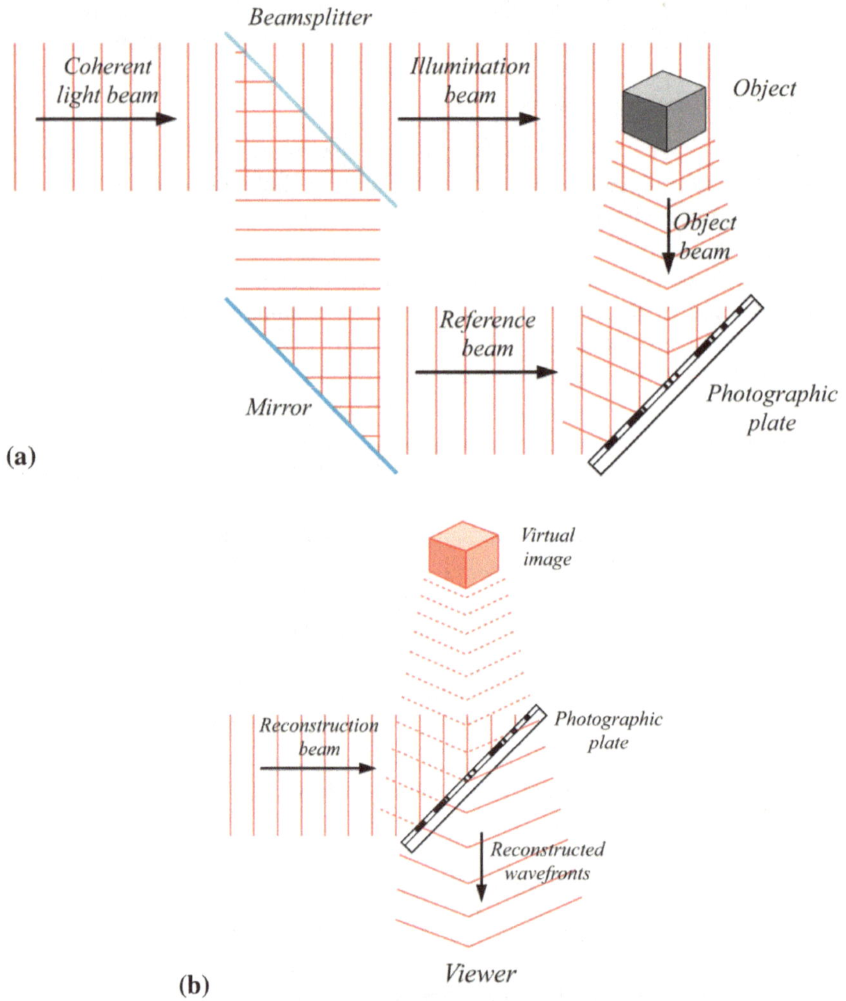

Fig. 10.3 Holographic image creation (**a**) and holographic image projection (**b**). Images created by DrBob at the English language Wikipedia, CC BY-SA 3.0, https://commons.wikimedia.org/w/index.php?curid=18103931 (**a**), https://commons.wikimedia.org/w/index.php?curid=18103933 (**b**)

another special case, typically done with a passive detector in the past. Both types of technologies benefit specifically from operation in the terahertz domain.

10.3.1 Technical Aspects of Radar Sensing

Radar sensing, also called radar detection, is based on the principle that when a signal with high frequency is transmitted, and subsequently intercepted by a target object, some amount of transmitted radiation will be reflected [4, 15]. If the reflected radiation is detected, signal processing can be deployed to determine several properties of the object. These include the distance of the target from the radar antenna, as well as its speed and size. A large amount of computing power is required for classifying radar targets; it is thus feasible to predict that the 6G feature of cloud computing would be one of the main drivers for radar sensing adoption as part of 6G. Some of the potential applications of radar in 6G include automated driving and the flying of drones, obstacle avoidance and gesture recognition. These will be discussed after the radar operation principle discussed next.

From the technology perspective, the radar system is similar to the communication system shown in Fig. 3.1, operating at the same frequency. There are some differences, however, the most important being that modulation of signals is not required, as there is no transmission of information. Secondly, the same antenna can be used for both transmitting and receiving signals, resulting in a monostatic radar. Figure 10.4 shows the basic operational principle and the building blocks of such a radar. To detect a target with radar cross-section σ at a distance R from the antenna, the radar needs to emit power that is sufficient to allow the attenuated, reflected wave to be detectable by the same system. Radar is described by the radar equation, which relates the transmitted power P_t, the antenna gain G, the received power P_r, wavelength λ and the target parameters [16]. This relationship is

$$P_r = \frac{P_t G^2 \lambda^2 \sigma}{(4\pi)^3 R^4}. \tag{10.1}$$

Note that that the received power decreases with the fourth power of the distance of the object from the radar, meaning that very powerful transmitters and very sensitive receivers have to be deployed to detect objects at reasonable distances. Given the

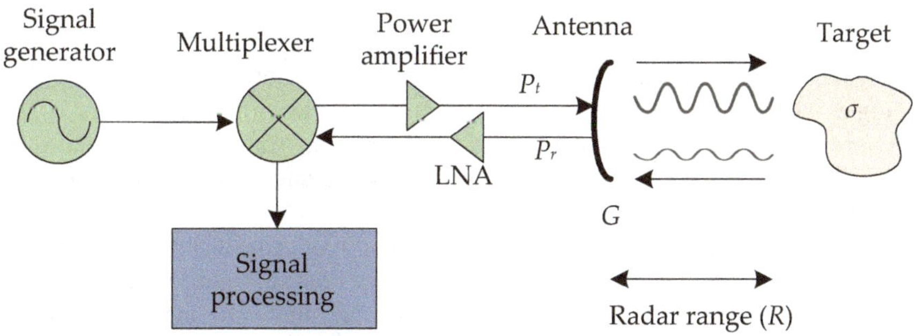

Fig. 10.4 The operating principle a of a monostatic radar. Adapted from [16]

maximum output power and the radar receiver sensitivity, the radar range can be worked out:

$$R_{\max} = \left(\frac{P_t G^2 \lambda^2 \sigma}{(4\pi)^3 P_{r,\min}} \right)^{\frac{1}{4}}. \tag{10.2}$$

From Eq. (10.2) it is clear that the radar range decreases with an increase in frequency (decrease in wavelength). However, there are many advantages that make up for this drawback [15], the most important being that because of the higher frequency of terahertz waves, higher radar bandwidth can be used, which yields better spatial resolution of detected objects. This, in turn, allows the radar to work with much smaller objects altogether. The radar resolution ΔR is related to the bandwidth B with relationship [17]

$$\Delta R = \frac{c}{2B}, \tag{10.3}$$

where c is the speed of light, as before. The higher frequency also implies higher directivity, allowing for the location of the target to be determined more closely. Another advantage comes from smaller radar antennas, when compared to radars at other frequencies, allowing integration in mobile devices and wearables.

Recently, a concept called MIMO radar has been proposed [18]. In this type of radar, antennas are arranged in a transmit array distributed over a wide geographical area with widely spaced elements, which allows for each antenna to see a different aspect of the target. This leads to better detection performance and improved parameter estimation. An alternative use of this so-called distributed MIMO radar is detection of multiple targets simultaneously (up to as many targets as there are MIMO radar antennas). For this, closely spaced antenna elements are needed. This implementation is called co-located MIMO radar. Hybrid distributed and co-located MIMO radars are also possible.

10.3.2 Services Associated with Radar Sensing

Many radar uses are envisioned for 6G. First, it is important to note that radar technologies will be integrated with mobile communication technologies. The aim of this is to provide all-round contextual information on the environment, from small scale to medium scale, to assist communications and communications security [19, 20].

The other services are more specific.

In automotive applications, terahertz radar can work together with millimeter-wave radar in safety features such as park distance control, to introduce additional safety measures. Terahertz radar could also add to the number of integrated automotive sensors and improve their capability [21], allowing for adaptive cruise control, as well as blind spot and lane change assistance in conventional vehicles. The potential of terahertz radar is even greater in self-driving vehicles, as it can provide fine resolution images even in bad weather conditions, such as fog, dust and snow. These images are traditionally obtained by cameras and lidars, which have compromised performance in adverse weather conditions. Terahertz radar can also provide insight into the motion parameters of various objects, for example pedestrians [22]. These automotive terahertz radar benefits can be extended to the context of avionics and that of drones, which have increased in popularity recently. In both automotive use and avionics, the higher degrees of freedom brought about by large-scale MIMO radar technology will be useful to deal with the continuously growing complexity of military and commercial scenarios [18].

Another emerging application of terahertz radar is gesture recognition, mentioned earlier in this chapter as a way of controlling 6G user devices. Gesture recognition may also be important for autonomous driving, where vehicles may need to understand hand gestures of traffic officers and even pedestrians and cyclists [23]. Research into gesture recognition started in the millimeter-wave domain some time ago [24], but because human fingers are, strictly speaking, very small targets, terahertz radar introduces the much needed improved resolution [25]. For successful gesture recognition, terahertz radar needs to measure the complete Doppler signature of the target. This is achieved by sensing the time delay, frequency and phase shift and the amplitude attenuation of the radar signal in real time, as shown in Fig. 10.5. The information is then interpreted as gestures.

Use of radar in the applications that were discussed are just some of the potential radar applications. As more advanced terahertz radar systems are prototyped over the next ten years, one can expect more and more potential applications to be identified.

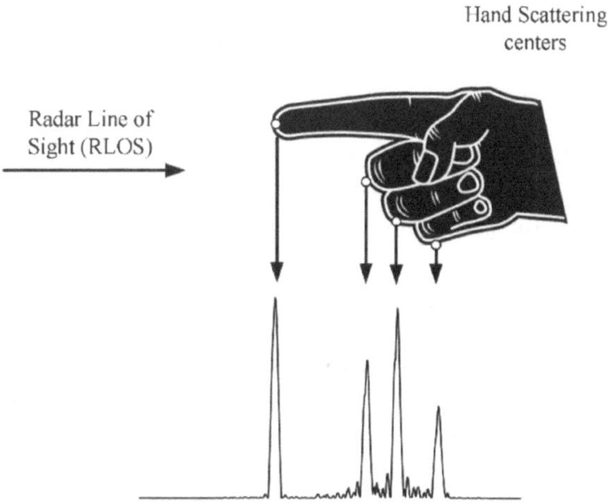

Fig. 10.5 Radar gesture recognition principle. Reproduced under Creative Commons Attribution License from [25]

10.4 Imaging and Sensing

As a standalone concept, terahertz imaging has been used in practice for quite some time. This was normally not achieved with the help of solid-state devices, which are capable of active amplification, but rather by using passive imaging methods, e.g. those using Schottky diodes and lasers [15]. Emerging inexpensive active imaging methods, however, can contribute to even wider adoption of imagining, paving the way to widespread adoption in 6G.

10.4.1 Technical Aspects of Terahertz Imaging

The most useful characteristic of terahertz imaging, or spectroscopy, is the fact that different materials exhibit different absorption spectra at terahertz frequencies, with some specific advantages attributed to radiation with submillimeter wavelengths. Terahertz wavelengths easily pass through clothes and cardboard, but are reflected by metal and plastic objects [15]. Furthermore, passive terahertz imagers running below 1 THz can provide 1 cm resolution even at a target distance up to 10 m. Lastly, terahertz radiation is not harmful to living tissue, since it is non-ionizing, as described in Chap. 9, nor is it destructive to other materials.

10.4.2 Services Associated with Imagining

All these considerations combined lead to numerous present and potential applications (or services in the 6G context) of terahertz imagining. Some will be mentioned here:

- In material research, terahertz radiation can be used to observe properties of matter.
- In environmental sensing, terahertz radiation can be deployed for gas detection in the process of rotational spectroscopy [2].
- In the avionics and space industry, damage to aircraft and spacecraft materials can be evaluated with terahertz waves.
- In food research, foreign particles and contaminants can be detected by terahertz radiation.
- The content of closed packaging can be explored in a non-destructive manner by terahertz imaging; this may be beneficial from both the security and logistics perspectives.
- In safety and security, terahertz radiation can be used in concealed weapon detection, where a real-time imaging capability adds to the speed of processing. (This is particularly important for busy settings, such as airports and train stations.)

- In medical applications, terahertz radiation can supplement and ultimately replace X-ray and ultrasound techniques, typically used for the purpose of establishing medical diagnoses.
- Imaging can also be used to increase the accuracy of outdoor positioning, as will be discussed shortly in Sect. 10.5.
- Similarly, gaining insight into the environment may assist in blockage detection and subsequent improvement of datalink reliability by deploying more dynamic operations [20].

It should also be mentioned that at terahertz frequencies, better image resolution can be achieved at reasonable hardware cost [26]. As an example, Fig. 10.6a shows the photograph of a packaged IC. Figure 10.6b shows an image of the same IC taken by an X-ray machine. Finally, Fig. 10.6c shows the process of generating an insightful terahertz image of the same IC, with the end result being an image of similar, if not better quality, as the X-ray image.

10.5 Localization

GPS technology has caused a revolution in the field in localization (positioning) of objects and navigation. However, there are at least three places where localization by technologies provided by 6G may be of assistance. The first is achieving more accurate outdoor positioning than is possible with the GPS alone, typically with centimeter-level accuracy. The second is the possibility of accurate indoor positioning, which needs accurate interpretation of the third dimension on top of the already existing 2D information. The third is the ability to introduce an additional three dimensions to localization: roll, pitch and yaw [18]. 6G promises to combine radio and VLC localization to get to six-dimensional centimeter-level accuracy, irrespective of whether the object or the user is to be found in an indoor or an outdoor setting.

10.5.1 Indoor Localization

VLC localization may prove to be the most adequate solution for indoor positioning. This concept was already discussed in Chap. 5, but a short summary will be presented here. The physics of light wave propagation dictates that a user can only communicate with each access point via either a direct link or a single reflection. Moreover, light does not propagate through the floors or walls of a building. Thus, triangulation that uses light waves (Fig. 5.24) is much more accurate than if radio waves, which may have multiple reflections or propagate through walls and floors, are used instead. This allows for the user to be located in a specific location on a single floor and in a room

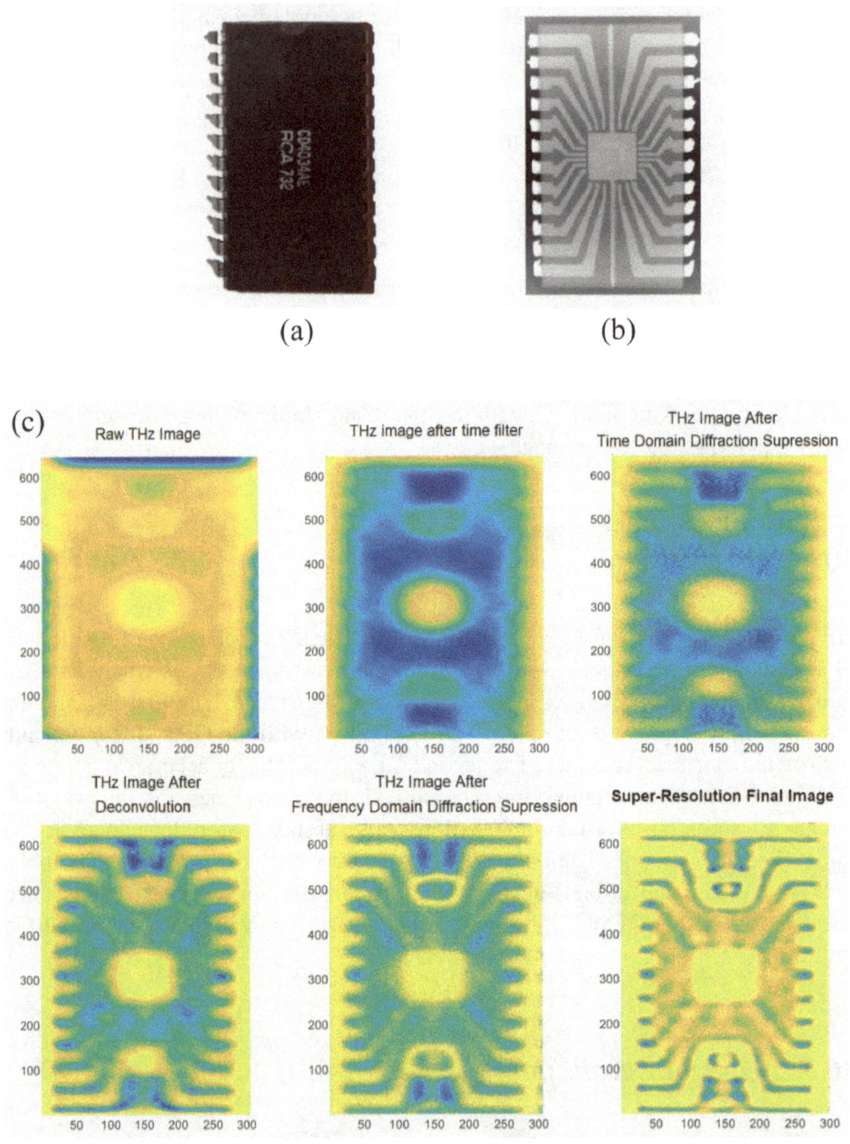

Fig. 10.6 Comparison between terahertz and X-ray imaging: **a** An IC (photograph), **b** The X-ray image of the same IC, **c** Process of generation of the terahertz image of the same IC. Reproduced under Creative Commons Attribution-Share Alike 4.0 International License from https://commons.wikimedia.org/wiki/File:XRay-Optical-THz-IC.gif

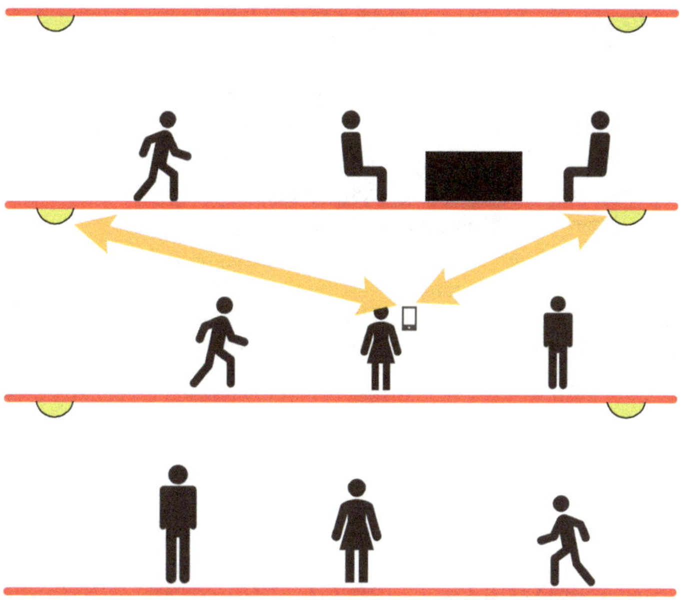

Fig. 10.7 Indoor 3D localization with VLC

of a building, which in practice gives insight into the third dimension, something like the situation illustrated in Fig. 10.7.

10.5.2 Outdoor Localization

For outdoor localization, terahertz imaging may be useful. One technique that has been proposed is called simultaneous localization and mapping (SLAM) [2, 27]. In SLAM, high-resolution images are provided by a terahertz imager (e.g. a steered radar). These are first used to compute a 3D image of the environment. Note that this requires a large amount of computing power; however, cloud computing of 6G can be used instead of computing by the localization terminal. Then, the accurate location of an object to be located in the real-time 3D image can be extracted by comparing its data with the data of the saved image of the environment without the said object present. One clear drawback of this method is the need to have prior knowledge of the environment.

10.5.3 Alternative Localization Methods

Some other alternative methods of positioning, which deploy terahertz or millimeter waves, have also been proposed. These include time-of-flight-based positioning, phase-based positioning, angle-based positioning, 3D millimeter-wave tracing, map-based localization and map-assisted localization using a single base station [27]. The last method is interesting because it uses the concept of multipath propagation, but it results only in candidate locations, which need to be screened using a different (less accurate) method. Another recently proposed localization method is localization with the aid of intelligent reflective surfaces [28], which were described in Chap. 7. Localization using MIMO arrays is also an alternative, where multiple antennas add to the special richness of the localized object.

Some of the problems with the techniques listed above include the difficulty of synchronization of terahertz waves, blockage, fading and noise. As a result, more research is needed before some of these techniques leave the confines of a laboratory.

10.5.4 Services Associated with Localization

Location-based services on their own are "nice to have" verticals for users, but knowledge of the accurate location of the user also has some practical advantages for the physical layer of the 6G network. The location of the user is an important consideration for beamforming. Accurate location can enhance beam recovery, avoid beam failure that may happen if beams are formed through prediction, and can also be used to optimize communication links further in general [11]. Other services of 6G may also benefit from accurate positioning. Good examples of these are autonomous vehicles and drones, which are the perfect platforms for active sensing with radar, and communications convergence [28], as will be described later in Sect. 10.6.4.

10.6 Other Verticals

The fast data rate and low latency of future wireless networks allow for the development of vertical industries that have been considered isolated concepts in the past. From Sect. 10.1, with communications, data transfer, sensing, imaging and localization already discussed, the focus can now shift to IoT and IoE, Industry 4.0 and Industry 5.0, VR, AR and XR, as well as to robotics and autonomous systems (e.g. autonomous driving).

10.6.1 Internet of Things and Internet of Everything

The IoT and IoE have been identified as playing the main role in achieving global SDGs and improving the quality of life. The wording "internet of things" is a quite good definition of the IoT concept on its own. Whereas the regular internet, the "internet of people", described earlier in Sect. 2.4.3, is responsible for connecting devices through which people interact (network) with one another (personal computers, cellphones, tablets), the IoT enables machines or objects ("things") to communicate with one another and with people as well [29]. In other words, in IoT, machines communicate on behalf of people, as opposed to people communicating via machines. If one pictures the mobile internet (enabled by 3G or newer wireless networks) as a 2D plane where one axis is space and the other one is time (that is, the internet allows humans to connect to a network from any place at any time), then IoT can be imagined as an additional axis that enables connection of any "thing" in addition to any human. This is illustrated graphically in Fig. 10.8. IoE is practically the same concept, with some researchers expanding IoT by defining four classes of connectivity: people, things, data and processes (such as those in manufacturing) [30]. The two concepts are often used interchangeably; in this book it will be assumed that IoE evolves from IoT.

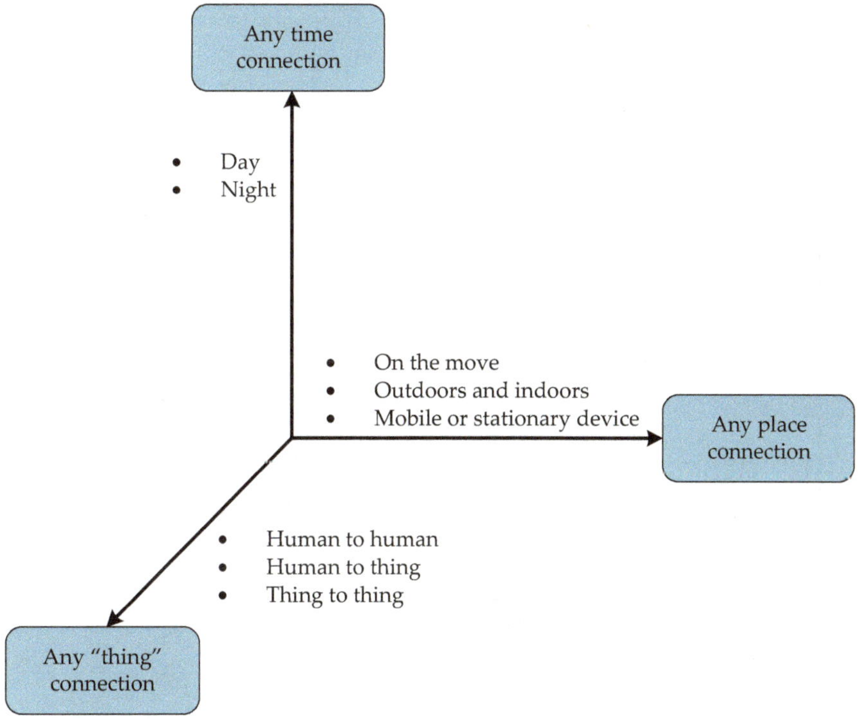

Fig. 10.8 3D representation of IoT. Adapted from [29]

Technical Aspects of Internet of Things

In practice, the IoT brings together two concepts: RFID of various devices with large-scale wireless networking [29]. Differently stated, it is the internet connectivity brought about by 3G, 4G, Wi-Fi and other (newer) wireless technologies that enable devices that have a unique communicable identifier to be connected to a single network. The RFID concept, which is considered pivotal for IoT, on its own has existed for almost 40 years already and has been widely used in logistics, pharmaceutical production, retailing, and supply chain management [31]. In the 1990s, the automatic identification and tracking of RFID technology were expanded into the concept of wireless sensor networks, which added a sensing capability and some intelligence to device connectivity on a small (localized scale). This eventually evolved into IoT in the late 2000s, with the evolution of networks that can handle the increased traffic load and technologies that support storage of large amounts of generated information. The complete evolution of IoT and IoE is illustrated in Fig. 10.9. The purpose of a potentially enormous amount and variety of data generated in the modern IoT is to provide new services to citizens, companies, and public administrations [32]. Various services, which include smart homes, smart cities, smart healthcare, smart businesses and smart industries among others, will be discussed towards the end of this section, but first, the architecture of the IoT will be discussed in some detail.

Today, IoT is considered to be built on the following three pillars [33]:

- Identifiability (anything identifies itself);
- Ability to communicate (anything communicates); and
- Ability to interact (all things interact either among themselves or with end-users or other entities in the network).

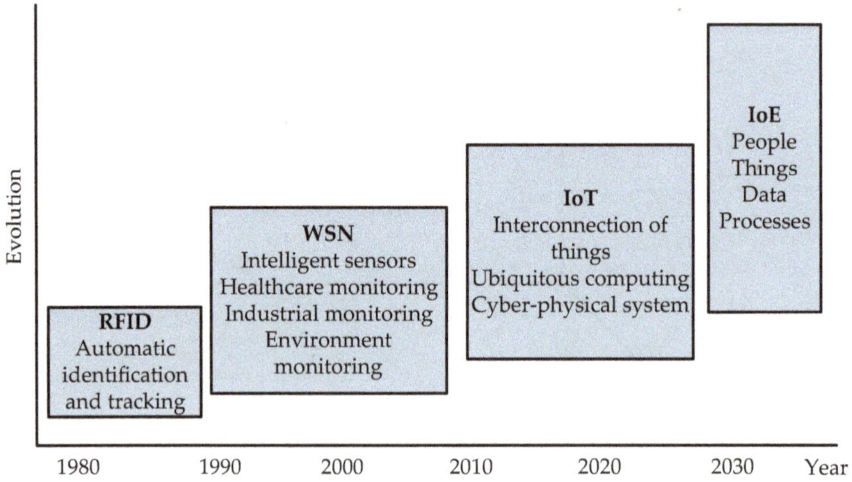

Fig. 10.9 Evolution of IoT and IoE. Adapted from [31]

For machines (things) to be considered fit for IoT, they need to possess certain qualities. These are:

- A physical body and a set of associated physical features (e.g., size, weight, shape, etc.);
- At least a minimum communication functionality, for example the ability to accept incoming messages and reply to them;
- A unique identifier, consisting of one name and one address, where the name is human readable and the address is machine-readable;
- Some basic computing capabilities, ranging from interpreting messages to performing complex computations, including service discovery and network management tasks; and
- Means to sense physical phenomena (e.g. temperature, light, pressure, sound intensity, pollution level, electromagnetic radiation level) or to trigger actions having an effect on physical reality, or both.

The last point above distinguishes between sensors and actuators; the other points in the list above basically define things as virtual personalities that are seamlessly integrated into the information network [31].

In IoT, the focus is on data and not on point-to-point communications, thus the IoT is a highly dynamic and radically distributed system [33]. As more and more machines fit this description, it is clear that the data traffic and the amount of information collected increase drastically, which makes IoT a perfect vertical for the imagined omni-present 6G network. 5G is already a step in the right direction for IoT, offering the ability to establish massive connections and providing standardized access methods to IoT [34], but the true potential will probably only be reached with 6G, with AI and edge computing enabling distributed decisions of the larger sensor arrays. Some researchers have coined the phrase "IoT meets AI" [35], whereas some others define "super IoT" [14], both of which can certainly be used in the context of 6G.

On the service level, IoT needs to support the following [33]:

- Device heterogeneity (support for devices with different computational communication capabilities);
- Scalability (ability to identify and address devices with a massive number of devices connected that increase over time);
- Ubiquitous data exchange through proximity wireless technologies (e.g. RF, terahertz communications, VLC);
- Energy efficiency (with a large number of devices energy consumption becomes problematic);
- Localization and tracking capabilities (for locating devices in the physical realm);
- Self-organizing capabilities (autonomous reaction to a wide range of different situations);
- Semantic interoperability and data management (for standardized data formatting); and
- Embedded privacy and security techniques.

Table 10.3 SOA of IoT

Layer	Description
Sensing layer	Hardware for identification, data acquisition and actuation
Networking layer	Basic networking (data transfer) support
Service layer	Creating and managing services that satisfy user needs
Interface layer	Interaction with users and the network

This is achieved by building IoT as a service-oriented architecture (SOA). A four-layer SOA is shown in Table 10.3 and consists of a sensing layer, networking layer, service layer and interface layer [31]. The description of each layer is given in the table itself.

Services Associated with Internet of Things

In this book, it is probably the types of services that IoT enables that will interest the reader. Thus, this section concludes with a short discussion on present and potential IoT services [31–34].

Smart homes or smart buildings. In this service, advanced IoT technologies are installed with the aim of saving resources such as electricity and water, and improving overall living (homes) or working (offices) conditions. For example, IoT can control lighting, air-conditioning and recycling, or assist in decreasing the overall carbon footprint of the building.

Smart cities. This concept deploys advanced communication infrastructure and novel services to bring about improvements in the quality of living on a city-wide scale, while also bringing down the operational costs of public administration. Services include smart traffic control, smart parking management, pollution control, waste management, noise monitoring, smart street lighting, city energy consumption monitoring, monitoring of the structural health of buildings and automation of public buildings, among others.

Smart environmental monitoring. IoT can be deployed to sense any range of environmental conditions, including temperature, wind, rainfall, river height, pollution, noise and radiation level. Environmental monitoring can be part of the smart city service, or it can be deployed in any other setting, for example in rural areas, or even in environments where human operator presence is not viable (remote areas, harsh environments, during natural disasters, etc.).

Smart security. IoT can provide the less invasive alternative to active security and surveillance that is typically achieved by operators monitoring situations via cameras or by in-person deployed security. This service of IoT can be applied to office buildings, factories, farms, car parks and other locations.

Smart businesses and industries. These include among others mining, farming, inventory, and logistics. In other words, IoT can assist any kind of industry that deals with the production and movement of goods. This service class will be illustrated by means of examples. For example, for general businesses, the RFID capability of IoT offers the capability of tracking goods through the supply chain, thus enhancing supply management systems. On the other hand, in mining, IoT can contribute to mining safety by sensing the environment or tracking miners underground. Lastly, in farming, IoT sensing can assist in quality management, operational efficiency and public food safety, while in cattle breeding, the RFID tracking capability can help in locating animals and the logistics associated with their movement.

Smart healthcare. Various wearable or implantable sensors can help patients and elderly people by monitoring their body vitals, such as temperature, blood pressure, heart rate, sugar level, breathing activity, etc. IoT can also help patients with logistics, diagnosis, therapy, recovery and medication management. Other sensors, such as accelerometers and proximity sensors, can help monitor their movement and issue warnings or assist in providing rapid response, for example in case of accidents (e.g. elderly person falling) or illness manifestation (e.g. patient fainting or having a heart attack). The overall effect of this approach results in improving patients' lifestyle by decreasing the frequency of doctor's visits or eliminating the need for in-house care. Whereas some of the smart healthcare ideas may seem unnecessary at present, it is a fact that the world's population is aging and that the way in which healthcare is rendered at present is not sustainable [36].

As seen in the (incomplete) list of services above, IoT has also found its place in various types of industrial environments. One could say that there is an unbreakable bond between IoT and Industry 4.0, in the sense that IoT has paved the way for the fourth industrial revolution [37]. Thus, it is fitting that Industry 4.0 is discussed next.

10.6.2 Industry 4.0 and Industry 5.0

Industry 4.0 is typically associated with sustainable manufacturing [38], and it is consequently a perfect match for achieving the SDGs as part of the 6G vision. As discussed in Chap. 1, the information and communication technologies of 5G are expected to promote the development of Industry 4.0, whereas the transition from 5G to 6G that brings about ubiquitous network access, low latency and high data throughput may further expand Industry 4.0 capabilities and allow for seamless transition into Industry 5.0.

The third industrial revolution introduced the so-called cyber systems into factories (digitalization with microelectronics and automation); however, in 21st-century manufacturing, the physical and cyber worlds remain separate. The paradigm shift of Industry 4.0 brings cyber and physical systems together into what can be considered cyber-physical systems. Industry 5.0 could bring about the co-existence of systems such as cyber-cyber, cyber-physical and physical-physical systems. According to a 6G white paper published by the 6G Flagship program in 2020 [39], Industry 5.0 will target producing customized and personalized products in mixed sensing/actuation/haptics scenarios, while Industry 4.0 focuses more on data collection and analytics. Industry 5.0 will also be a much more interactive concept than Industry 4.0. However, Industry 5.0 is still some time away from becoming a reality, whereas Industry 4.0 has been around for some time but is still under development.

The concept of Industry 4.0 was conceived in 2011 by the German government as Industrie 4.0, an initiative to transform industrial manufacturing.

In this chapter, the focus will remain on Industry 4.0, as the concept of Industry 5.0 is still only an interesting thought. Some examples of Industry 4.0 include interconnecting machines, robotics in industries such as agriculture and high-precision manufacturing.

Technical Details of Industry 4.0

The premise of Industry 4.0 is advanced digitalization within factories, while combining internet technologies and future-oriented technologies in the field of "smart" objects [40]. The term "Industry 4.0" is practically synonymous with the fourth industrial revolution, but the naming was specifically coined to point out the digital component by alluding to software versioning. With the society changes that the 21st century introduced, a number of social, economic and political factors emerged that are regarded as triggers of this recent industrialization paradigm shift. These include:

- Requirements for short innovation and product development cycles brought about by the fast pace of 21st century living;
- Change in the buyers and sellers' market that led to individualization on demand;
- Need for flexibility in manufacturing;
- Need for decentralization of industry; and
- Need for resource efficiency brought about by shortages or an increase in the prices of various resources.

At the same time, industries experienced a push from emerging technologies, including a further increase in mechanization and automation, digitalization and networking, as well as miniaturization. Availability and use of connectivity (internet and IoT), integration of technical processes and business processes in companies, as well as digital mapping and virtualization of the real world also contributed to this push [41].

Today, it is understood that Industry 4.0 consists of these fundamental concepts [40–42]:

- *Smart factory.* In a smart factory, manufacturing is equipped with sensors, actuators, and autonomous systems. In practice this leads to "smart" industrial production and "smart" products (e.g. products that are able to communicate with the (IoT) network, for example to provide their location or their health state).
- *Cyber-physical systems.* In Industry 4.0, physical and cyber systems will merge. This happens when physical and digital representation cannot be differentiated in a reasonable way anymore. That implies that autonomous systems make their own decisions, typically based on machine learning algorithms, real-time data capture, analytics results and results of successful historical decisions.
- *Self-organization.* Manufacturing now happens on its own, without central control.
- *New systems in distribution and procurement.* This means that distribution and procurement are increasingly individualized.

- *New systems in the development of products and services.* Similar to the point made above, product and service development will be individualized.
- *Adaptation to human needs.* Whereas traditionally, human needs were adapted to existing manufactured products, in Industry 4.0, manufacturing can adapt to human needs. In other words, Industry 4.0 promotes massive customization of manufactured products.
- *Corporate social responsibility.* This point indicates that Industry 4.0 is led by ideas of sustainability and resource efficiency. The allocation of resources (including products, materials, energy and water) should be realized in a more efficient way, which is enabled by intelligent cross-linked value creation modules [38].

Naturally, with more efficient manufacturing, the cost of production is also supposed to decrease. According to Rojko [41], Industry 4.0 should bring production costs down by 10–30%, logistic costs by 10–30%, and quality management costs by 10–20%.

The system integration of Industry 4.0 is essentially achieved in three dimensions [38, 43]:

1. *Horizontal integration across the entire value creation network.* This is, in essence, cross-company and company-internal intelligent cross-linking and digitalization.
2. *Vertical integration and networked manufacturing systems.* This is intelligent cross-linking and digitalization on the different aggregation and hierarchical levels of a value creation module. These levels include manufacturing stations, manufacturing cells, lines and factories, as well as marketing, sales and technology development.
3. *End-to-end engineering across the entire product life cycle.* This entails intelligent cross-linking and digitalization throughout all phases of a product life cycle (starting with raw material sourcing, manufacturing itself, product use and ending with product decommissioning).

The last dimension in the above list practically describes that the Industry 4.0 concept is not limited to only direct manufacturing, but rather includes a complete value chain from providers to customers, with all enterprise's business functions and services in between [41]. The system integration of Industry 4.0 is shown in Fig. 10.10.

Challenges Associated with Industry 4.0

A number of challenges are associated with Industry 4.0, requiring further research into this discipline before any major adoption steps. These include (i) challenges with world-wide standardization, (i) technical challenges and (iii) challenges with information security and privacy protection [37].

Where standardization is concerned, initial standardization efforts have been reported, but currently, there is no universally adopted single standard that describes Industry 4.0. Potential standard candidates include (1) the open-source standard

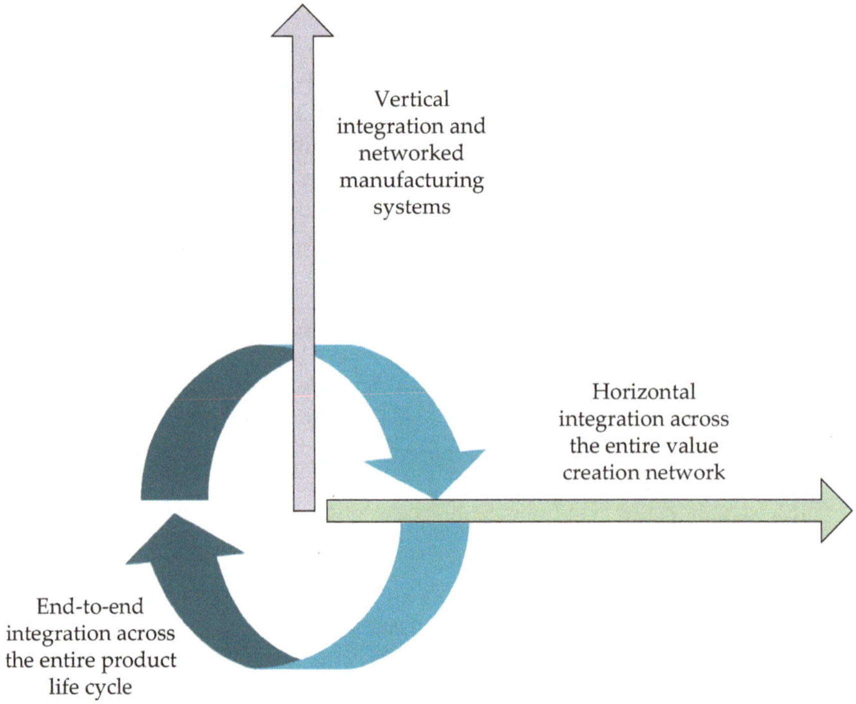

Fig. 10.10 System integration of Industry 4.0. Adapted from [43]

called the Industrial Internet Reference Architecture (IIRA) and (2) industrial IoT (IIoT). A future unified standard will probably be based on merging IIRA and IIoT with other ideas.

There are still numerous technical challenges. First, in many manufacturing companies, technology infrastructure is not ready to support the digital transformation that Industry 4.0 requires. Second, the issue with scalability remains. Like IoT itself, Industry 4.0 must support connection of an increasing number of physical objects that are connected to the manufacturing network. Third, the data analytics form the bare bones of Industry 4.0, and unless it can be processed efficiently, it holds little value. Lastly, any IoT-related challenges, such as delays in transmission, inadequate computing efficiency, limited collaboration ability between different IoT nodes and lack of cross-integration of IoT with legacy systems may also be troublesome. It is hoped that 6G has improved data transfer and edge computing capabilities that might solve many of the technical issues listed above.

The last challenge associated with Industry 4.0 is the question of information security. Recently occurring cases of cyber-physical attacks have warned industries of the possible dangers and threats that Industry 4.0 may face [44]. Thus, it is recommended that proper security should be viewed in equal measure along with production efficiency and productivity and should therefore be in the focus of future Industry 4.0 research.

10.6.3 Multi-sensory Extended Reality and Tactile Internet

Virtual, augmented, mixed and extended realities are terms used to describe several related technologies, the potential of which will finally be fully empowered in 6G [1]:

- VR places users into a virtual environment that is controlled by a set of rules. The user is isolated from the physical world, with the perception of most of their senses replaced by simulated perception [45]. VR provides visual, auditory and even tactile stimuli. Users communicate with the VR environment by using their regular motor skills [46].
- AR is a concept in which physical and synthetic concepts exist simultaneously. The aim of AR is to supplement the physical world through a virtual layer. This can be achieved, for example, by the user wearing glasses that have overlay displays describing objects seen by the user, based on the information captured by various sensors (cameras, microphones, gyroscopes, GPS). Another example of AR is telepresence, where users interact with people who are not physically in the room.
- MR is the term used to describe the experience resulting from combining both VR and AR.
- More recently, the term XR was coined as a way to describe all three types of realities under one term (with "X" standing for not only "extended" but also "any").

Multi-Sensory Extended Reality in 6G

For immersive interaction with XR, high-fidelity multi-sensory interaction is required, and stable and seamless user coordination is a must. This is currently not supported, particularly not in mobile environments such as 4G and the initial 5G releases. Wearables and mobile devices used for VR and XR require extremely capable graphics processing units (GPUs) and thus, at least desktop computers or fast servers are needed to render the information required. Thus, the information needs to be transferred to dedicated computers or servers, but the delay in information transmission needs to be less than 100 ms for a fully immersive experience. 5G seems to fall just short of making this possible. However, with the promise of high data rates and low-latency data transmissions, 6G will allow computing to be offloaded to GPUs on the cloud (or cloud edge) and thus allow for the required quality of multi-sensory XR experience.

Note that, as already stated in Chap. 1, development and deployment of XR, on its own, will require multidisciplinary collaboration between engineering and social disciplines, because it is not trivial to develop a technology that plays with human senses, cognition and psychology.

Tactile Internet

Enabled by 6G, XR may also evolve into a new concept, called the tactile internet [35, 46, 47] or internet of senses [39]. Haptic control of the virtual or augmented environment may grow outside the XR realm and allow for anything to be controlled remotely via the tactile internet. For example, the real-time control of interconnected nodes or machines of IoT and Industry 4.0 may be achieved. This includes robots used in the manufacturing industry, robots used for exploration, or robots used for emergency response in difficult to access or harsh environments. Even the systems used for tele-diagnosis or remote surgery will benefit from the tactile internet. Other uses of tactile internet may include entertainment and education.

With the tactile internet, the so-called multi-modal sensory experience becomes realistic. In this concept, tactile feedback adds to the visual and auditory feedback that the user can already experience with the regular internet [46].

10.6.4 Connected Robotics and Autonomous Systems

In Chap. 1, CRAS was defined as a wide term and includes technologies such as drone-delivery systems, autonomous cars and other vehicles, as well as autonomous robotics [1]. All of these services require high reliability systems (BER of the order of 10^{-9} [9]) and accessibility everywhere, granting the ability to provide high bandwidths and data rates as well as low latency (1 ms or less), to perform in a satisfactory manner.

Autonomous Driving

One of the most talked about concepts of CRAS is autonomous driving. Autonomous vehicles are paramount for ecologically sustainable transportation in the future [3]. Besides, autonomous vehicles are an interesting example of a vertical of 6G, because they use numerous interrelated systems, including radars, lidars and cameras, to observe their environment and plot a course of their movement, which are combined with GPS, DSRC and V2X systems that are used to communicate with other objects, such as other vehicles on the road and traffic lights [19]. Ultra-reliable V2X communication on its own is becoming increasingly important [48]. The major obstacle in establishing vehicular communication is that both the transmitter and the receiver are moving at great speeds. If terahertz or VLC communication is established, however, the required transmission can occur at such a fast rate that the relative position of vehicles and channel properties change at a much lower rate, causing the transmitter and receiver to appear practically stationary over the short period of time required for the exchange of information [6]. By being able to facilitate this communication, 6G will greatly simplify vehicular communication. in addition, 6G AI and machine learning with edge computing are ideal tools for processing information on the environment, road conditions, potential hazards and information on other vehicles, people and animals, collected via all different means, necessary for safe driving and avoiding potential collisions [19]. It is also possible that automated driving will allow vehicles to move at speeds that would not be possible with human drivers.

The level of vehicular automation is gradually increasing as technology evolves. This trend is expected to continue [49]. As illustrated in Fig. 10.11, six levels of vehicular automation can be identified. Automation level 0 means the complete absence of automation (true for most low-end vehicles sold at present). In levels 1 and 2, the driver is still in control of the vehicle, and the technology works to assist the driver. In levels 3 and 4, it is the automated system that has full control of the vehicle, but a driver is present and is able to intervene. In level 5, which is the last level of automation (that is still some time away), a driver may not even need to be present in the vehicle.

Autonomous Drones

CRAS also introduces the idea of autonomous drones. Examples of uses of drones include search and rescue missions (drones are deployed faster than manned vehicles), drones for mailing and delivery (this allows for faster delivery of parcels within the carrying capacity of a drone) and marine drones (such as those used in underwater exploration missions) [46]. In applications involving marine drones, 3D connectivity via the air-ground-underwater network becomes essential. Note that drones themselves may actually be one of the main enablers of ubiquitous access, providing 6G coverage by serving as airborne base stations [2, 46].

Autonomous Robotics

Autonomous robotics is a concept of CRAS that has already been inherently discussed, as most of it can be applied to the concept of Industry 4.0. However, there are naturally many other, typically commercial, uses of robots outside manufacturing plants. For example, automated lawnmowers and vacuum cleaners are by no means unheard of even at present. In 6G, these can be enhanced by improved connectivity (such as talking to other robots performing the same or similar tasks), or making decisions based on data collected by various sensors or imaging (in a similar manner to that in which an autonomous vehicle or a drone would operate, that is, by communicating with the edge). Other uses may include robots used in the hospitality industry, healthcare, warehousing or package delivery [50].

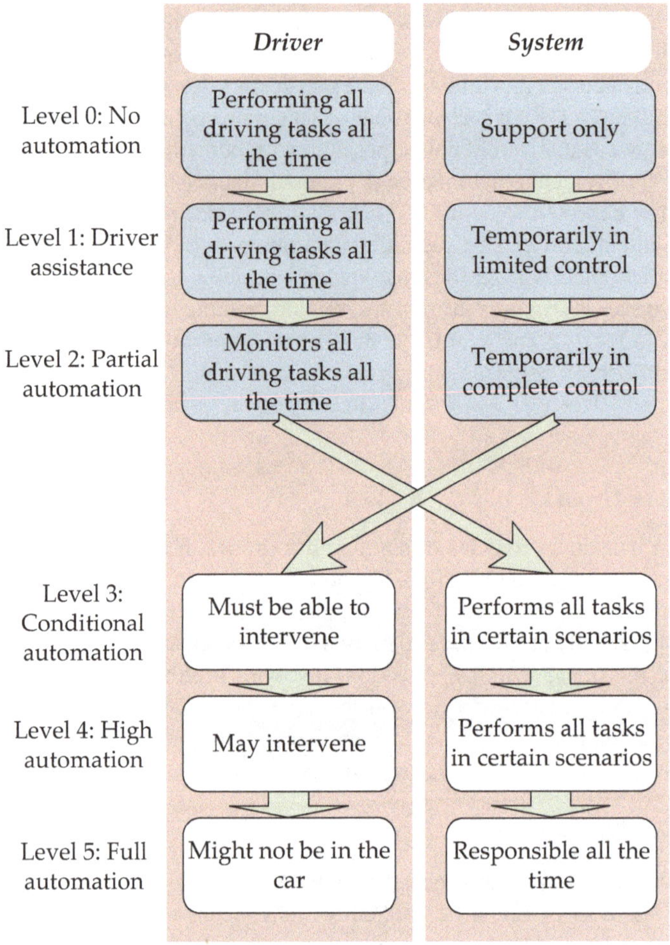

Fig. 10.11 Six levels of automation in autonomous driving. From [49]

10.7 Concluding Remarks

With the discussion on vertical industries of 6G carried out in this chapter, this book is brought to a close. Over the course of ten chapters, we discussed the vision of 6G, the history of communications as a whole, wireless communications and the internet, as well as introducing the existing and emerging technologies that will define 6G, the potential problems when it comes to wide adoption of this future concept, and the envisioned services of this network.

6G is still at least ten years from becoming a reality if everything develops according to the plan. Certainly, from the historical perspective, over the past 40 years or so, every ten years a new wireless network generation was introduced, and with 5G deployment around 2020, it is expected that 6G will come to life in 2030. However, there are many obstacles to be overcome before this concept will be ready commercially. Some can be predicted and researchers can work to overcome them before they arise; others may come completely unexpectedly (think about the 2020 Covid-19 pandemic that has practically brought the world to a standstill for at least a year at the time of writing, resulting in many technology-related expositions, symposia and other events being canceled or converted to virtual events). Nevertheless, the fusion of verticals that 6G will bring about is thrilling, as the goal of many of them is to improve the quality of life of both rich and poor, while connecting the East and the West and merging the real world with the virtual world. We, as the authors of the book, are excited that we can make at least this small contribution to this cause that we call "the leap to 6G".

References

1. Saad W, Bennis M, Chen M (2019) A vision of 6G wireless systems: applications, trends, technologies, and open research problems. IEEE Netw 1–9
2. Sarieddeen H, Saeed N, Al-Naffouri TY, Alouini M-S (2020) Next generation terahertz communications: a rendezvous of sensing, imaging, and localization. IEEE Commun Mag 58:69–75
3. Latva-aho M, Lppanen K (eds) Key drivers and research challenges for 6G ubiquitous wireless intelligence (white paper). 6G Vissions 1. Oulu: Univerisity of Oulu
4. Božanić M, Sinha S (2020) Millimeter-wave integrated circuits: methodologies for research. Design Innov, Springer Nature
5. Saeedkia D (2013) Handbook of terahertz technology for imaging, sensing and communications. Elsevier
6. Mumtaz S, Jornet JM, Aulin J, Gerstacker WH, Dong X, Ai B (2017) Terahertz communication for vehicular networks. IEEE Trans Veh Technol 66:5617–25
7. Kürner T, Priebe S (2014) Towards THz communications—status in research, standardization and regulation. J Infrared Milli Terahz Waves 35:53–62
8. Akyildiz IF, Jornet JM, Han C (2014) Terahertz band: next frontier for wireless communications. Phys Commun 12:16–32
9. Rajatheva N, Atzeni I, Bjornson E, Bourdoux A, Buzzi S, Dore J-B et al (2020) White paper on broadband connectivity in 6G. arXiv:200414247 [eess] [Internet]. [cited 2020 Sep 16]. http://arxiv.org/abs/2004.14247
10. Tajima T, Kosugi T, Song H-J, Hamada H, El Moutaouakil A, Sugiyama H et al (2016) Terahertz MMICs and antenna-in-package technology at 300 GHz for KIOSK download system. J Infrared Milli Terahz Waves 37:1213–24
11. Chen Y, Zhu P, He G, Yan X, Baligh H, Wu J From connected people, connected things, to connected intelligence. In: 2020 2nd 6G Wireless Summit (6G SUMMIT), pp 1–7
12. Lucente M (2011) The first 20 years of holographic video—and the next 20. In: SMPTE 2nd annual international conference on stereoscopic 3D for media and entertainment, pp 21–3
13. Huang C, Hu S, Alexandropoulos GC, Zappone A, Yuen C, Zhang R et al (2020) Holographic MIMO surfaces for 6G wireless networks: opportunities, challenges, and trends. IEEE Wirel Commun 1–8
14. Zhang L, Liang Y-C, Niyato D (2019) 6G Visions: Mobile ultra-broadband, super internet-of-things, and artificial intelligence. China Commun 16:1–14
15. Božanić M, Sinha S (2019) Emerging transistor technologies capable of terahertz amplification: a way to re-engineer terahertz radar sensors. Sensors 19:2454

16. Pozar DM (2012) Microwave engineering. Wiley
17. du Preez J, Sinha S (2017) Millimeter-wave power amplifiers. Springer
18. Björnson E, Sanguinetti L, Wymeersch H, Hoydis J, Marzetta TL (2019) Massive MIMO is a reality—what is next? Five promising research directions for antenna arrays. Digit Signal Proc 94:3–20
19. Tariq F, Khandaker M, Wong K-K, Imran M, Bennis M, Debbah M A speculative study on 6G. IEEE Wirel Commun 27:118–25
20. Matinmikko-Blue M, Yrjölä S, Ahokangas P (2020) Spectrum management in the 6G Era: the role of regulation and spectrum sharing. In: 2020 2nd 6G wireless summit (6G SUMMIT), pp 1–5
21. Choi J, Va V, Gonzalez-Prelcic N, Daniels R, Bhat CR, Heath RW (2016) Millimeter-wave vehicular communication to support massive automotive sensing. IEEE Commun Mag 54:160–7
22. Jasteh D, Hoare EG, Cherniakov M, Gashinova M (2016) Experimental low-terahertz radar image analysis for automotive terrain sensing. IEEE Geosci Remote Sens Lett 13:490–4
23. Weaver C (2020) The new driver's ED: game developers teach Cruise's autonomous vehicles to understand gestures made by people on the street. IEEE Spectr 57:32–7
24. Lien J, Gillian N, Karagozler ME, Amihood P, Schwesig C, Olson E et al (2016) Soli: ubiquitous gesture sensing with millimeter wave radar. ACM Trans Graph 35:142:1–142:19
25. Zhou Z, Cao Z, Pi Y (2018) Dynamic gesture recognition with a terahertz radar based on range profile sequences and doppler signatures. Sensors 18:10
26. Nuessler D, Schubert M, Kose S, Pohl N (2017) Swifter security scanning: millimeter-wave imaging with spin. IEEE Microwave Mag 18:70–8
27. Rappaport TS, Xing Y, Kanhere O, Ju S, Madanayake A, Mandal S et al (2019) Wireless communications and applications above 100 GHz: opportunities and challenges for 6G and beyond. IEEE Access 7:78729–57
28. Bourdoux A, Barreto AN, van Liempd B, de Lima C, Dardari D, Belot D et al (2020) 6G white paper on localization and sensing. arXiv:200601779 [cs, eess] [Internet]. Accessed 16 Sep 2020. http://arxiv.org/abs/2006.01779
29. Lu T, Wang N (2010) Future internet: the internet of things. In: 2010 3rd International conference on advanced computer theory and engineering (ICACTE), pp V5-376–V5-380
30. Miraz MH, Ali M, Excell PS, Picking R (2015) A review on Internet of Things (IoT), Internet of Everything (IoE) and Internet of Nano Things (IoNT). In: 2015 Internet technologies and applications (ITA), pp 219–24
31. Xu LD, He W, Li S (2014) Internet of things in industries: a survey. IEEE Trans Industr Inf 10:2233–43
32. Zanella A, Bui N, Castellani A, Vangelista L, Zorzi M (2014) Internet of things for smart cities. IEEE Internet Things J 1:22–32
33. Miorandi D, Sicari S, De Pellegrini F, Chlamtac I (2012) Internet of things: vision, applications and research challenges. Ad Hoc Netw 10:1497–516
34. Li S, Xu LD, Zhao S (2018) 5G Internet of Things: a survey. J Indus Inf Integr 10:1–9
35. Nawaz SJ, Sharma SK, Wyne S, Patwary MN, Asaduzzaman Md (2019) Quantum machine learning for 6G communication networks: state-of-the-art and vision for the future. IEEE Access 7:46317–46350
36. Mucchi L, Jayousi S, Caputo S, Paoletti E, Zoppi P, Geli S et al (2020) How 6G technology can change the future wireless healthcare. In: 2020 2nd 6G wireless summit (6G SUMMIT), pp 1–6
37. Xu LD, Xu EL, Li L (2018) Industry 4.0: state of the art and future trends. Int J Product Res 56:2941–2962. Taylor & Francis
38. Stock T, Seliger G (2016) Opportunities of sustainable manufacturing in industry 4.0. Proc CIRP 40:536–541
39. Mahmood NH, Böcker S, Munari A, Clazzer F, Moerman I, Mikhaylov K et al (2020) White paper on critical and massive machine type communication towards 6G. arXiv:200414146 [cs, eess] [Internet]. Accessed 5 Oct 2020. http://arxiv.org/abs/2004.14146

40. Lasi H, Fettke P, Kemper H-G, Feld T, Hoffmann M (2014) Industry 4.0. Bus Inf Syst Eng 6:239–42
41. Rojko A (2017) Industry 4.0 concept: background and overview. Int J Interact Mob Technol (iJIM) 11:77–90
42. Lu Y (2017) Industry 4.0: a survey on technologies, applications and open research issues. J Indus Inf Integr 6:1–10
43. Vaidya S, Ambad P (2018) Bhosle S. Industry 4.0—a glimpse. Proc Manuf 20:233–238
44. Prinsloo J, Sinha S, von Solms B (2019) A review of industry 4.0 manufacturing process security risks. Appl Sci 9:5105. Multidisciplinary Digital Publishing Institute
45. Braud T, Bijarbooneh FH, Chatzopoulos D, Hui P (2017) Future networking challenges: the case of mobile augmented reality. In: 2017 IEEE 37th International conference on distributed computing systems (ICDCS), pp 1796–1807
46. Bariah L, Mohjazi L, Muhaidat S, Sofotasios PC, Kurt GK, Yanikomeroglu H et al (2020) A prospective look: key enabling technologies, applications and open research topics in 6G networks. arXiv:200406049 [eess] [Internet]. Accessed 14 Sep 2020. http://arxiv.org/abs/2004.06049
47. Wikström G, Peisa J, Rugeland P, Johansson N, Parkvall S, Girnyk M, et al (2020) Challenges and technologies for 6G. In: 2020 2nd 6G wireless summit (6G SUMMIT), pp 1–5
48. Yuan Z, Ma Y, Hu Y, Li W (2020) High-efficiency full-duplex V2V communication. In: 2020 2nd 6G wireless summit (6G SUMMIT), pp 1–5
49. The road to 5G in automotive - with infographic to download (2669436) [Internet]. Accessed 29 Sept 2020. https://event.on24.com/eventRegistration/console/EventConsoleApollo.jsp?&eventid=2669436&sessionid=1&username=&partnerref=&format=fhvideo1&mobile=&flashsupportedmobiledevice=&helpcenter=&key=A8991B66ACA1D28ADB3ACAEF13C096C4&newConsole=true&nxChe=true&text_language_id=en&playerwidth=748&playerheight=526&eventuserid=359879989&contenttype=A&mediametricsessionid=307679236&mediametricid=3761811&usercd=359879989&mode=launch
50. Viswanathan H, Mogensen PE (2020) Communications in the 6G Era. IEEE Access 8:57063–74